Palgrave Studies in Technical and Vocational Education and Training

Series Editorss

This series explores the field of Technical and Vocational Education and Training (TVET), emphasising the various dimensions involved in research, policymaking and practices of teaching and learning. The series illustrates the diverse tensions, challenges and research problems in various national systems of TVET, as well as cross-national aspects.

Contributions to the series will enhance knowledge and insights within the field based on empirical research and theoretical/conceptual analyses. The editors welcome contributions from different TVET contexts worldwide, as well as cross-national studies, and the series is designed to stimulate intellectual debate and influence practices in different levels of TVET enactment.

Unique to the series is an understanding of TVET as a complex field, where research problems and new knowledge are located and interact at macro, meso and micro levels, consequently demanding a multidisciplinary, theoretically and methodologically diverse research approach. The role of TVET in sustainable development will be given particular attention.

Eli Smeplass · Chris Zirkle
Editors

Opportunity Structures in Vocational Education and Training

National Studies and International Discussions

palgrave
macmillan

Editors
Eli Smeplass
Department of Teacher Education
Norwegian University of Science
and Technology (NTNU)
Trondheim, Norway

Chris Zirkle
College of Education and Human
Ecology
Ohio State Univerity
Columbus, OH, USA

ISSN 2948-1678 ISSN 2948-1686 (electronic)
Palgrave Studies in Technical and Vocational Education and Training
ISBN 978-3-032-04409-9 ISBN 978-3-032-04410-5 (eBook)
https://doi.org/10.1007/978-3-032-04410-5

This work was supported by the Norwegian University of Science and Technology.

Cover credit: filo/gettyimages

This Palgrave Macmillan imprint is published by the registered company Springer Nature Switzerland AG
The registered company address is: Gewerbestrasse 11, 6330 Cham, Switzerland

Preface

This project grew out of a shared curiosity and commitment to exploring the possibilities and strengths of vocational education and training (VET). As scholars in the field, we have long been aware of the critical role VET plays in shaping opportunities for both youth and adults across the globe. Yet, we have also been struck by how much of the academic literature leans toward a problem-oriented approach, often overlooking the positive contributions and everyday successes of VET systems.

The idea for this book was born in conversations between the two of us, as we compared the structures of VET in Ohio and Norway. Despite differences in history, governance, and labor markets, we were fascinated by the similarities that emerged at a human level. Teachers, students, and apprentices in both contexts work tirelessly to build competence, deliver quality services and products, and support the development of professional identities, just transitions, and realistic career trajectories. These parallels convinced us that there is value in examining how systems, no matter how different, can be understood through shared conceptual tools.

At the same time, we recognized a challenge. VET is deeply embedded in local and national labor markets, making it difficult to identify common ground across countries. This is precisely why we believe it is necessary to take responsibility for building international dialogue. VET research has much to gain from comparative perspectives, not only to highlight systemic challenges, but also to learn from each other's strengths and innovations. The concept of *opportunity structures* offered us a way forward—a lens through which diverse cases could be connected while still honoring their specific social and economic contexts. This volume is our attempt to contribute to that international conversation.

We are deeply grateful to our colleagues from around the world who embraced this idea and contributed their expertise. Their chapters illustrate the richness of VET as it unfolds in very different societal structures and systems, while also showing the potential of opportunity structures as a shared framework. It has been an inspiring process to work alongside such dedicated scholars, and we have learned a great deal from their perspectives.

We would also like to thank our reviewers for their careful feedback, the series editors for their support, and the excellent team at Palgrave Macmillan and Springer Nature who guided this project to completion. Most of all, we extend our appreciation to the chapter authors, whose openness, flexibility, and insight made this book possible.

Trondheim, Norway — Eli Smeplass
Columbus, OH, USA — Chris Zirkle
September 2025

About This Book

Opportunity Structures in Vocational Education and Training is an edited volume that brings a fresh perspective to the global discussion on vocational education and training (VET). Too often, VET is framed in problem-oriented terms, overshadowing its potential to foster social mobility, economic development, and lifelong learning. This book challenges that narrative by highlighting the strengths of VET systems worldwide, emphasizing their role in building skills, creating career pathways, and supporting regional economies.

Edited by Eli Smeplass and Chris Zirkle, this volume unites leading international scholars to explore how opportunity structures in VET shape individuals' educational and career trajectories. Through comparative perspectives and case studies, the book examines positive aspects of VET systems and presents examples of how they can function as engines of inclusion and economic development, while also adapting to changing labor markets.

Contents

Notes on Contributors

Dr. Stephen Billett is Professor of Adult and Vocational Education at Griffith University, Brisbane, Australia and Australian Research Council Future Fellow. Following an earlier work life in garment manufacturing, he has subsequently worked as a vocational educator, educational administrator, teacher educator, professional development practitioner and policy developer in the Australian vocational education system and teacher and researcher at Griffith University. He is a Fulbright scholar, national teaching fellow, recipient of honorary doctorates from Jyväskylä University (Finland), University of Geneva (Switzerland) and University West (Sweden), and is elected Fellow of the Academy of Social Sciences of Australia. He is the founding editor of the journal *Vocations and Learning: Studies in professional and vocational education*, and the Professional and Practice based Learning book series.

Prof. Dr. John Chakamba is a Senior lecturer specializing in the built environment in the Department of Art Design and Technology Education at the University of Zimbabwe. With a deep passion for technical and vocational education, Dr. Chakamba has dedicated his career to advancing technical teacher training, aiming to equip future educators with the skills necessary for success in the ever-evolving world of technology and design. His work emphasizes the importance of practical learning and the development of a skilled workforce, which he believes is critical to the growth and innovation of the nation.

Prof. Dr. Erika Gericke heads the Chair of School and Teaching Development at Vocational Colleges at the University of Siegen. She studied sociology, vocational education and training and English language and literature (in Germany and Wales). In her doctorate, she examined the interaction between the vocational training (VET) system and biographical vocational orientations/behavioral patterns in Germany and England. Among other things, she is currently investigating VET cultures from a comparative perspective. Her main areas of work include international comparative VET research, socio-pedagogically oriented VET and digitalization in VET.

Petteri Hansen has a PhD in pedagogy and works as an Associate professor in pedagogy at the Norwegian University of Science and Technology (NTNU). His research field is comparative educational policy studies with an emphasis on Nordic countries and issues related to the sociology of knowledge and time. His theoretical interests are related to sociological systems theory.

Hannes Hautz is Assistant Professor at the Department of Organization and Learning, University of Innsbruck (Austria). His research interests focus on teacher professionalism, educational governance, learning analytics, and post-structuralist approaches in VET. He received his PhD in Business Education from the University of Innsbruck in 2020 with the thesis 'Subjectivation processes of VET teachers in the quality dispositive'. His research has been published for example in the British Journal of Sociology of Education, in Education + Training, and in the Journal of Vocational Education & Training. In 2021, he received the Austrian VET Research Award.

Elisabeth Hovdhaugen is a Senior researcher at NIFU (Nordic Institute for Studies in Innovation, Research, and Education) and holds a PhD in sociology. Her research bridges upper secondary education and higher education, primarily focusing on the student perspective. Recent research on upper secondary education includes a paper on pathways to higher education for students coming from vocational education and a paper on students taking one of more gap years after upper secondary, and how this impacts their progression in higher education.

Dr. Seong Ji Jeong is an Assistant professor in the Workforce Education and Development in the College of Education at the Pennsylvania State University. Her scholarly focus centers on understanding how education and training can unlock human potential in a rigid labor market structure, particularly for individuals navigating life transitions under vulnerable situations. Her research interests include school-to-work transition, career trajectories, and the aging workforce. Prior to joining Penn State, she was a postdoctoral scholar at Ohio State University, where she evaluated state-wide workforce development programs using state longitudinal administrative data. She got her PhD at Seoul National University, Korea.

Christian Helms Jørgensen is Professor emeritus at Roskilde University, Denmark. His main research field is adult and vocational education, learning in working life, school-to-work transitions of young people, integration of refugees through vocational education, education policy studies, students' drop-out from education and comparative studies of vocational education especially in the Nordic Countries.

Dr. Junmin Li is a Senior researcher at the UNESCO Chair on TVET and Competence Development for the Future of Work at the TU Dortmund University. She has a wide range of research experience in the field of international comparative VET, with a focus on Asia, Europe and the Anglo-Saxon regions. She is one of the chairs of the European Emerging Researchers Network (EERN-VETNET) and the JVET Early Career Researcher Network and a reviewer for several international journals. Previously, she worked as a researcher at the Chair of Business Education and International VET Research at the University of Cologne from 2012 to 2024. Her particular interests lie in the areas of policy learning, design-based research, teaching and learning and work-based learning.

Rønnaug H. Lyckander is a Senior researcher at NIFU (Nordic Institute for Studies in Innovation, Research and Education) and holds a PhD in Educational Sciences with a specialization in Teacher Education. Her work centers on vocational education and training (VET) within upper secondary education, alongside exploring teacher competence and professional development. Recently, her research has focused on how VET can address the needs of newly arrived immigrants, including a review of European studies on the role of VET in integrating immigrants into the workforce.

Fernando Marhuenda-Fluixá is a full university Professor at the Department of Didactics and School Organization, University of Valencia (Spain). He lectures in the Degree of Social Education and the Master of

Social Economy, and supervises doctoral dissertations in Social Economy (non-profit organizations) and Education. He is coordinator of the multidisciplinary research group (didactics and school organization, sociology, developmental psychology and social psychology) Transicions, with cooperation agreements with the Spanish Association of Second Chance Schools and the Spanish Federation of Work Integration Enterprises https://produccioncientifica.uv.es/grupos/40381/detalle.

His research interests are in vocational education and training, transitions from education into employment with a particular focus upon vulnerable people.

Dr. Marcel Martsch is a Research associate at the the Chair of School and Teaching Development at Vocational Colleges at the University of Siegen. He studied psychology and focuses on the design, implementation, and evaluation of vocational training. His development and research work is not limited to Germany but extends globally. This includes, for example, the design of competency-based curricula, modular (practice-oriented, needs-based) training concepts, Train the Trainer courses, as well as doctoral programs for interested target countries. Moreover, he curiously follows all developments and insights related to vocational education and incorporates them into university courses for teachers at vocational schools.

Dr. Davison M. Mupinga is a Professor in career and technical education at Kent State University in Ohio, USA. He holds graduate degrees in technical and vocational education and training (TVET) and has extensive experience in program development, technical and vocational education policies, teacher development, and program evaluation. His research interests are in international TVET, technical teacher development, and integrating technology into teaching and learning. He was the editor of International Journal of Vocational Education and Training and has published several journal articles on TVET. He has consulted on TVET in Kenya, Lesotho, Uganda, Nigeria, USA, and Zimbabwe.

Dr. Filiz Oskay is PhD candidate in the Department of Educational Studies at The Ohio State University. She also holds a position with the university's Office of Equity, Diversity, and Inclusion, where she contributes to advancing institutional initiatives aimed at fostering equity and social justice. Her research centers on social justice in education, with particular attention to recognition theory, Culturally Responsive School

Leadership (CRSL), and the role of Career and Technical Education (CTE) in addressing opportunity gaps and promoting educational equity.

Davinia Palomares-Montero holds PhD in Sociology from the University of Valencia. She is a tenured professor in the Department of Didactics and Scholar Organization in Faculty of Philosophy and Sciences of Education at University of Valencia, Spain. She is a member of the research group 'Transitions to the Working World in At-risk Populations' (TRANSICIONS). She is interested in the study of higher education curriculum design, vocational training, and the analysis of multidisciplinary innovation for social change.

Prof. Dr. Matthias Pilz is Full Professor of Business Education and International VET Research at the University of Cologne and Director of the German Research Center for Comparative Vocational Education and Training (G.R.E.A.T.). Prior to becoming an academic, he worked as a teacher at a Business College in Hannover and was an advisor for European Union education projects at the district government of Hannover. He is Editor in Chief of the International Journal of Training and Development. His research interests are in international comparative research in VET, transitions from education to employment, and teaching and learning.

Dr. Jay Plasman PhD is an Assistant professor of Workforce Development and Education at The Ohio State University. He also serves as the Director of Research at the Dennis Learning Center where he focuses on postsecondary success and transitions into the workforce. Broadly, his work focuses on education policy with a particular emphasis college and career readiness.

Dr. Katrin J. Posch worked for several years at the Dean's Office of the WiSo Faculty at the University of Cologne before taking over the leadership of the credit transfer center. Before working at the university, she worked in a bank. Her work on an Erasmus+ project prompted her to delve deeper into the topic of refugee integration, which resulted in her doctoral thesis.

Ekaterina Schlüter is a Research associate at the Chair of Business Education and International VET Research at the University of Cologne in Germany. Her research interests focus on the quality of vocational education and training in the German and European context as well as on international comparative research in vocational education and training.

Julián Bell Sebastián The doctoral research of Sebastián focuses on the relationship between vocational education and migration, exploring pathways and challenges for students with migrant backgrounds. His research interests include vocational education and training (VET), mobility, and migration. He is a member of the research group Culture, Diversity, and Development (CUDIDE) and collaborates with the Transitions between Training and Employment in Contexts of Social Vulnerability (TRANSICIONS) group.

Antti Seitamaa is a PhD candidate at the University of Helsinki. His work focuses on education policy and politics from both national and comparative perspectives. Prior to entering academia, he worked as an advocate on behalf of VET students' rights in the national education policy field. His doctoral work has been supported by the Emil Aaltonen Foundation.

Asgeir Skålholt is a senior researcher at NIFU (Nordic Institute for Studies in Innovation, Research, and Education). His work centers on understanding transition within and from upper secondary education, with a special focus on vocational education and training (VET). His work is mainly quantitative and recently includes studies who investigate how socio-economic background is mitigated in vocational education transitions and how economic shocks affects chance and type of transition within vocational education.

Dr. Eli Smeplass is an Associate professor at the Department of Teacher Education at the Norwegian University of Science and Technology (NTNU). She was a Fulbright Scholar at The Ohio State University in 2024, conducting a comparative study of Norwegian and American vocational education and training (VET) systems. With a background in sociology, Smeplass defended her PhD in 2018 on Norwegian teacher training within an organizational and political framework. She has since expanded her research in this field, publishing internationally on vocational education, teacher training, and related policy issues. She is currently engaged in several international research projects on vocational education and vocational teacher training, contributing to the field with a broad and interdisciplinary approach.

Dr. Daniel Láscarez Smith is a Professor at the National Technical University of Costa Rica and currently serves as the Director of the TVET Research Department. His research focuses on the structural dimensions of Technical and Vocational Education and Training (TVET), including governance, learning systems, education policies, and the engagement of social actors. Dr. Láscarez is an active member of several research networks specializing in the sociology of labor in Latin America, with a particular emphasis on labor transitions, productive configurations, and Global South theories. His work contributes to advance critical perspectives on workforce development, education-to-employment pathways, and the role of vocational education in shaping equitable labor markets.

Dr. Chris Zirkle is a full Professor at the Norwegian University of Science and Technology (NTNU) and an Associate Professor in Workforce Development and Education at the College of Education and Human Ecology, The Ohio State University. With four decades of experience in vocational-technical education, Dr. Zirkle has taught at the middle school, high school, community college, and university levels. He also has administrative experience as a high school building administrator and holds current Ohio teaching and administrative licenses in multiple fields. Dr. Zirkle's research focuses on vocational-technical education, and he has published widely in national and international journals, as well as contributed book chapters on vocational education in the U.S., Europe, and Asia. In the fall of 2021, he was a Fulbright Scholar at Haaga-Helia University of Applied Sciences in Helsinki, Finland. He has also served as a visiting scholar at universities in China, Germany, Japan, and Taiwan.

List of Figures

List of Tables

CHAPTER 1

Opportunity Structure as a Core Concept for Vocational Education and Training

Eli Smeplass and Chris Zirkle

Opportunity Structures as a Concept Tool

Meeting with colleagues and scholars from around the world, it is always fascinating to learn how various nations and regions approach learning and training for those preparing to enter the workforce. In many countries, vocational educational tracks are well-established systems designed to attract and support young people by offering training programs that build confidence, maturity, and both technical and practical skills. These programs incorporate hands-on tasks and learning activities specifically tailored to foster human development, employment, and productivity.

E. Smeplass (✉)
Department of Teacher Education, Norwegian University of Science and Technology, Trondheim, Norway
e-mail: eli.smeplass@ntnu.no

C. Zirkle
College of Education & Human Ecology, Ohio State University, Columbus, OH, USA
e-mail: zirkle.6@osu.edu

E. Smeplass and C. Zirkle (eds.), *Opportunity Structures in Vocational Education and Training*, Palgrave Studies in Technical and Vocational Education and Training,
https://doi.org/10.1007/978-3-032-04410-5_1

Additionally, many systems extend opportunities for training or retraining to adults, preparing them for new, engaging, and value-creating roles.

Researchers in this field often work within the frameworks of their national education systems, engaging with context-specific definitions, challenges, and policy concerns. Their work is typically embedded in national research communities and communicated in their native languages, with a focus on addressing local inefficiencies and reform needs. Because VET is deeply intertwined with the historical development, design, and governance of each country's broader education system, much of the research remains domestically oriented. While this contributes significantly to national knowledge and policy development, it can also mean that successful practices and institutional strengths are not always communicated to the international research community. Consequently, cross-national discussions may overlook these positive aspects, limiting opportunities for shared learning and collaboration.

Several attempts have been made to define and understand developments in the field of VET[1]—particularly through comparative approaches that develop new analytical tools for capturing its complexities and variations across different contexts (Cedefop, 2017). These approaches seek to identify patterns and trends that transcend national differences, shedding light on how structural conditions, policy frameworks, and institutional arrangements shape opportunities for individuals and economies. Strengthening the international research community is essential for advancing knowledge, fostering collaboration, and broadening

[1] In various contexts, the terminology for Vocational Education and Training (VET) varies, even when referring to similar systems. While terms like Technical Vocational Education and Training (TVET) and Career and Technical Education (CTE) are widely used in English-speaking contexts, most non-English-speaking countries have their own unique terms that reflect local traditions and systems. For example, in Norway, the term yrkesfag is used, which roughly translates to "professional subjects." However, yrkesfag has a distinct meaning in the Norwegian context, referring specifically to educational tracks that lead to a journeyman's or craftsman's certificate. Such specificity is not unique to Norway; many countries have their own terms rooted in their cultural and linguistic traditions. These terms often encapsulate nuances unique to the country's vocational education system, making direct translation or comparison challenging. In this book, the editors have chosen to use the term vocational education and training, or VET, as it is a broad and neutral term commonly recognized in international discourse. However, chapter authors were encouraged to use the terminology relevant to their context, provided they explain its meaning and implications within their study. Recognizing these linguistic and contextual nuances is crucial for understanding both the shared and unique aspects of VET systems across the globe.

perspectives on vocational learning. Comparative approaches play a crucial role in identifying overarching patterns and facilitating cross-national learning about the diverse ways in which VET contributes to economic and social development.

However, while these insights are valuable, they should not be the sole framework for understanding the field. Focusing solely on transnational trends can risk overlooking the unique strengths embedded in national and regional contexts. Many well-functioning systems already provide structured pathways that support learners and contribute to economic and social stability, yet these strengths are often taken for granted and under-communicated in academic discourse. A prevailing problem-oriented perspective—where scholars are trained to critically assess inefficiencies, disparities, and reform needs—can inadvertently obscure the existing benefits and achievements of VET. This imbalance may reinforce the notion that vocational education is a field in perpetual crisis rather than one that consistently provides viable opportunities for individuals and societies.

To foster a more balanced perspective, there is a need for a conceptual tool that facilitates both comparative analysis and recognition of national strengths—a tool that captures the diverse ways vocational learning functions while emphasizing its potential. In this book, we propose the concept of *opportunity structures* as a framework for investigating how education and training systems create pathways for individuals, linking them to meaningful work and broader social participation. By focusing on the mechanisms that enable access, progression, and skill development, the concept of opportunity structures offers a means to articulate the strengths of different systems without reducing them to mere challenges. This perspective shifts attention toward how vocational education actively contributes to personal and economic growth.

Many of the strengths of VET are well-known and inherently positive—even if they are sometimes taken for granted. Modern individuals participate in complex economic systems, where specialized jobs allow us to exchange time and effort for goods and services we cannot produce on our own. This specialization enables people to engage in professions that are integral to society, creating a dynamic network in which various roles support and depend on one another—much like the components of a living organism. The benefits of this system are manifold: we gain access to resources and services through systematic cooperation, and the interdependence of specialized roles fosters a resilient and interconnected society.

As theorists have noted—from Tönnies' distinction between *Gemeinschaft* and *Gesellschaft* to contemporary discussions on isomorphism in education—these patterns of specialization and exchange are foundational to modern social and economic life.

Vocational education and training reflect this dual nature: they are shaped by highly context-specific factors while also sharing notable commonalities. These include practical training components, professions that are central to everyday life and work, robust professional communities, and the very opportunity structures that empower students and lifelong learners. When we meet with colleagues and discuss the importance of vocational education, there is a broad consensus that these pathways are vital and beneficial. Yet, much of the research tends to focus predominantly on challenges and systemic deficiencies. Why is this the case?

In parallel, several scholars have explored potential future directions for VET, considering its evolving role in response to economic, technological, and societal changes. McGrath et al. (2019) discussed VET's function within the learning economy and its contribution to sustainable development, highlighting the need for systems that are adaptable to shifting labor market demands. Moodie (2008) examines the relationship between VET and higher education, suggesting a growing convergence that may lead to more integrated forms of tertiary education. Similarly, Wheelahan (2012) and Young (2008) debated the nature of vocational knowledge, questioning whether VET curricula should prioritize occupation-specific competencies or broader educational capabilities that prepare learners for a dynamic and uncertain future. These discussions contribute to ongoing debates about the direction of VET and provide insights into how systems may evolve to balance workforce preparation with broader educational objectives.

Vocational education and training play a vital role in educational systems worldwide (Clarke & Winch, 2007; Maclean et al., 2009). Yet, it is often viewed through a lens of challenges and problems (Clarke et al., 2021; Mutohhari et al., 2021). Fewer publications acknowledge the obvious importance and success of many of these educational pathways (Mirabel et al., 2022; Pavlova, 2008). While highlighting problems is essential for improvement, it is equally crucial to explore the positive aspects and potential of VET—namely, to provide people with skills and certifications that preserve craftsmanship, enable social mobility, foster labor market inclusion, offer alternatives to academic tracks, and provide lifelong training and development opportunities (Mouzakitis, 2010; Palmer, 2006). A key to VET systems' success lies in aligning the interests of multiple actors, as collaboration is intrinsic to vocational education (Arinaitwe, 2021; McGrath, 2012). Our book project seeks to challenge problematizing narratives by examining VET as a catalyst for possibilities and opportunities across the globe. We seek to spotlight the importance of VET in diverse contexts, documenting its significance for learners, organizations, and other stakeholders.

VET is grounded in multiple disciplines and theoretical perspectives. While a definitive global theory of VET does not exist, its goals and objectives are informed by frameworks from economics, sociology, education, and philosophy—as well as by ideas associated with adult learning, career development, human capital, inclusion, and collaboration. VET is further shaped by external forces such as labor market demands, political influences, and environmental and sustainability concerns. Another focus of this book is also to question the persistent emphasis on VET challenges while also highlighting its deep connections to society. We recognize the positive impact of VET in nurturing highly skilled individuals with promising careers (Goldin & Katz, 2009; Hall & Mirvis, 1995). By adopting an exploratory approach, this project encourages critical thinking about the positive facets of VET across various settings worldwide, documenting its potential, core functions, innovative practices, and inspiring cases.

Scholars in the field recognize that VET offers diverse learning pathways tailored to individuals with varied backgrounds and aspirations. From traditional classroom-based training to online courses, virtual/augmented reality learning, apprenticeships, and work-based training, VET provides multiple entry points and progression routes. Through close collaboration with stakeholders, VET programs ensure relevance and responsiveness to contemporary needs. Therefore, our goal is to communicate VET's potential to empower learners, drive economic development, and create opportunities for youth and adults alike.

Recognizing Opportunity Structures in VET

At the heart of this book is the concept of *opportunity structures*, which we define as the organizational frameworks within education, training, and the workforce that open pathways, fostering accessibility and progression for numerous individuals within society. At its core, these structures are composed of policies, practices, and collaborative networks that together create a positive ecosystem for learners to navigate their educational and career trajectories. In VET, these structures facilitate the development of professionalized educational and training routes that not only lead to formal certifications but also open new career and employment opportunities for entrants.

Recognizing the need for a broader international dialogue on VET, we launched this project with a call for contributions that would showcase both the diversity and commonalities of opportunity structures in different national contexts. We sought to bring together scholars from a range of educational and economic systems to explore how vocational learning functions as a pathway to social mobility, employment, and lifelong learning. The positive response to this initiative reinforced the relevance of opportunity structures as a unifying concept for understanding VET's role in different societies.

The expansion of mass education over the past two centuries has transformed the ways in which knowledge and skills are transferred through structured educational pathways (Boli et al., 1985). While general and higher education often follow standardized models, VET remains deeply shaped by historical, social, and economic factors unique to each country (Gessler et al., 2021). Understanding how *opportunity structures* operate within these diverse systems requires perspectives that acknowledge their national and regional characteristics while identifying broader trends that transcend borders.

To build this perspective, we invited researchers to examine the policies, practices, and institutional frameworks that shape vocational learning and career trajectories. Their contributions reflect a wide array of systems, from well-established apprenticeship traditions to newer models of vocational education responding to economic and technological change. This book brings together scholars from Austria, Australia, Costa Rica, Denmark, England, Finland, Germany, South Korea, Norway, Spain, the United States, and Zimbabwe to explore how opportunity structures function in different educational and labor market settings.

The evolution of VET in these contexts has followed distinct trajectories. In Europe, vocational education has roots in medieval guilds that regulated training and professional standards (Wollschlager & Reuter-Kumpmann, 2004). In the United States, vocational education was formalized through federal policy in 1917 (Gordon & Schultz, 2020), expanding into workforce retraining and career and technical education

(Zirkle & Martin, 2012). In Asia, the development of VET has been diverse, and in many nations emphasizing training institutions and on-the-job training rather than formal vocational schools, with employers playing a crucial role in skill development (Tilak, 2003). Meanwhile, on the African continent, the growth of VET has been significantly influenced by the interplay of economic development and colonial legacies, facing challenges such as fragmentation, inequality, and underfunding, yet it remains a pivotal strategy for enhancing youth employment (McGrath et al., 2020; African Union, 2018). Across these contexts, *opportunity structures* can contribute to highlighting how individuals access, navigate, and benefit from vocational education.

Sociological research provides valuable frameworks for understanding these structures. *Opportunity structures*, as explored by Cloward and Ohlin (1960), emphasized how economic and political conditions shape individuals' access to education and career pathways. Similarly, Bæck (2019) pointed out how deliberations at multiple levels influence resource allocation in education and training. Applying these perspectives to VET allows us to examine not only how vocational education adapts to labor market demands but also how it actively contributes to social mobility, workforce development, and economic resilience.

The chapters in this book highlight a diversity of opportunity structures in VET, but they also invite deeper reflection on the forces that shape these structures and their implications for learners, institutions, and societies. Underlying each contribution is a set of broader, more fundamental questions: How do we define opportunity in vocational education? What determines whether an educational system truly offers meaningful pathways for social mobility, inclusion, and professional development? And how do different nations balance the tension between tailoring VET to their unique social, economic, and historical contexts while also learning from one another? Perhaps one of the most pressing questions we must ask is: How do we ensure that the strengths of different VET systems are recognized and shared beyond their national borders? Despite growing international collaboration, vocational education research remains heavily shaped by local policy concerns and institutional frameworks, often limiting the exchange of best practices. How can scholars and practitioners work toward a more open and engaged international research community—one that does not merely compare systems but actively contributes to their mutual enrichment?

Taken together, these chapters serve as a reminder that vocational education should not be defined solely by its challenges but also by its essential contributions to individuals and society. Too often, research on VET remains confined within national boundaries, engaging primarily with local policy debates while missing opportunities to participate in broader international discussions on effective practices and successful models. In the first section, authors examine their respective VET systems within their national or regional contexts, demonstrating how these structures operate and why their cases hold relevance for a global audience. Through their selected topics, they illustrate how opportunity structures function in practice, offering insights into policies, institutional frameworks, and pathways that enable access, progression, and success. The second section shifts toward a broader theoretical lens, presenting conceptual frameworks that provide a foundation for cross-national perspectives on the future of vocational education. By expanding this discourse, we emphasize that VET is not merely a response to labor market needs but a vital educational pathway—one that fosters individual empowerment, social mobility, and economic sustainability. Recognizing and sharing these insights across borders strengthens our collective understanding of vocational education's evolving role in shaping both careers and societies.

Moving forward, improving how we share knowledge on *opportunity structures* in the international research community is essential. Despite growing collaboration, VET research is still shaped by national institutions, funding structures, and policy agendas that often limit cross-national exchange. This book is a step toward bridging these gaps, encouraging scholars to articulate what works in their systems and contribute to a more balanced and constructive global discussion. By strengthening the international research community, we can deepen our understanding of vocational education as an evolving field that plays a crucial role in shaping the future of work and learning worldwide.

Chapters in the Book

This book is divided into two parts, each offering a distinct yet complementary perspective on *opportunity structures* in VET.

Part I

The first section in the book, *Contextualizing Opportunity Structures*, brings together a series of empirical case studies that explore how various national and regional systems organize access to vocational education and enable career progression. Each chapter examines a particular topic of local significance, shedding light on the ways opportunity structures emerge, evolve, and address challenges within their specific contexts. By presenting these national and state-level experiences, authors offer valuable insights into the diverse ways through which vocational education systems can foster skills development, expand employment prospects, and contribute to social mobility.

Chapter 2 by Daniel Láscarez Smith examines the implementation of dual apprenticeships in Costa Rica as a response to youth unemployment, critically assessing how opportunity structures emerge through policy transfer. While dual training has been successful in Europe, its adaptation in Costa Rica has faced challenges, including weak employer engagement, labor market mismatches, and political resistance. By tracing historical developments and labor market data, Láscarez Smith reveals the gap between policy ambitions and actual employment outcomes. This case raises important questions about how vocational education reforms can genuinely expand opportunity structures and contribute to social mobility in diverse economic contexts.

Chapter 3 by Rønnaug H. Lyckander, Asgeir Skålholt, and Elisabeth Hovdhaugen examines shifts in Norway's VET system to improve access for adult learners. The chapter traces how reforms since the 1994 restructuring of upper secondary education initially limited opportunities for adults but have, over the last decade, expanded rights and flexible pathways. By analyzing policy changes and participation trends, the authors highlight how opportunity structures in adult VET have been reshaped through modular programs, workplace-based training, and targeted policies for immigrants. They show how these reforms reflect a broader shift toward lifelong learning but also raise questions about structural barriers, institutional coordination, and the integration of underrepresented groups. The Norwegian example of adult students in VET offers insights into how VET can evolve to better support adults in changing labor markets while balancing flexibility with quality and equity.

Chapter 4 by Chris Zirkle traces over a century of vocational education development in the U.S. state of Ohio, highlighting how a well-structured and articulated career and technical education (CTE) system has expanded opportunity structures for learners of all ages. The Ohio model provides multiple entry and exit points, beginning in middle school, continuing through high schools, career centers, and community colleges, and extending into adult technical training and apprenticeships. The chapter illustrates how Ohio's CTE system is deeply connected to the state's evolving labor market, particularly in agriculture, manufacturing, and healthcare. By fostering strong institutional cooperation and clear pathways between education levels and employment, Ohio has created a system that balances workforce preparation with postsecondary opportunities. This case underscores how vocational education can be structured to offer lifelong learning opportunities while addressing economic and social needs.

Seong Ji Jeong's Chap. 5 explores South Korea's Meister High School (MHS) initiative as a targeted response to the long-standing challenges of vocational education in a credential-oriented society. Designed to directly link secondary education with industry needs, MHS schools operate with extensive autonomy, integrating flexible curricula, employer partnerships, and state support to enhance the prestige and effectiveness of vocational education. The chapter highlights how the program not only boosts graduate employability but also establishes a structured pathway for lifelong learning through the *Work First, Study Later* approach. However, sustaining this model requires greater industry participation, improved job quality, and strategies to retain skilled workers in regional labor markets. The South Korean experience offers valuable insights into how state-led vocational education reforms can create meaningful career pathways while maintaining adaptability in an evolving economic landscape.

Antti Seitamaa and Petteri Hansen critically examine the evolving role of VET in Finland in Chap. 6, through the opportunity structures lens, revealing an ongoing discursive struggle between educational equality and economic utility. The authors trace how Finnish VET, historically positioned as a mechanism for both social mobility and national development, has increasingly been shaped by European integration and neoliberal labor market priorities. While the system has maintained high participation rates and formal pathways to higher education, the expansion of work-based learning and competency-based assessment reflects a shift away from its earlier egalitarian ethos. The chapter argues that Finnish VET is now primarily structured as an adaptive mechanism for labor market responsiveness, particularly for marginalized groups, raising questions about the extent to which it still serves as a vehicle for democratic citizenship and lifelong learning. By interrogating how policymakers and experts navigate these tensions, this chapter offers insights into the political and ideological forces that shape vocational education reforms in an era of economic precarity and Europeanization.

In Chap. 7, Katrin J. Posch, Junmin Li, and Matthias Pilz examine how German VET institutions can function as opportunity structures for refugees, focusing on school principals' acceptance of quality indicators. While barriers such as resource limitations, funding constraints and policy misalignment persist, the chapter highlights how VET can facilitate refugee integration through language learning, support systems, and employer engagement. School principals view some tailored measures—such as specialized teacher training and cultural education—as essential, while others are seen as beyond their control. The chapter argues that VET's potential for social mobility depends on resources, policy coordination, and balancing educational and employment goals. Through this example, the authors demonstrate how VET can foster inclusion while exposing the limits of school-level decision-making within existing governance structures.

Chapter 8 by Filiz Oskay and Jay Plasman explores the evolution of VET in the United States, analyzing its historical development, policy shifts, and role in expanding opportunity structures. The authors examine how VET, once stigmatized as a secondary track for less academically inclined students, has transformed into an integrated pathway that bridges academic and technical education. Federal policies, particularly the Perkins legislation, have played a key role in aligning VET with broader educational and labor market goals, emphasizing career readiness and social mobility. The chapter highlights the benefits of VET participation, including improved school engagement, postsecondary transitions, and labor market outcomes, particularly for marginalized groups. The authors suggest that strengthening VET requires ongoing efforts to enhance program accessibility, align training with labor market needs, and foster lifelong learning.

Chapter 9 by Davinia Palomares-Montero explores Spain's Second Chance Schools (SCS) as an opportunity structure for young people who have left education without qualifications. These schools provide vocational training, academic support, and social integration measures to help students re-engage with education and improve their employability. While Spain has made progress in reducing early school leaving, dropout rates remain among the highest in Europe, particularly affecting disadvantaged youth. SCS aim to address these challenges by offering flexible training programs, work-based learning opportunities, and individualized support tailored to students' needs. However, access to formal qualifications varies by region, and funding inconsistencies create barriers to long-term sustainability. The chapter highlights how SCS function as opportunity structures by expanding educational and employment pathways, particularly for marginalized youth. At the same time, it underscores ongoing challenges related to program retention, gender disparities in training opportunities, and the need for stronger policy recognition. Through this case, the chapter illustrates the potential of alternative education programs to foster inclusion and social mobility within Spain's VET system.

Chapter 10 by Davison M. Mupinga and John Chakamba examines the benefits, challenges, and opportunities of VET in Zimbabwe, particularly in high schools. VET programs play a crucial role in equipping students with industry-relevant skills, increasing employability, and fostering entrepreneurship, especially amid the country's economic challenges. However, these programs face significant obstacles, including inadequate funding, outdated equipment, a shortage of qualified instructors, and persistent societal stigma. Despite these challenges, VET remains a vital opportunity structure for Zimbabwean youth, particularly those from disadvantaged backgrounds. The chapter explores strategies to improve VET's effectiveness, such as strengthening education-industry partnerships, expanding work-based learning, and increasing policy support to enhance sustainability and relevance. Ultimately, the discussion underscores VET's potential to bridge the skills gap, drive economic development, and provide employment pathways in Zimbabwe's evolving labor market.

Part II

The second section, *Conceptualizing VET Success Between Individual Needs and Societal Development*, shifts the focus from single-case analyses to broader, cross-national and theoretical considerations. The chapters in this section examine some of the overarching forces shaping vocational education—how it addresses societal demands, fosters innovation, and supports lifelong learning. By examining how VET systems balance individual development with wider economic and social goals, chapters in this collection illuminate the role and functions of vocational education in building more equitable, adaptable, and sustainable societies.

Chapter 11 by Stephen Billett offers a thoughtful reflection on what VET should aim to achieve in today's societies. Rather than treating VET as merely a means to fill labor market shortages, Billett argues that it should be seen as a system that enables individuals to find meaningful work, adapt to shifting occupational demands, and contribute to workplace innovation. A key concern in the chapter is the low status of VET, which deters young people from seeing it as a viable or desirable option. Billett emphasizes the importance of actively engaging students and their families early on, ensuring they understand the opportunities VET can provide. The chapter also considers the broader function of VET in helping individuals identify careers that align with their interests and skills, rather than being pushed into training programs without a sense of purpose. Beyond initial education, Billett makes a strong case for vocational learning as a lifelong process. He highlights how continuing education and training (CET) is crucial for workers navigating new technologies, changing job roles, and the increasing expectation that employees contribute to workplace innovation. Rather than treating innovation as something driven solely by management, the chapter illustrates how VET can prepare workers to shape their own work environments, making them active participants in economic and technological change. This chapter invites readers to reconsider VET as more than just a supply-side response to labor market needs. Instead, Billett presents it as a system that fosters agency, inclusion, and long-term employability. Readers interested in policy development, education reform, and workforce planning will find valuable insights here, as the chapter challenges conventional thinking about the role of vocational education in modern societies.

In Chap. 12, Eli Smeplass examine Norway's VET system as a strategic *investment in youth*, balancing education, labor market needs, and social inclusion. Rooted in Norway's welfare state, VET provides structured pathways that support career mobility, ensuring that students gain both theoretical knowledge and practical skills through a flexible system that allows transitions between vocational and academic tracks. A key feature of the Norwegian model is the tripartite collaboration between the state, employers, and unions, which keeps VET relevant while securing student rights, training salaries, and employer subsidies. This integration reflects Norway's broader social and economic policies, particularly its 'work-first principle', which promotes high labor market participation while maintaining strong social protections. The chapter highlights how Norway's VET system reduces the risks associated with educational choices by offering financial support, legal guarantees, and a structured apprenticeship system. It provides insights into how vocational education can function as both an economic strategy and a tool for social mobility, making it an essential read for those interested in education policy and workforce development.

In Chap. 13, Stephen Billett challenges the prevailing top-down approach to VET, arguing instead for a model that prioritizes local capacities and engagement. While national agencies and industry bodies set overarching goals, the chapter contends that the real potential of VET can only be realized through localized responses that address specific community needs, economic contexts, and opportunities for innovation. Billett builds his case on a three-decade-long research program, demonstrating that local social, educational, and administrative infrastructure is essential for VET to be effective. He emphasizes that communities, businesses, and educators at the local level—not just national policymakers—should shape VET systems to ensure they are relevant, adaptable, and capable of fostering both employment and social development. This perspective aligns with the broader theme of "opportunity structures" in the volume, highlighting how education and training must be embedded within the unique social and economic realities of different regions. By questioning the dominance of centralized mandates and advocating for a more participatory, community-driven approach, this chapter offers an important critique of conventional VET policy. The chapter describes how vocational education can be more effectively designed to meet both individual aspirations and local labor market demands, offering insights into how systems can foster sustainable employment and regional economic resilience.

In Chap. 14, Junmin Li, Ekaterina Schlüter, Hannes Hautz, Julián Bell Sebastián, and Fernando Marhuenda-Fluixá examine how quality management (QM) systems in VET reflect different governance structures and the implications for opportunity structures. Comparing Austria, Germany, and Spain, the chapter explores how centralized, federalized, and market-influenced governance regimes shape QM practices and, in turn, affect pathways for learners. By linking QM approaches to opportunity structures, the chapter highlights a critical tension: efforts to ensure quality through standardized metrics and external evaluations may inadvertently limit local flexibility and constrain school-level discretion. This is particularly evident in non-formal VET, especially in Spain, where QM often functions more as a marketing tool than as a means of enhancing educational quality. The authors argue that while QM can help maintain transparency and comparability, it does not necessarily addresses the conditions required to expand meaningful pathways for learners. Instead, it tends to stabilize existing governance arrangements and may risk reproducing rather than transforming opportunity structures. Through this comparative analysis, the authors provide insights into how governance arrangements influence the design and impact of QM systems in VET. This chapter is concerned with how quality assurance can be aligned more closely with expanding access, equity, and meaningful career opportunities in diverse educational contexts.

In the last contribution of this book, Chapter 15, Christian Helms Jørgensen examines the central role of vocations as an opportunity structure in VET and working life. By comparing vocations with professions, Jørgensen highlights how vocations uniquely combine practical skills, occupational cultures, and institutional frameworks to connect education and work. These connections enable VET to serve as a bridge for young people transitioning from school to employment while offering a cohesive identity and meaning that many learners find motivating. The chapter contextualizes vocations within three different VET regimes—market-led, state-led, and collective skill formation—and argues that the collective regime best supports strong vocational identities, industry-wide standards, and broad skill portability. In this model, vocations play a central role in aligning the needs of the labor market with educational curricula, while also facilitating ongoing negotiation among stakeholders to maintain relevance and quality. Jørgensen reflects on how the institutional and personal dimensions of vocations make them resilient to changing labor market demands and evolving educational policies. However, he also acknowledges tensions within vocational systems—such as balancing broad, transferable skills with specialized training, and addressing structural barriers like gender inequalities and the academic-vocational divide. These challenges underscore the complexity of maintaining vocations as viable opportunity structures in the face of rapid societal and economic changes. Ultimately, Jørgensen's chapter reaffirms the enduring value of vocations in providing meaningful pathways into work, fostering social mobility, and promoting economic and social cohesion. By examining the historical, institutional, and personal aspects of vocations, Jørgensen invites readers to consider their continued relevance and potential as a key component of modern VET systems.

In the concluding chapter, *Looking Backwards and Forwards: Concluding Remarks* (Chap. 16), we synthesize key insights from the volume and reflect on the role of VET within global opportunity structures. The chapter revisits central themes, emphasizing how VET serves as a platform for lifelong learning, social mobility, and economic development. It underscores the importance of balancing critical analysis with recognizing VET's strengths, advocating for a research agenda that addresses systemic challenges *and* highlights successful practices. Looking ahead, we discuss how VET systems must continue adapting to technological change, labor market transformations, and evolving policy frameworks. The chapter concludes with a call for increased international collaboration and theoretical development, emphasizing the need for shared learning and deeper engagement across diverse VET contexts.

How to Engage with the Collection

Each chapter in this book offers a perspective on how VET functions within a broader system. By reading each chapter on its own terms, readers can gain insight into the different ways VET supports learning and career development around the world. The diversity of these accounts reflects not only the specific contexts from which they emerge but also the shared goal of improving VET systems to meet the evolving needs of society.

A central theme of this collection is the concept of opportunity structures—organizational systems, policies, and networks that open pathways to personal and professional growth. The contributing authors, each deeply familiar with their respective systems, demonstrate how these structures foster skills development, connect learners to industry, and prepare individuals for meaningful careers. The chapters highlight VET's core strengths: hands-on training, close industry partnerships, and a focus on creating practical and lasting opportunities. By examining the ways opportunity structures function in practice, readers are invited to consider how VET contributes to individual empowerment but also to community and innovation. The collection encourages us to envision a future in which VET plays an increasingly vital role. Whether through exploring specific case studies or reflecting on the broader theoretical discussions, the collection challenges conventional thinking and offers new ways to conceptualize and strengthen vocational education.

We encourage readers to take a moment to think about how the insights presented here might inform their own research or professional practice. Scholars can consider how to connect more closely with other scholars and practitioners in the field to build a shared understanding of VET's potential. The book demonstrates that while VET systems vary widely, some fundamental qualities are consistent and ready for further exploration. In short, this collection is more than a set of academic contributions. It is an invitation to rethink how we approach VET and recognize its profound importance in shaping education and work. The chapters contained here offer a starting point for new ideas, fresh collaborations, and a renewed focus on the positive aspects of vocational education. We hope you find inspiration within these pages and that they help catalyze new inquiry, innovation, and advocacy in VET research.

References

African Union. (2018, October 22). *Continental strategy for technical and vocational educational and training (TVET) to foster youth employment*. Retrieved from https://au.int/en/documents/20181022/continental-strategy-technical-and-vocational-educational-and-training-tvet

Arinaitwe, D. (2021). Practices and strategies for enhancing learning through collaboration between vocational teacher training institutions and workplaces. *Empirical Research in Vocational Education and Training, 13*, 1–22.

Bæck, U. D. K. (2019). Spatial maneuvering in education. Educational experiences and local opportunity structures among rural youth in Norway. *Nordic Journal of Comparative and International Education, 3*(3), 61–74. https://doi.org/10.7577/njcie.3274

Boli, J., Ramirez, F. O., & Meyer, J. W. (1985). Explaining the origins and expansion of mass education. *Comparative Education Review, 29*(2), 145–170.

Cedefop. (2017). *The changing nature and role of vocational education and training in Europe* (Conceptions of vocational education and training: An analytical framework) (Vol. 1). Cedefop. https://doi.org/10.2801/532605

Clarke, L., & Winch, C. (Eds.). (2007). *Vocational education: International approaches, developments and systems*. Routledge.

Clarke, L., Westerhuis, A., & Winch, C. (2021). Comparative VET European research since the 1980s: Accommodating changes in VET systems and labour markets. *Journal of Vocational Education & Training, 73*(2), 295–315.

Cloward, R. A., & Ohlin, L. E. (1960). *Delinquency and opportunity: A study of delinquent gangs*. Routledge.

Gessler, M., Bohlinger, S., & Zlatkin-Troitschanskaia, O. (2021). International vocational education and training research: An introduction to the special issue. *International Journal for Research in Vocational Education and Training, 8*(4), 1–15.

Goldin, C., & Katz, L. F. (2009). *The race between education and technology*. Harvard University Press.

Gordon, H. R., & Schultz, D. (2020). *The history and growth of career and technical education in America*. Waveland Press.

Hall, D. T., & Mirvis, P. H. (1995). The new career contract: Developing the whole person at midlife and beyond. *Journal of Vocational Behavior, 47*(3), 269–289.

Maclean, R., Wilson, D., & Chinien, C. A. (2009). *International handbook of education for the changing world of work: Bridging academic and vocational learning* (Vol. 1). Springer Netherlands.

McGrath, S. (2012). Vocational education and training for development: A policy in need of a theory? *International Journal of Educational Development, 32*(5), 623–631. https://doi.org/10.1016/j.ijedudev.2011.12.001

McGrath, S., Mulder, M., Papier, J., & Suart, R. (2019). *Handbook of vocational education and training: Developments in the changing world of work*. Springer.

McGrath, S., Ramsarup, P., Zeelen, J., Wedekind, V., Allais, S., Lotz-Sisitka, H., Monk, D., Openjuru, G., & Russon, J. A. (2020). Vocational education and training for African development: A literature review. *Journal of Vocational Education & Training, 72*(4), 465–487. https://doi.org/10.1080/13636820.2019.167996

Mirabel, K. T., Sewagegn, A. A., & Diale, B. M. (2022). Technical vocational education and Training (TVET) to empower secondary school learners from low socioeconomic backgrounds. *Journal of Technical Education and Training, 14*(1), 138–149.

Moodie, G. (2008). *From vocational to higher education: An international perspective*. McGraw-Hill/Open University Press.

Mouzakitis, G. S. (2010). The role of vocational education and training curricula in economic development. *Procedia—Social and Behavioral Sciences, 2*(2), 3914–3920.

Mutohhari, F., Sutiman, S., Nurtanto, M., Kholifah, N., & Samsudin, A. (2021). Difficulties in implementing 21st century skills competence in vocational education learning. *International Journal of Evaluation and Research in Education, 10*(4), 1229–1236.

Palmer, R. (2006). Beyond the basics: Balancing education and training systems in developing countries. *Journal for Education in International Development, 2*(1), 1–21.

Pavlova, M. (2008). *Technology and vocational education for sustainable development: Empowering individuals for the future* (Vol. 10). Springer Science & Business Media.

Tilak, J. B. (2003). Vocational education and training in Asia. In *International handbook of educational research in the Asia-Pacific region: Part one* (pp. 673–686). Springer.

Wheelahan, L. (2012). *Why knowledge matters in curriculum: A social realist argument*. Routledge.

Wollschlager, N., & Reuter-Kumpmann, H. (2004). From divergence to convergence: A history of vocational education and training in Europe. *European Journal of Vocational Training, 32*, 6–17.

Young, M. (2008). *Bringing knowledge back in: From social constructivism to social realism in the sociology of education*. Routledge.

Zirkle, C., & Martin, L. (2012). Challenges and opportunities for technical and vocational education and training (TVET) in the United States. In M. Pilz (Ed.), *The future of vocational education and training in a changing world* (pp. 9–24). Springer.

PART I

Contextualizing Opportunity Structures

CHAPTER 2

Challenges and Contradictions Between the Reduction of Youth Unemployment and the Implementation of Dual Apprenticeship in Costa Rica

Daniel Láscarez Smith

Introduction

The analysis of the dual apprenticeship[1] system in Costa Rica represents an exceptional opportunity to explore the complex dynamics of educational policy transfer from the Global North to the Global South. This process not only involves the adoption of a successful model in European contexts, such as Germany, but also raises fundamental questions about

[1] It is a work-based learning system in which participants learn, in a systematic and controlled manner, a set of skills and knowledge that will enable them to strengthen their career paths and ensure the provision of a well-qualified labor force.

D. L. Smith (✉)
TVET Research Department, National Technical University of Costa Rica, Alajuela, Costa Rica
e-mail: dlascarez@utn.ac.cr; dlascarez04@gmail.com

E. Smeplass and C. Zirkle (eds.), *Opportunity Structures in Vocational Education and Training*, Palgrave Studies in Technical and Vocational Education and Training,
https://doi.org/10.1007/978-3-032-04410-5_2

the capacity of these systems to adjust to profoundly different realities. In sociological terms, the study of this transfer reveals structural and cultural tensions that manifest themselves in the interaction between foreign institutional models and local configurations, challenging universal notions of efficiency and adaptability.

The specificity of the Costa Rican context adds another analytical dimension to this phenomenon. Costa Rica, with its economic structure in transition, its marked legacy of neoliberal policies and its position as a middle-income country in Latin America, presents a scenario where unique labor, cultural, and structural characteristics converge. The implementation of the dual system in this environment not only allows us to assess the material conditions that facilitate or hinder its adoption, but also to understand how national development narratives influence the reconfiguration of its objectives and strategies. In this sense, Costa Rica acts as a mirror of the global challenges surrounding the construction of education and training systems that respond to both the demands of capital and the needs of social inclusion.

Finally, the Costa Rican case highlights the possibility of building opportunity structures that have a significant impact on social and labor inclusion, provided that the successes and failures of these transfers are evaluated in advance. From a critical perspective, the dual system has the potential to operate as a social mobility device, especially for vulnerable young populations. However, it also faces the risk of reproducing structural inequalities if its implementation is not accompanied by a thorough understanding of the socio-economic context in which it is inserted. The analysis of this experience, therefore, is fundamental to identify the conditions under which this model can be an engine of inclusion, rather than a perpetuator of exclusion.

Historical Construction of Dual Apprenticeships: A Brief Overview

The implementation of the dual apprenticeship system in Costa Rica responds to a complex interaction between specific economic and political transformations and the need to articulate organizational frameworks that generate structures of opportunity for young people in a context marked by challenges of social equity and adaptation to the global labor market, especially for the reduction of youth unemployment. These structures, understood as frameworks that integrate policies, practices and collaboration networks, are fundamental for configuring educational itineraries and for generating new possibilities for employment and social mobility.

During the 1970s and 1980s, within the framework of the structural adjustment programs derived from the fiscal crisis (Raventós, 1995), neoliberal elites promoted the adoption of dual apprenticeship as part of an economic-educational strategy aimed at providing the business sector with a more qualified labor force and as a labor policy. This model was part of a structural change that sought to diversify the Costa Rican economy, improve competitiveness in globalized markets and respond to the transformations in the labor market resulting from the transition from the import substitution industrialization model to a neoliberal paradigm (Díaz Arias, 2019).

Since the 1990s, the opening of markets and the attraction of foreign direct investment consolidated this transition and positioned dual apprenticeships as a strategic tool to reconfigure the Technical and Vocational Education and Training (TVET) system. This perception was reinforced after the global financial crisis of 2008, when dual apprenticeship became more relevant as a policy response to youth unemployment, consolidating itself as a key strategy in the institutionalization of TVET in Costa Rica. However, limitations in the articulation between educational, business and social actors hindered the creation of a fully inclusive ecosystem, adapted to the specificities of the Costa Rican context.

This chapter analyzes, from a historical perspective, the impact of dual apprenticeship on the reduction of youth unemployment in Costa Rica, problematizing the tensions between the promises of labor inclusion that underpinned its adoption and the results obtained after decades of implementation. Based on a critical approach, we explore the structural, institutional, and political conditions that have shaped the opportunity structures associated with this model, evaluating its ability to balance the demands of the labor market with the principles of social equity. In this way, it seeks to contribute to the debate on the design and effectiveness of technical training systems in similar contexts.

Method

This study adopts an analytical-critical methodology to examine the relationship between the implementation of dual apprenticeship and its impact on the reduction of youth unemployment in Costa Rica, integrating quantitative and qualitative approaches in a rigorous interpretative framework. The analysis is based on an exhaustive review of academic literature and institutional documents, complemented by an empirical study of official data, which allows us to comprehensively address the historical, structural and discursive dynamics associated with this educational policy.

The quantitative basis of the study is constructed from official data from the National Institute of Statistics and Census (INEC) and other government sources, covering the period 2010–2023. This time range was strategically selected to capture the effects of the consolidation of dual learning policies in the country and to guarantee the validity and comparability of the figures, given the use of unified methodological criteria. These data made it possible to identify historical patterns and trends in youth unemployment, providing a solid empirical context for evaluating the impact of these policies.

From a qualitative approach, the analysis is complemented by a critical review of key documents, including reports of the Dual Apprenticeship Advisory and Promotion Commission, legislative projects such as Law No. 9728, and speeches of political and business actors and international organizations such as the The Organization for Economic Co-operation and Development (OECD).

Overview of Youth Unemployment in Costa Rica[2]

Youth unemployment has become a structural problem in the sense that its solution depends on multiple productive, political, educational, public and private investment factors, which has aggravated the social problem of inequality and poverty. Several authors, such as Jiménez (2020), state that unemployment arises as a result of a combination of structural factors, among which the economic development model adopted by the country stands out.

[2] The concept of youth unemployment refers to the situation in which people aged 15–24 are actively seeking employment, but are unable to find it.

This model has generated adverse effects on the least qualified sectors of the labor force. In this context, one of the main challenges is that policies aimed at economic growth and integration into global capital markets are part of a political agenda that has generated tensions in the fiscal, educational and social spheres. This suggests a misalignment between economic development strategies and educational policies designed to improve employability, reflecting a disconnect between economic priorities and the formation of human capital adapted to the demands of the labor market.

The working population of Costa Rica in 2024 was 2,157,476, of which 1,335,668 are men and 821,808 women. The tertiary sector, i.e., the service sector, is where most people work, both young and old. Costa Rica has experienced a persistent problem of youth unemployment, with rates consistently higher than the national average. Historically the average general unemployment rate for this 15-year period is approximately 10.2%. In contrast, the average youth unemployment rate over the last 15 years is 33.15%.

As shown in Fig. 2.1 between 2010 and 2024, the youth unemployment rate in Costa Rica has followed a fluctuating pattern, shaped by structural challenges and significant economic events. This trend reflects the persistent difficulties young people face in integrating into the labor market and their vulnerability to changes in the economic environment.

During the early years of the period (2010–2014), youth unemployment remained relatively stable, with rates ranging between 20% and 25%. While this stability may appear positive, it highlights a structural issue: the limited access young people have to quality employment opportunities, likely due to gaps between the skills required by the market and those available in the youth population.

From 2015 onward, unemployment rates began to increase slightly, reaching their highest pre-pandemic point in 2019, at nearly 30%. This rise could be attributed to an economic slowdown that negatively impacted job creation, particularly in sectors where young people traditionally find opportunities, such as commerce, tourism, and services.

The impact of the pandemic in 2020 was devastating, with a sharp rise in youth unemployment surpassing 45%. This marked the most critical point of the period analyzed, caused by economic shutdowns and a general contraction of the labor market. Young people, who are traditionally employed in more precarious or informal roles, were particularly affected, facing mass layoffs and a drastic reduction in job offers.

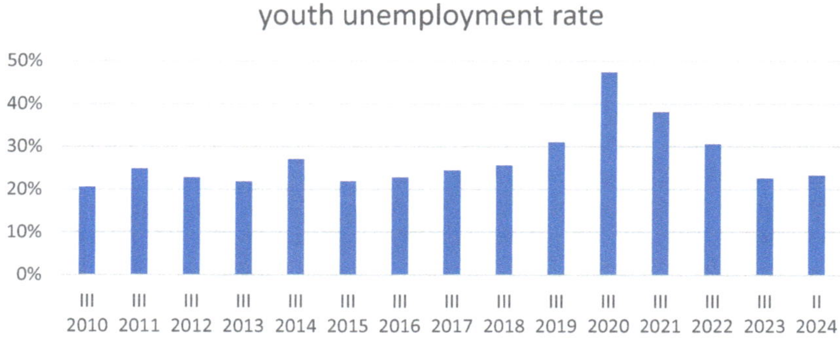

Note: INEC-Costa Rica. Encuesta Continua de Empleo, 2024.

Fig. 2.1 Ratio of youth unemployment to the labor force aged 15–24 years (2010–2024). (**Note**: INEC-Costa Rica. Encuesta Continua de Empleo 2024)

In the post-pandemic years, a gradual recovery is observed. In 2021 and 2022, youth unemployment rates began to decline, albeit unevenly, reflecting partial economic recovery. However, by 2023 and 2024, the rate appears to have stabilized again at around 20%, similar to the levels observed at the beginning of the period. This suggests that structural challenges remain unresolved.

If the historical data on youth unemployment since 2010 is segregated according to economic sector, it can be seen that the tertiary sector is the sector with the highest youth unemployment. This sector is precisely the one with the greatest dynamism and the greatest changes in occupational profiles thanks to the growth of the free trade zone regime.

Table 2.1 highlights the distribution of youth unemployment across three key economic sectors in Costa Rica: primary, secondary, and services. The primary sector, which includes agriculture, livestock farming, fishing, forestry, and mining, has consistently had the lowest share of unemployed youth. This reflects both a declining interest among young workers in these industries and structural shifts, such as increased mechanization and the sector's reliance on seasonal, often precarious employment. Meanwhile, the secondary sector, encompassing manufacturing, construction, and industrial production, has experienced fluctuations in youth unemployment over the years. While some jobs in this sector require specialized technical training, others remain low-skilled and vulnerable to economic cycles that impact industrial activity and infrastructure investment. However, the services sector dominates youth unemployment figures, consistently accounting for the highest percentage of unemployed individuals aged 15–24. This is probably related to the fact that the service sector requires higher levels of qualification and bilingual skills, which young people, who need to work, do not have or have not yet achieved. As Costa Rica's most dynamic and largest employment provider, this sector includes commerce, tourism, education, healthcare, transportation, technology, and finance.

Although the overall participation of men and women in unemployment figures seems balanced, the analysis reveals significant differences in specific years, shaped by structural, situational, and global factors (see Fig. 2.2).

At the beginning of the period, in 2010, the proportion of unemployment between men and women was relatively similar. However, as the decade progressed, important fluctuations emerged. For instance, in 2016, women faced significantly higher unemployment levels than men,

Table 2.1 Youth unemployment by economic sector in Costa Rica (2010–2024): trends in the 15–24 age group

Quarter	*Primary sector (%)*	*Secondary sector (%)*	*Services sector (%)*
IV 2010	14.44	17.55	67.99
IV 2011	10.63	22.08	66.67
IV 2012	11.86	17.73	70.50
IV 2013	11.54	16.94	71.34
IV 2014	14.21	17.87	67.60
IV 2015	12.22	19.07	68.72
IV 2016	11.60	19.43	68.34
IV 2017	13.50	23.99	62.18
IV 2018	13.86	18.17	65.58
IV 2019	13.47	20.29	66.20
IV 2020	18.09	20.46	61.52
IV 2021	13.71	17.78	68.45
IV 2022	10.24	21.33	67.91
IV 2023	9.41	23.33	64.00
III 2024	9.00	20.99	70.10

Note: Instituto Nacional de Estadística y Censos (2024). Encuesta Continua de Empleo: Población joven 15–24 años según indicadores de actividad 2010–2024. Instituto Nacional de Estadística y Censos. https://inec.cr/estadisticas-fuentes/encuestas/encuesta-continua-empleo?page=3

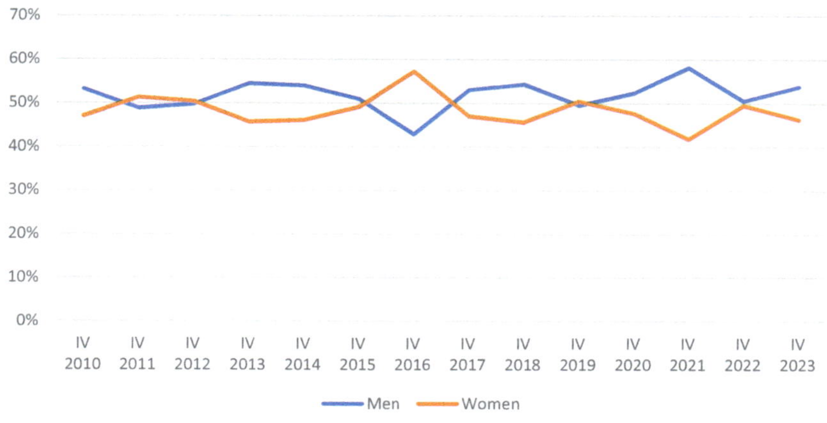

Note: Encuesta Continua de Empleo, INEC, 2024. https://inec.cr/estadisticas-fuentes/encuestas/encuesta-continua-empleo?page=3

Fig. 2.2 Youth unemployment rate by gender in Costa Rica (2010–2024). (**Note**: Encuesta Continua de Empleo, INEC, 2024. https://inec.cr/estadisticas-fuentes/encuestas/encuesta-continua-empleo?page=3)

which might be related to changes in the most affected economic sectors or lower absorption of female labor during that time.

The onset of the pandemic in 2020 marked a turning point in the labor market. Both men and women experienced a sharp rise in unemployment, reflecting economic restrictions and the downturn in key sectors. However, women appeared to recover more quickly in subsequent years, possibly due to their greater presence in sectors such as services, healthcare, and education, which showed resilience or were essential during and after the health crisis.

In the final years of the analyzed period (2023–2024), a general recovery in unemployment levels is evident. The gap between men and women narrows once again, with a slight advantage in male unemployment rates by 2023. This suggests that, while structural challenges persist, the labor market has achieved some post-pandemic stability.

This trajectory highlights how macroeconomic and social events influence men and women differently, depending on their sectoral participation and the dynamics of the labor market. While progress toward greater equality is evident, historical fluctuations emphasize the need for public policies that consider the specific circumstances of each gender, fostering a more equitable and resilient labor market in the face of future challenges.

The analysis of the educational profile of unemployed youth between 2010 and 2024 reveals a significant structural shift in unemployment distribution by educational level. In 2010, more than 50% of unemployed youth were concentrated in the groups with primary education or less and incomplete secondary education. By 2024, this proportion has shifted toward groups with completed secondary education and university qualifications, who now account for over 50% of the unemployed (Table 2.2).

From a statistical perspective, this indicates a redistribution of unemployment incidence toward higher educational levels. Unemployment among youth with completed secondary education has shown a sustained increase, reaching critical levels in recent years. This suggests that this level of education, historically considered a competitive advantage, no longer guarantees employment. This phenomenon points to a structural

Table 2.2 Educational profile of unemployed youth 2010–2024

Quarter	*Primary school completed or less (%)*	*Incomplete secondary (%)*	*Complete secondary (%)*	*University graduate with or without degree (%)*
IV 2010	29.57	38.71	20.79	10.77
IV 2011	20.46	45.69	19.10	14.41
IV 2012	22.61	41.04	19.83	15.61
IV 2013	23.22	46.17	19.18	11.52
IV 2014	23.97	43.84	18.51	13.68
IV 2015	17.30	40.37	26.51	15.65
IV 2016	20.67	39.37	23.85	16.06
IV 2017	22.12	47.76	21.04	9.08
IV 2018	12.14	51.30	18.99	17.58
IV 2019	9.89	47.53	24.59	18.00
IV 2020	7.60	45.95	30.87	15.58
IV 2021	7.55	39.58	32.64	20.19
IV 2022	5.43	36.53	38.50	19.91
IV 2023	11.28	21.77	55.76	10.60
III 2024	10.81	26.56	44.67	17.93

Note: Encuesta Continua de Empleo, INEC, 2024. https://inec.cr/estadisticas-fuentes/encuestas/encuesta-continua-empleo?page=3

mismatch in the labor market, where economic growth has not been accompanied by the creation of sufficient skilled jobs to absorb the more educated population.

The impact of the 2020 pandemic was reflected in a sharp increase in unemployment across all educational levels, exacerbating pre-existing vulnerabilities in the labor market. Although the subsequent economic recovery has reduced unemployment rates among lower educational levels, this recovery has been uneven, with persistent challenges in the labor absorption of highly educated workers. This could indicate dynamics of overqualification or a labor market that is not fully aligned with the capabilities of the available workforce.

Costa Rica's State of the Nation Program (PEN, 2022, 2024) has outlined a series of strategies to address the persistent issue of unemployment, emphasizing the need for a comprehensive and inclusive approach. A key proposal is the creation of a National Employment Policy that integrates gender, territorial, sectoral, and age-based perspectives. This policy should transcend political cycles and be built through tripartite dialogue among the state, employers, and workers to ensure continuity and effectiveness. To stimulate job creation, the plan also advocates for subsidies and temporary tax incentives for companies with strong growth potential, encouraging them to hire individuals facing labor disadvantages, particularly young people and women. Additionally, the establishment of a governing body dedicated to supporting national enterprises and generating employment opportunities is proposed, granting it the authority to design, implement, and evaluate labor market interventions.

A crucial component of these strategies is the strengthening of educational and vocational training systems to enhance employability. Immediate action is required to address the ongoing crisis in Costa Rica's public education system, as education and access to decent work are fundamental pathways to social mobility. Gender disparities in the labor market also require targeted policies to facilitate women's workforce participation, including a reallocation of household responsibilities. Strengthening the National Care Network, as established under Law 9220, and encouraging workplaces to accommodate family needs are essential steps, as the lack of early childhood care remains a significant barrier preventing many mothers from entering the labor market.

Finally, the PEN (2022) emphasizes the need to consolidate dual vocational education and training as a mechanism for bridging the gap between education and employment. A more inclusive dual education model, adapted to regional economic dynamics, could optimize local resources and expand opportunities for marginalized groups, particularly young people. By addressing structural labor market challenges through a combination of educational reform, employment incentives, and gender-sensitive policies, Costa Rica can move toward a more equitable and sustainable employment landscape.

This last point is relevant in the context of this chapter. It is proposed to consolidate an inclusive dual apprenticeships system adapted to the needs of the different territories. There are several points here that are useful for analysis: a dual apprenticeships system must make sense in the context of productive dynamics and employment policies and be part of an economic and social development strategy, but, first, in Costa Rica, the economy is growing but does not generate jobs, second, participation in the labor market has been reduced, people are no longer looking for formal work, and third, there is no employment policy. The result is a high concentration of wealth in a few hands and high levels of poverty.

Brief History of the Implementation of Dual Apprenticeships in Costa Rica

The implementation of dual apprenticeships in the Costa Rican context began decisively in the early 1980s when an agreement was signed between two vocational training institutions in Costa Rica and the Federal Republic of Germany. The first pilot plan that was developed from the interest of the executive branch was carried out in 1994–1996 in the field of mechanical automotive training (Mittmann, 2001). From 1980 to 2015, various pilot schemes were implemented at different qualification levels offered by the National Institute of Apprenticeship, the Ministry of Public Education and public and private universities in Costa Rica (Láscarez Smith & Baumann, 2020).

In 2015, Costa Rica and Germany signed a bilateral cooperation agreement in different areas, including the implementation of dual apprenticeships with technical support from the German Federal Ministry of Education and Research. Since then, various political and business groups have formulated at least four different bills to try to establish a Costa Rican-style dual apprenticeships model (Láscarez Smith, 2024a).

This context of draft legislation was one of the most important periods of social conflict in the field of education in Costa Rica in the last 10 years. Teachers' unions, student groups and political and academic organizations opposed the business proposal to create a dual apprenticeships system, arguing labor exploitation, cheap labor for transnational sectors and the loss of students' and workers' rights. On the other hand, neoliberal business groups and politicians defended the need for dual apprenticeships in the context of high unemployment and the high proportion of young people who neither study nor work in Costa Rica.

In 2017, the MEP (Ministerio de Educación Pública[3]) asked the ILO to set up a tripartite social dialogue between employers, trade unions and the government in order to reduce the high social conflict and to reach a consensus on the principles of the Costa Rican dual apprenticeships system. The tripartite dialogue lasted approximately 7 months and ended without firm agreements for various political reasons. In 2019, after four years of intense debate, the first dual apprenticeships law (Law No. 9728) was passed. Since the passing of this law, commissions have been established and agreements between companies and training institutions started to be signed to officially implement dual apprenticeships.

The early development of dual apprenticeships in Costa Rica demonstrates the gradual evolution of a system that sought to bridge the gap between classroom learning and the needs of industry. This historical background provides valuable insights into the origins of dual apprenticeships in the country, shedding light on the principles and practices that have shaped its implementation, such as social dialogue, regulation and associated standards, as well as institutionalization and its social, economic and political effects.

Dual apprenticeships are currently implemented in the following careers in the MEP[4]: Organization of food and beverage service operations; Accommodation business operations; Precision mechanics; Installation and maintenance of industrial electrical systems; Kitchen management and production; Industrial electronics; Electromechanics and Web development.

At INA (Instituto Nacional de Aprendizaje[5]), the dual training offer is as follows: Service provision and guest care; Operation of production processes; Food and beverage service operations; Kitchen assistance

[3] Ministry of Public Education.

[4] https://educacionyformaciondual.cr/

[5] National Learning Institute.

operations; Industrial maintenance; Graphic printing; Quality management; Electromechanics; Graphic design; Advanced cooking; Point of sale management and Repair of light vehicle systems.[6]

General Educational Outcomes of the Dual Apprenticeship System

Despite being positioned as a strategy to address youth unemployment, the dual apprenticeship system in Costa Rica has yet to achieve significant scale or demonstrate measurable outcomes. In 2023, the initial enrollment in dual education programs comprised 245 students, distributed as follows: 109 men enrolled in INA programs and 27 in MEP programs. The initial participation of women in dual education programs was 100 in INA and 9 in MEP. By the end of 2023, 169 students were actively participating in dual TVET programs offered by MEP and INA, of whom 90 were men and 79 were women. These figures reflect a modest increase in participation, nearing 200 students, but still highlight the limited scale and reach of the dual apprenticeship system in Costa Rica (Comisión Asesora y Promotora de la Educación y Formación Técnica Profesional Dual, 2023[7]).

This limited participation highlights two key issues. First, the dual apprenticeship system, as currently implemented, is far from being a mainstream educational model in Costa Rica. Second, the small number of participants prevents any meaningful analysis of its impact on reducing youth unemployment. Moreover, the problem is not only that the total number of participants remains modest, but also that the quality of the school-based learning system, job creation, and the qualification levels required by the labor market must be thoroughly analyzed. The absence of these complementary evaluations makes it difficult to fully understand the model's true potential.

Furthermore, these figures illustrate the inherent difficulty of transferring international educational models to local contexts. The development of dual apprenticeships in Costa Rica represents a decades-long process of negotiation and adaptation among various social actors, including policymakers, educators, and the business sector. This prolonged trajectory

[6] https://educacionyformaciondual.cr/

[7] Consultative and Promotional Commission for Dual Technical Vocational Education and Training, 2023.

emphasizes the complexity of aligning global educational frameworks with local labor market dynamics and social equity goals. In this sense, the configuration of the economic growth model and employment policy must be analyzed as a complex framework in which individuals, based on their educational level, ethnicity, gender, migrant status, and other conditions, should be targeted through specific policy programs that address their unique realities and needs.

Ultimately, while the dual apprenticeship model holds potential as a tool for improving employability, its current scale and implementation in Costa Rica are insufficient to address the structural causes of youth unemployment. A more comprehensive strategy is essential, integrating dual apprenticeships with broader educational reforms, labor policies, and economic development plans to create meaningful opportunities for young people in the labor market.

Political and Business Arguments for Dual Apprenticeships

From a comparative point of view, in countries such as Germany, dual apprenticeships have been a strategy that has developed in a cultural context, which has evolved from multiple and complex socio-economic processes, and has been related to multiple objectives including the need to train a more qualified workforce and thus meet the demands of a growing economy[8] (Greinert, 1995; Gessler & Howe, 2013). However, in the Costa Rican context the beginnings of dual apprenticeships are not culturally bound, as the Spanish colonial heritage framed the existence of a school-based apprenticeship system, mostly inspired by Salesian pedagogical principles (Láscarez Smith, 2024b).

In this sense, the implementation of dual apprenticeships is justified as an option, according to neoliberal politicians and businessmen, in the face of the failure of the import substitution model that generated poverty, unemployment and inequality. These three issues have been the main political arguments in favor of dual apprenticeships since the 1980s and especially the 1990s (Guzmán, 1992).

[8] "Today's dual vocational training system in Germany is not the result of conscious planning and development, but rather arose within a complex historical process to form a connected whole" (Greinert, 1995, p. 19). (ver el texto de Gessler 2013).

However, as Costa Rica has built a school-based apprenticeship system (and still is), the attempt to introduce dual structures has become a real challenge that has limited the expected positive effects on employment and inequality reduction. In particular, this is true because it is those school-based educational structures (curriculum, study programs, institutional cultures, business cultures and policies) that have been attempted to dualise.

The functioning and relevance of the dual apprenticeships system is currently being questioned, especially because poverty and unemployment have increased steadily and because the relationship between the production system, dual apprenticeships and the reduction of unemployment in Costa Rica has not been established. This is because there are actually few technical careers under this modality and because the proportion of apprentices is practically insignificant.

Despite the progress made, the implementation of dual apprenticeships in Costa Rica has faced challenges, such as the need to increase the number of participating enterprises and to ensure the sustainability of the programs. Efforts have also been made to broaden the scope of dual apprenticeships beyond the already highly employable technical fields and to promote their adoption among small and medium-sized enterprises.

In the Costa Rican context, the implementation of dual apprenticeship programs reveals a prolonged, four-decade endeavor hindered by substantial structural constraints and limited explanatory frameworks available to researchers examining this field. A primary challenge is rooted in the institutional inertia of established educational entities and other influential social actors who regard dual education as a potential disruptor to the established educational paradigm. This perception underscores a broader resistance to structural change, stemming from concerns that dual apprenticeship models could undermine the traditional educational hegemony and shift the balance of educational influence toward more industry-oriented, skill-based training approaches.

Dual Apprenticeships and Youth Unemployment: Between Rhetoric and Evidence

Unemployment is a multifaceted socio-economic phenomenon that not only impacts individuals directly, depriving them of income and stability, but also generates knock-on effects at the societal level, including weakening the social fabric, reducing aggregate demand and increasing state dependency. Lack of employment affects the ability of individuals to participate fully in the economy and undermines collective well-being, intensifying structural inequalities and limiting the potential for macro-level economic development.

Education is commonly conceived as a strategic tool in the reduction of unemployment, as it is assumed to equip people with the skills and knowledge indispensable for integration into the labor market. However, this relationship turns out to be less linear than traditional models assume, revealing itself, in sociological and economic terms, as a necessary but not sufficient condition, especially in Latin American contexts. In these scenarios, structural factors, such as labor market segmentation and inequalities in access to quality jobs, complicate the transition of individuals from the education system to effective participation in formal employment.

Theoretical approaches that attempt to explain the phenomenon of unemployment are diverse, although often reductionist, as they fail to comprehensively address the different levels at which the problem manifests itself. A predominant approach to explaining unemployment is that of human capital theories, which argue that youth unemployment is mainly due to a lack of skills and experience required by the labor market. This approach places the responsibility for unemployment on the individual, conceptualizing it as a consequence of personal decisions and actions, while minimizing, or even ignoring, the structural factors and contradictions inherent in capitalist development (Sandoval & Hernández, 2018).

In this sense, Bourdieu's (1984) analysis of inequalities in cultural and economic capital is fundamental to understanding the mechanisms of social exclusion and unequal access to labor opportunities.

Harvey (2005) highlights that capitalism systematically produces "surplus populations", a phenomenon that aligns with Marx's concept of the "industrial reserve army" (Marx, 1867, p. 791). According to Marx, this reserve workforce emerges as a result of capitalist accumulation, which continuously expels and reabsorbs workers depending on economic cycles, thus intensifying structural inequalities. This dynamic disproportionately affects youth employment by creating unstable labor conditions and limiting access to stable jobs. In contrast, human capital theory provides a reductive explanation of youth unemployment, as it largely attributes employability to individual skill acquisition, overlooking systemic factors such as labor market segmentation, job insecurity, and restricted social mobility (Castel, 1999). These structural constraints shape the persistent difficulties young people face in securing stable employment within capitalist economies.

This one-dimensional view contrasts with critical approaches that stress the interaction between economic, social and cultural capital, and highlight how these forms of capital are unevenly distributed, thus conditioning individuals' labor and life trajectories (Bourdieu, 1986).

Hinkelammert (2003) expands this analysis with a radical critique of the capitalist system, arguing that capitalist logic not only generates material inequalities but also an "instrumental reason" that subordinates human life to profitability and accumulates wealth at the expense of collective well-being. For Hinkelammert, unemployment serves a functional role within the system, maximizing profit rates by reducing labor costs; however, this logic also produces "disposable populations"—individuals deemed superfluous in economic calculations, reflecting a structural dehumanization inherent to global capitalism.

Meanwhile, decolonial theories emphasize that global capitalism reproduces a dependency relationship between the Global North and South. Both Quijano (2000) and Hinkelammert argue that the system imposes conditions of underdevelopment on the South through mechanisms that extract cheap resources and labor, while obstructing the transfer of productive capital and technology. This unequal flow reinforces the South's exclusion from the benefits of industrial and technological development, perpetuating conditions of precarious labor and underemployment instead of creating quality jobs. In this sense, unemployment is not merely a lack of opportunity but a manifestation of an accumulation logic that systematically marginalizes and dispossesses global peripheries.

In the Costa Rican context, high general and youth unemployment has been promoted as a key justification for adopting dual apprenticeships. This initiative is partly fuelled by the successful publicity and prestige that Germany has been able to link between its dual apprenticeships model and its low youth unemployment rates (Lassnigg, 2015; Láscarez Smith & Schmees, 2021). However, this adoption also responds to a flow of ideas and policies from the global North that have a political-ideological background and, when transposed, do not always thoroughly consider the structural particularities of the global South.

From a critical political economy perspective, dual apprenticeships can be seen as an approach that, although it has proven benefits in certain contexts, poses challenges of adaptation in peripheral economic realities where unemployment has complex causes. Beyond the symbolic value associated with the German system, local actors have faced challenges in developing comprehensive explanations of the structural causes of unemployment, which raises the need for solutions that address the specific problems of the Costa Rican labor market.

We have to follow the argument of myth and brand. At the level of the policy rhetoric we can easily see that the established relationship between the German Dual System (GDS) and youth unemployment lacks substance, and is based on very superficial empirical correspondences and analogies that do not stand a deeper questioning. If we take the notion of a myth literally, which somehow refers to a kind of deeper reflection of complex, sometimes mysterious issues of life or world that are not easy to understand, then the reasoning behind the GDS and employment is rather a trivialisation of this concept (Lassnigg, 2015, p. 13).[9]

Lassnig observes that educational models and forms of qualification are traditionally interconnected with occupational development, providing a direct and concrete pathway between education and labor market needs. However, he concludes that political influences have introduced distortions within apprenticeship systems, decoupling them from the specific economic and productive contexts they aim to serve. Consequently, Lassnig argues that the relationship between the dual apprenticeship system and unemployment in Germany is not as straightforward as often portrayed. He contends that more holistic approaches are required—ones that integrate employment policy, economic development strategies, work culture, and the broader social context—into analyses of apprenticeship outcomes and labor market dynamics.

The ideologies and theories about education have always somehow interacted with work and employment, and to some point they were also explicitly applied to learning for work and occupations. Overall, these interactions and interrelations were highly contingent, and in varying degrees organizations and institutions emerged that typically started to cover some parts of the overall fields, and through time have spread more broadly across the fields of work and occupations. The diversity of the world of work and occupations, and its historical development,

[9] A central contribution of Lassnig is on the treatment of the concept of myth and the social representations that emerge in the process of political and scientific argumentation on dual apprenticeships: is there a political discourse and a scientific discourse on the myth of the benefits of dual apprenticeships?

has been accompanied by varied learning practices and vocational education systems. However, as Lassnigg (2015) argues, these frameworks are not neutral or universally applicable; rather, they emerge from specific socio-economic and political contexts. Vocational education systems have been shaped by scattered and diverse policy influences, meaning that their interrelation with work practices is neither uniform nor easily transferable across different labor markets.

In the study by Láscarez Smith and Schmees (2021, 2023) on the arguments and political positions of actors involved in Technical and Vocational Education and Training (TVET) in support of implementing dual training in Costa Rica, it is revealed that youth unemployment is often attributed, by business leaders and proponents of the 2019 legislation, to individual factors or to the limitations of educational institutions in capturing and adequately responding to business sector demands. However, a critical analysis suggests that youth unemployment should be understood as a symptom of contradictions inherent in the productive structure, the mode of production, and prevailing institutional approaches in the country. Effectively addressing this issue requires comprehensive political interventions that go beyond mere educational adaptation, involving a structural reconfiguration that enables better alignment among economic development, employment policies, and social inclusion.

For example, Brenner (2006) argues that economic crises, derived from the internal contradictions of capitalism, generate fluctuations in employment that disproportionately affect the most vulnerable sectors, such as young people. This should include intersectional approaches that explain poverty, inequality of certain social groups through a focus on social class reproduction, ethnic, gender, or vulnerable group disadvantage.

In addition, segmented labor market theory (Fernández-Huerga, 2010) provides a critical understanding by suggesting that youth unemployment is perpetuated not only by economic cycles, but also by regulatory rigidity and structural discrimination in labor markets. Fernández-Huerga (2010) points out that this theory tends to encompass a range of approaches, quite diverse in their origins and content, which began to emerge in the late 1960s driven by dissatisfaction with the neoclassical explanation of the labor market. Piore and Doeringer (1971) argue that labor markets are segmented into 'primary' and 'secondary' sectors, with young people, especially those from marginalized groups, largely confined to the secondary sector, characterized by precarious employment, low mobility and limited career development opportunities. This segmentation and the institutional barriers that underpin it reinforce the exclusion of young people from the formal labor market, thus perpetuating their economic vulnerability, or what in Marxism is called social class reproduction.

Conclusions

Since 2010, the youth unemployment rate in Costa Rica has fluctuated between 22% and 39%. Although efforts have been made during this period to implement dual apprenticeships, a robust evaluation of their impact on youth unemployment has not been achieved due to the lack of system expansion and limited implementation. This chapter demonstrates that while the dual model is promising, it cannot be considered the sole solution to the issue of youth unemployment in the country. However, it can contribute as part of a broader approach to employability, decent work, and equitable development.

The analysis highlights the complexity of the relationship between education and employment. While education is a fundamental tool for improving job opportunities, its effectiveness largely depends on its alignment with labor market demands and the ability of educational systems to adapt to global transformations. The Costa Rican experience suggests that the implementation of dual apprenticeships requires a more systemic approach that considers not only expanding access but also improving the quality and relevance of technical education, as well as labor policies that foster the creation of quality jobs.

Despite four decades of efforts to implement dual apprenticeships in Costa Rica, the debates have lacked an adequate systemic understanding, limiting their potential impact. The main limitations include resistance from traditional educational actors, the disconnect between political objectives and the real needs of the labor market, and the low participation of small and medium-sized enterprises in the dual system. Furthermore, the current approach has, to some extent, reproduced structural inequalities by failing to effectively integrate the most vulnerable sectors of the youth population.

Looking ahead, it is crucial for Costa Rica to adopt a more comprehensive approach to address youth unemployment. This includes:

- Developing an inclusive dual apprenticeship system that considers territorial and sectoral particularities, with a strong emphasis on the participation of small and medium-sized enterprises.
- Implementing a national employment policy that transcends governmental periods, with a focus on tripartite dialogue (state, employers, and workers) and the generation of decent jobs.
- Improving alignment between education and the labor market, promoting training programs that match the most in-demand occupational profiles and the needs of emerging economic sectors.
- Strengthening gender equity and social inclusion through policies that facilitate the participation of women and youth from vulnerable populations in the labor market.

As Lassnigg (2015) argues, the relationship between dual apprenticeships and employment outcomes is often oversimplified in political discourse, lacking the necessary empirical depth to justify the expectations placed on this system. As a general conclusion, we would like to state that while dual apprenticeships are a promising strategy, no solid connection has been demonstrated between this policy and the reduction of youth unemployment in Costa Rica. Finally, the implementation of dual apprenticeships must be part of a broader socio-economic development strategy that addresses structural inequalities and strengthens the country's educational and productive capacities. Only through a holistic and contextualized approach can the impact of dual apprenticeships on reducing youth unemployment in Costa Rica be maximized.

References

Bourdieu, P. (1984). *Distinction: A social critique of the judgement of taste.* Routledge.

Bourdieu, P. (1986). The forms of capital. In J. G. Richardson (Ed.), *Handbook of theory and research for the sociology of education* (pp. 241–258). Greenwood Press.

Brenner, R. (2006). *The economics of global turbulence: The advanced capitalist economies from long boom to long downturn, 1945–2005.* Verso.

Castel, R. (1999). La metamorfosis de la cuestión social. Crónica de un asalariado.

Comisión Asesora y Promotora de la Educación y Formación Técnica Profesional Dual. (2023). *III Informe del Estado de la Educación y Formación Técnica*

Profesional Dual en Costa Rica. Ministerio de Educación Pública, III Informe Anual Estado EFTP Dual—Costa Rica 2023 | Plataforma de comunicación de Educación y Formación Técnica Dual en Costa Rica.

Díaz Arias, D. (2019). *Historia del neoliberalismo en Costa Rica: La aparición en la contienda electoral, 1977–1978/David G. Díaz Arias* (1st ed.). Vicerrectoría de Investigación: Centro de Investigaciones Históricas de América Central.

EFTP. (2024). *Oferta educativa dual*. EFTP, Plataforma de comunicación de Educación y Formación Técnica Dual en Costa Rica.

Fernández-Huerga, E. (2010). La teoría de la segmentación del mercado de trabajo: Enfoques, situación actual y perspectivas de futuro. *Investigación Económica, 69*(273), 115–150. Retrieved from http://www.scielo.org.mx/scielo.php?script=sci_arttext&pid=S0185-16672010000300004&lng=es&tlng=es

Gessler, M., & Howe, F. (2013). The German Dual Vocational Training System—The origin of the current architecture. *Bulletin of Institute of Technology and Vocational Education Nagoya University, 10*, 17–28.

Greinert, W. D. (1995). *Das deutsche System der Berufsausbildung. Geschichte, Organisation, Perspektiven* (2nd ed.). Nomos.

Guzmán, M. (Ed.). (1992). *Posibilidades para la implementación en Costa Rica del sistema dual de formación profesional 1991*. Memoria.

Harvey, D. (2005). *A brief history of neoliberalism*. Oxford University Press.

Hinkelammert, F. (2003). *El huracán de la globalización: La paradoja de la globalización económica*. Editorial DEI.

INEC. (2024). *Encuesta Continua de Empleo (ECE)*. INEC. Retrieved from https://bit.ly/3NkOn6e

Láscarez Smith, D. (2024a). Political participation of business organizations in the construction of the technical vocational education and training system in Costa Rica between 1980 and 2021. *International Journal of Vocational Education Studies, 1*(1), 57–70.

Láscarez Smith, D. (2024b). *Political participation of business organizations in the construction of the technical vocational education and training system in Costa Rica between 1980 and 2021*. Doctoral Thesis. https://doi.org/10.48693/410

Láscarez Smith, D., & Baumann, F. (2020). Costa Rica: Berufsbildung im Wandel. In F. A. Baumann, D. Frommberger, M. Gessler, L. Holle, L. Krichewsky-Wegener, S. Peters, & J. Vossiek (Eds.), *Berufliche Bildung in Lateinamerika und Subsahara-Afrika. Entwicklungsstand und Herausforderungen dualer Strukturansätze*. Springer.

Láscarez Smith, D., & Schmees, J. (2021). *Los conceptos del sector empresarial costarricense sobre la transferencia de la formación dual alemana*. https://doi.org/10.15359/abra.43-66.1

Lascarez Smith, D., & Schmees, J. (2023). Union perspectives on dual apprenticeship in Costa Rica: Experiences during the tripartite dialogue. *Revista ABRA, 43*(66), 1–21. https://doi.org/10.15359/abra.43-66.1

Lassnigg, L. (2015). The political branding of apprenticeship into the "dual system": Reflections about exporting the myth of employment transition. In A. Heikkinen & L. Lassnigg (Eds.), *Myths and brands in vocational education* (1st ed., pp. 78–98). Cambridge Scholars Publishing.

Marx, K. (1867). *El capital, tomo i: El proceso de producción del capital*. Siglo xxi. Editores.

Mittmann, F. (2001). *La educación dual en Costa Rica. Proyecto piloto mecánica automática del Colegio Vocacional Monseñor Sanabria 1996–1999*. INA.

Mora Jiménez, H. (2020). *La crisis del empleo en Costa Rica*. Friedrich Ebert Stiftung. Retrieved from https://library.fes.de/pdf-files/bueros/fesamcentral/16399.pdf

PEN. (2022). *Capítulo 07: Debates para el desarrollo: Rutas de ejecución para la resolución de las dificultades de generar más y mejores oportunidades de empleo fuera de las Zonas Francas y el GAM [2022]*. Programa Estado Nación.

PEN. (2024). ¿Por qué es importante generar más y mejores oportunidades de trabajo en Costa Rica? Retrieved from https://estadonacion.or.cr/dia-trabajo-en-costa-rica-2024/

Piore, M. J., & Doeringer, P. B. (1971). *Internal labor markets and manpower analysis*. U.S. Government Printing Office.

Quijano, A. (2000). Colonialidad del poder, eurocentrismo y América Latina. *Revista Internacional de Ciencias Sociales, 167*.

Raventós, C. (1995). *The construction of an order: Structural adjustment in Costa Rica (1985–1995)*. Tesis doctoral, New School for Social Research de New York.

Sandoval, J., & Hernández, G. (2018). Crítica a la teoría del capital humano, educación y desarrollo socioeconómico. *Revista Ensayos Pedagógicos, XIII*(2). https://doi.org/10.15359/rep.13-2.7

CHAPTER 3

Improving Adult Students Access to Vocational Education and Training in Norway

Rønnaug H. Lyckander, Asgeir Skålholt, and Elisabeth Hovdhaugen

Introduction

In today's society, completing upper secondary education, thus being educated beyond compulsory schooling, is increasingly more important for individuals to be able to access the labour market (OECD, 2023; Müller & Gangl, 2003). There is both international and Norwegian evidence that completing upper secondary education opens up many more labour market opportunities (Bratsberg et al., 2020).

The finding that completion of an upper secondary qualification is vital for labour market inclusion was one of the reasons for the Norwegian government to launch the 'Completion reform' (*Fullføringsreformen*), a

R. H. Lyckander (✉) · A. Skålholt · E. Hovdhaugen
Nordic Institute for Studies in Innovation, Research and Education, Oslo, Norway
e-mail: ronnaug.lyckander@nifu.no

E. Smeplass and C. Zirkle (eds.), *Opportunity Structures in Vocational Education and Training*, Palgrave Studies in Technical and Vocational Education and Training,
https://doi.org/10.1007/978-3-032-04410-5_3

white paper published in 2021 addressing adults' opportunities in upper secondary education (Ministry of Education and Research, 2021). This white paper concludes a longstanding national discussion on the importance of adults' access to upper secondary education, and particularly to vocational education and training (referred to as VET from now on). The backdrop to this is that in 1994, Norway implemented a comprehensive reform of upper secondary education, called Reform94. One feature of Reform94 was that students who had completed 10 years of compulsory schooling had a statutory right of admission to upper secondary education, to one of their three programme choices (NOU 1991:4; Ministry of Education, Research and Church Affairs, 1992). Hence, this change only applied to students 16–19 years old and was thus a way to prioritise youth access to upper secondary education. In practice, this meant that from 1994, adult students, particularly those over 20 years of age, were not necessarily prioritised in VET education.

Prior to the reform, older youths and adults had filled most of the apprenticeships, but after the introduction of Reform94, applicants who had completed the second year of their vocational programme were given priority (Høst et al., 2008). This radically changed the situation for adult students in VET, and in the first few years after the implementation of Reform94, the proportion of adults in the VET system dropped. Hence, for adult VET students as a group, opportunities to gain a trade certificate were drastically reduced. In this chapter we focus on the opportunity structures for adult VET students in Norway from the onset of the major reform in 1994 to measures introduced in the latest reform, implemented in 2024 (*Fullføringsreformen*, Ministry of Education and Research, 2021).

The overarching research question this chapter is addressing is *How have the opportunities for adults to complete a VET qualification changed since the implementation of Reform94?* This question covers the entire period, but as most of the measures taken to improve opportunities for adult students have been introduced during the last decade, we will also address issues about *What measures have been taken to improve opportunities for adults to complete a VET qualification?* and *How has participation of adults in VET changed during this time?* As Reform94 largely ignored adult students who needed to get an upper secondary qualification, the aim of the chapter is to document how the Norwegian educational system has spent the last couple of decades moving in a direction in which adult students have more rights and more opportunities.

Methodological Approach

To answer our research questions, we conducted a document analysis of white papers, official Norwegian reports, and recommendations to parliament. In addition to this, we analysed administrative data, to show the proportion of adults in VET, how this has changed over time, and the proportion of immigrants as part of the number of adults in VET. The statistics presented are based on our own analysis of administrative data from Statistics Norway, covering the school years 1998/1999 until 2022/2023. These data come from a database covering the entire population, in which individual educational paths can be tracked over time (Vangen, 2007). Hence, we used these data to document how the proportion of adults (25 years and older) has changed over time, and the share of students with immigrant background who are part of the adult student group.

To define 'adult students', we distinguished between students who are under and over 25 years old. The reason for choosing 25 years as a dividing point is that while there has been changes in how many years the statutory right to upper secondary education is valid for, it has not been extended beyond age 24. We included immigrant status, defined as having immigrated to Norway after the age of six, and thus were born abroad to parents born abroad. The reasoning for this was to include only those who has not done all their compulsory education in Norway, as the aim is to investigate the proportion of relatively recently arrived immigrants in the data.

Upper Secondary Education in Norway

Upper secondary education in Norway consists of an academic track and a vocational track. Programmes in the academic track last 3 years, while the main model within the vocational track commonly is 2 years of schooling followed by 2 years of apprenticeship. VET is thus an integrated part of Norway's upper secondary education system. However, as this paper will display, there are several variants to this 'normal structure' that is also used to gain a VET qualification, particularly for adults. Interestingly, the most used is not formal education, but an option for documentation of qualifications for those with at least 5 years of relevant work experience. Among adults, more than 60% of those taking the trade certificate, takes it after 5 years of relevant experience. In this paper, however, we focus on those who are either in formal education or in training.

Earlier Research

In the Norwegian context, education for adults in upper secondary school is an under-researched field (Ministry of Education and Research, 2016). In the studies that exist, questions of participation and barriers to participation in various pathways have been explored. Three pathways are commonly used to shorten the path to a formal VET qualification for adults. These are regular admission into a programme, reduced programme duration, and qualification without a mandatory programme (Kis & Windisch, 2018). The alternative pathway in VET, the experience-based practice candidate scheme (a qualification without a mandatory programme), represents an important 'second chance' for adults without formal VET qualifications (Tønder & Aspøy, 2017), especially in the health care sector or in technical vocations (Bratsberg et al., 2017). Bratsberg et al. (2017) indicate that the social background of these candidates resembles young people at risk of dropping out of upper secondary education. Thus, this pathway helps to reduce the social gradient in education. It is also associated with economic returns in the form of higher earnings (Bratsberg et al., 2020). However, this alternative model requires at least 5 years of full-time relevant work experience, disqualifying many adult workers in part-time jobs or those with varied experience (Ministry of Education and Research, 2016).

Sterri et al. (2015) have investigated how regional education authorities administer questions about who has the right to undertake adult education, focusing on the groups without a statutory right of access to upper secondary education. They find that this group is heterogeneous, with a range of challenges and support needs, and the regional authorities have different practices in their assessments of admittance.

One way of making VET more accessible to adult learners is to recognise their competence. In a competence reform in 1999, the right for adults to get their formal, non-formal and informal competence assessed, was established (Ministry of Education, Research and Church Affairs, 1998). A key change of the reform is that more adults undergo a validation of their prior learning before starting formal education. This, combined with flexible arrangements introduced by the Competence Reform, has significantly shortened the time required for individuals to achieve a trade certification, particularly in school-based vocational education, with health care workers being the largest programme (Høst, 2008).

Immigrants are overrepresented in adult VET and the proportion of immigrants is rising. Sterri et al. (2015) find that immigrants meet additional barriers in Norwegian VET, because courses are not adjusted to their needs. For example, immigrants have had no right to language training as part of VET for adults. Nevertheless, obtaining formal qualifications in the country of settlement, such as by undertaking upper secondary education or occupational-specific qualifications, has the potential to increase labour market participation among immigrants and young refugees (Geven & Spörlein, 2023; Manhica et al., 2019). In contrast, the evidence of the benefits of "work-first policies", that emphasise a quick entry to the labour market, is more limited and with mixed results (Andersson Joona, 2019; Bredgaard & Thomsen, 2018) and may not lead to inclusion in the labour market in the longer term (Arendt, 2022; Arendt & Bolvig, 2020). Consequently, an ongoing discussion revolves around whether integration measures should prioritise improving immigrants' skills through educational activities to improve long-term employment or early labour market entry. In Norwegian policy, the focus has shifted to a focus on formal education and skills development (Arbo, 2022). Recent reforms have strengthened individual rights to education, emphasising the need for an integrated approach that considers both basic education and lifelong learning, tailored to different age groups and life circumstances.

Theoretical Framework

In this paper, we use the concept of *transition systems* to describe changes in the Norwegian VET system for adults. While many researchers have tried to characterise countries by models or typologies of VET systems (Busemeyer & Trampusch, 2012; Pilz, 2016) or transition arrangements (Walther, 2006), such typologies have been criticised for being limited and for glossing over important differences between countries that are perceived to be similar (Jørgensen et al., 2019; Raffe, 2008). For instance, based on a study of transition policies for young people in Sweden, Finland and Denmark, Jørgensen et al. (2019) question a categorisation of a common Nordic *universalistic regime*.

Transition systems is a concept used to study a country's national transition patterns between education and employment (Raffe, 2008). According to Raffe (2008), the concept can be used to investigate the institutional and structural factors that shape education and training in a nation or between nations. Raffe argues that the factors that shape transitions are broader than education and training, and therefore includes the organisation of labour markets as well as contextual features such as social welfare systems and family structures. Raffe defines a transition system as 'the relatively enduring features of a country's institutional and structural arrangements which shape transition processes and outcomes' (Smyth et al., 2001, p. 19). For adults in Norwegian VET, two features have characterised the transition system, a workplace-based route and the ordinary dual route, combining school-based and company-based training as apprentices (Tønder & Aspøy, 2017).

To study the changing opportunities for adult learners in VET, we will discuss two out of the six key ingredients of successful transition systems that have been identified in several policy analysis cases, and are viewed as universally applicable (OECD, 2000). These are *well-organised pathways* and *widespread opportunities for combining workplace experience with education*. Firstly, *well-organised pathways* can be characterised by a clearly defined qualification framework, linking the route of education, the content of the programmes, the qualification and the destination to which the qualification provides entry (work or further study). Secondly, these pathways provide access to education by being open to different participants (OECD, 2000). In Europe, restructuring VET pathways in more flexible ways, aim to meet the challenge of accommodating the growing diversity of students' needs as well as a shortage of skilled labour (Pilz et al., 2018). A range of trends can be observed, such as the broadening of vocational programmes, and more combinations of school-based and work-based learning (OECD, 2000). We describe the latter point in more detail later.

One specific feature of making pathways more flexible, is by introducing VET modularisation. Many European countries have already introduced modular forms of VET (CEDEFOP, 2015) to offer students greater flexibility and choice, allowing them to customise their learning pathways in terms of course options, sequencing, and pace of completion (see e.g. Pilz et al., 2018). Modular programmes can be targeted at different student groups. For instance, Finland offers a variety of modular pre-vocational and alternative VET programmes aimed at unemployed youth, students with additional learning needs, immigrants, and individuals with disabilities (Pilz et al., 2018). In Norway, modular programmes for adults represent a completely new pathway.

Pathway reform must carefully balance competing objectives and inherent tensions. While flexible pathways can offer students more autonomy in shaping their educational journeys, it may come at the expense of maintaining high quality standards, reducing transparency in educational options and qualifications, and greater inequalities (OECD, 2000). Moreover, navigating the most suitable pathways can become a complex challenge for individuals, necessitating guidance and support systems. Modularisation, while promoting flexibility, may inadvertently encourage individuals to exit education with only partial skills or qualifications, limiting their employability.

Widespread opportunities for combining workplace experience with education is another key feature of successful transition systems (OECD, 2000) and there has been a growing political interest in promoting apprenticeships and work-based learning in VET (Jørgensen, 2015). Workplace experience can be organised in various ways, with apprenticeships being the most well-known. Typically, the workplace is responsible for recruiting apprentices, who gradually take on more productive work as their skills develop. Research finds that the best transition outcomes in labour market terms are achieved in countries where a substantial number of young people enter work through apprenticeships (Blommaert et al., 2020).

Education and work can also be combined through school-organised workplace experiences, which can range from short-term placements to longer internships. These programmes serve several purposes: they help students acquire the knowledge and skills required for specific occupations as well as develop personal and social competencies; provide opportunities to practise skills learned in school within a real work environment and assist in career exploration and planning (OECD, 2000).

To sum up, through the concept of transition systems (Raffe, 2008), we analyse policy changes in Norway aimed at including more adult learners in VET. This paper focuses on two key aspects of successful transition systems: (1) well-organised pathways and (2) widespread opportunities for combining work and education.

Changes in Access to VET for Adults Over Time

In this section, we address our first research question, and shed light on how the opportunities for adults to complete a VET-qualification have changed since the implementation of Reform94. The right to access upper secondary education, and thus to access VET for those who had not completed an upper secondary qualification, was not granted to adults over age 25 until 2008, after being proposed in a white paper on lifelong learning (Ministry of Education and Research, 2006). Hence, from this point onwards, adults over 25 had rights similar to those introduced for youths aged 16–19 over 20 years earlier in Reform94. In that reform, the statutory right to education was linked to birth year (starting in 1978, the birth year of the first cohort affected by the reform), which left those who did not manage to complete an upper secondary qualification within the statutory period without the right of access. As will be shown later, adults still entered the system, but that was at the discretion of the regional education authorities, if there were spaces available in the VET education system. In addition, adult employees with long work experience still could obtain experience-based trade certification, thus affecting the numbers of skilled workers (Bratsberg et al., 2017; Høst, 2008).

This anomaly, that students over 25 years of age without formal qualifications had fewer educational opportunities than their younger counterparts, had been pointed out in several white papers preceding the white paper on lifelong learning (Ministry of Education and Research, 2006). The issue was first raised in a white paper on upper secondary education, just a few years after the implementation of Reform94 (Ministry of Education, Research and Church Affairs, 1999), but this did not lead to any changes. About 5 years later, a Norwegian official report made the same remark, and explicitly suggested that the age limit for adults' right to access upper secondary education be set at 25 years (NOU 2003:16). However, this suggestion was not directly followed, but rather only set to be evaluated (Ministry of Education and Research, 2004). Consequently, the formal right for adults over 25 years of age to access upper secondary education was not implemented until 2008.

However, these policy changes did not have many direct implications for adult participation in VET. Even if adults over age 25 did not have the right to access upper secondary education, they still did, and they were particularly salient in VET.

As will be shown below, 2016 constituted a shift in educational policies focused on adults. That year, a white paper on adult learners was published, focusing on educational opportunities as a basis for stable employment opportunities (Ministry of Education and Research, 2016). The white paper highlights the importance for adults with limited formal qualifications to get the opportunity to participate in education, and that this lack of formal qualification may be key in explaining their limited labour market participation. Thus, the main goal of the white paper was to promote skills development which would enable long-term workforce participation. Consequently, this white paper brought discussions on education for adults closer to discussions on employment and the need for individuals to meet the demands of the labour market.

Share of Adults Pursuing a VET-Education

Evaluations of the implementation of Reform 94 showed that the implementation of a statutory right to 3 years of upper secondary education led to a drop in share of adults applying for and accessing upper secondary education (Støren et al., 1998). This drop was linked to the initial capacity in the system, and the fact that many adults who did not have the statutory right to access upper secondary education were not offered a spot. However, as shown in the evaluation, was this a temporary drop in the proportion of adults in the system. Just a couple of years later, the proportion of applicants without a statutory right to access had picked up again (Støren et al., 1998).

To answer our research question of how participation of adults in VET has changed during the past three decades, we present our analyses of administrative statistical data. Figure 3.1 displays the proportion adults over age 25 constitute of all students in VET programmes, both at school and in various forms of apprenticeships. The figure shows that adults are a minority in the system, accounting for about 15% during the first 10 years of the period studied. From 2008–2009 until 2016–2017, we saw a reduction in proportion, which could be linked to there being fewer adults who have not completed upper secondary education, and thus indicate a reduction in demand. However, during the last 6 years there has been an increase in adult participation in VET, which may be linked both to changes in policy, and to changes in who participates in VET as an adult.

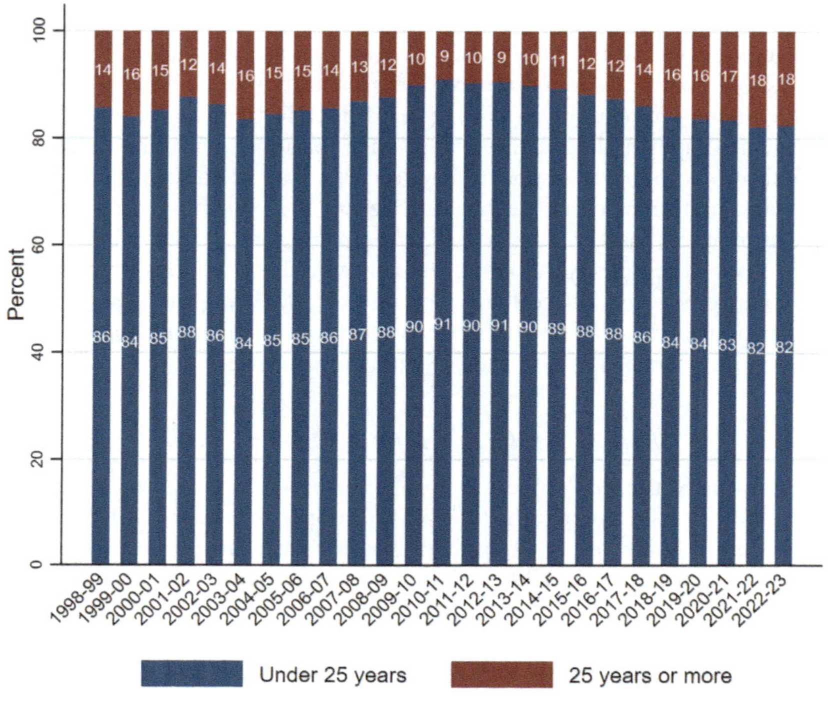

Fig. 3.1 Proportion over and under 25 years old participating in Norwegian VET, 1998–1999 to 2022–2023

Figure 3.2 may contribute to explaining the increase in share of adults over 25 in Norwegian VET. In the period we are looking at the share of immigrants as part of all adults over 25 in VET has increased from 6% to 50%, and the increase has been particularly strong over the past decade, as it has doubled in this period. This may indicate that the group of students taking VET as adults over 25 is changing, and thus potentially is serving other purposes than VET education for adults over 25 did at the start of the period.

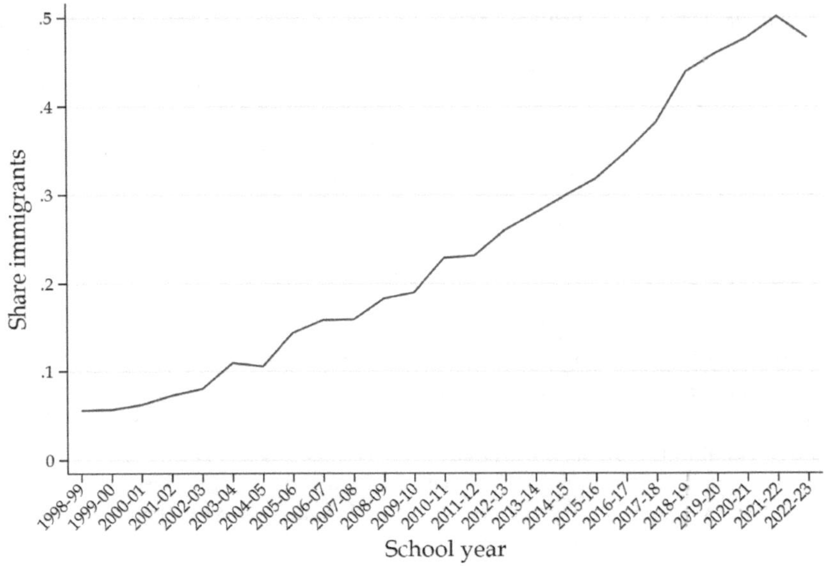

Fig. 3.2 Share of immigrants among adults (+25) in Norwegian VET, 1998–1999 to 2022–2023

Significant Shifts in Policy Measures Targeting Adults in VET

The white paper on adult learning in 2016 introduces a new initiative to develop a comprehensive policy for adult education that spans multiple sectors, including education, employment, social inclusion, and immigration (Ministry of Education and Research, 2016). The focus is at-risk adults, such as those with limited formal education, inadequate basic skills, and immigrants. The policy is guided by two key principles: (1) providing individualised and flexible learning options tailored to adults' needs, and (2) fostering cross-sector collaboration. In the most recent reform on upper secondary education, on how more students can complete their education (*Fullføringsreformen*), adult learners' needs are again addressed, and new measures for participation were introduced (Ministry of Education and Research, 2021).

In the following section, we explore the measures aimed at improving opportunities for adults in VET, organised under two main themes:

increased participation and *flexible pathways*. This analysis addresses the research question of what measures have been implemented to support adults in completing a VET qualification. Key questions include how well these measures address existing challenges and incorporate effective elements from transition systems, ultimately helping adults complete VET qualifications and secure stable employment.

Measures Supporting Increased Participation of Adult Learners in VET

Based on suggestions described in the white paper on adult learners (Ministry of Education and Research, 2016), in 2017 the Norwegian parliament passed legislation extending the right to upper secondary education to two new groups: young adults aged 19–24 and adults whose upper secondary education, completed abroad, had not been recognised in Norway. The prioritisation of youth in earlier reforms, described in the introduction, had lasting effects on adults' access to education. This new legislation helps to close the gap between the educational rights of young people (aged 16–21) and adults (aged 25 and above).

The extent to which this extension of educational rights will lead to increased participation of adults remains uncertain. Even before the new legislation, education authorities could admit adult students without formal rights, and many regional education authorities did so, albeit to varying degrees (Sterri et al., 2015). While this extension may not be the most significant factor in boosting participation, it is an important policy achievement, as it guarantees universal access to education without age restrictions, removing a key structural barrier. Furthermore, it reduces ambiguity in local admission decisions, ensuring more consistent access to VET. Sterri et al. (2015) highlight how regional education authorities previously adopted either restrictive or more liberal approaches in their admission processes.

Equally significant was the policy measure of granting immigrants with upper secondary qualifications from abroad the right to access upper secondary education in Norway. Many immigrants were unable to exercise this right during their youth, either due to limited Norwegian language skills or because they were not residing in Norway at the appropriate age (Sterri et al., 2015). Under the previous regulation, this group faced an unfair situation, as their only option for obtaining upper secondary qualifications was through costly private institutions. This even led to contradictory outcomes, with some immigrants misrepresenting their educational background and formal qualifications to gain access to education (Dæhlen et al., 2013). Additionally, the process of determining eligibility for upper secondary education was lengthy and complex (Dæhlen et al., 2013).

The most recent reform aimed at increasing upper secondary education completion rates (Ministry of Education and Research, 2021), extends the right to education to a new group of adult students, namely those who previously failed or dropped out. Now, under the reform, students are guaranteed the right to continue their education until they obtain a qualification, regardless of how long it takes. This ensures that students without a qualification can no longer lose their entitlement to complete upper secondary education.

Moreover, the same white paper introduces a *vocational requalification right (rekvalifiseringsrett)* (Ministry of Education and Research, 2021). It argues that, unlike students in higher education who have the flexibility to pursue additional degrees or study programmes, students in upper secondary education with a vocational qualification, face unfair restrictions in upper secondary education when seeking to obtain an additional qualification. The *requalification right* allows adults who have already completed upper secondary education to obtain a new vocational qualification. This is particularly relevant for those seeking to change careers, or to enhance their skills to meet new demands in the labour market. This right grant adults' access to upper secondary education in order to obtain a vocational trade certificate, without having to pay for the education.

In Table 3.1, we provide an overview of the rights that have been granted to adult learners over the past 10 years.

Table 3.1 Overview of new rights granted to adults in VET the last 10 years

Granted right	*When*
Right for adults aged 19–24 to access upper secondary education	2017
Right to access upper secondary education for immigrants with this qualification completed outside of Norway	2017
Right to complete upper secondary education (Fullføringsrett)	2024
Vocational requalification right (Rekvalifiseringsrett)	2024

Measures Aimed at Introducing More Flexible Pathways in VET

The 2015 white paper on adult learners introduced three key measures designed to create more flexible educational pathways tailored to the specific needs of adult students, with the aim of potentially reducing the time required for training (Ministry of Education and Research, 2016). One of the main initiatives was the introduction of a new workplace-based pathway called *Trade Certificate on the Job* (*Fagbrev på jobb*). While the existing experience-based route had proven successful for adult workers, accounting for one-third of all annual trade certificates, it was not fully adapted to the demands of the modern labour market, in which part-time and short-term employment are increasingly common. To address this, a new pathway was introduced in 2018, requiring only 1 year of work experience, making it more accessible to workers in these employment situations.

Eligible candidates for the *Trade Certificate on the Job* remain employed during the training period, which allows them to remain in paid employment—a significant advantage of the scheme. Employers participating in the programme must be certified as approved training companies and receive an apprenticeship grant. The candidate's training plan is developed in collaboration with the company and the regional education authorities, ensuring that the training meets all required standards.

The second measure focuses on pilot testing a modularised programme for adults in 13 selected trades. A key principle established by the government is that adult learners should only undergo the training necessary to complete their trade certificate successfully (Ministry of Education and Research, 2016). To achieve this, the programme mandates a recognition of prior learning, allowing participants to begin their training in the appropriate module based on their existing skills and experience. The modules are organised flexibly, enabling concurrent study without requiring a set sequence or time period. Each module concludes with a certificate of competency.

The initial module, common to all participants, includes competency goals in the Norwegian language and social studies relevant to their vocational field. Training is predominantly workplace-based, with agreements between the regional education authorities and the training company outlining each party's responsibilities. To complete the programme, participants must pass the standard trade examination, but they must first succeed in a written exam to qualify for the final trade test.

The evaluation of the modular VET programme indicates that this pilot programme has been successful in creating effective pathways for adult learners and attracting new groups into vocational education (Dahle et al., 2024). On average, participants spent 2 years in training before completing the trade examination, with 50% of the nearly 1000 participants recruited from the labour market administration (NAV) and the introductory programme for newly arrived immigrants. However, the higher dropout rates among those recruited from the introductory programme suggest that this group may need additional support. One of the major challenges in modular VET remains ensuring income security for participants during their training. Preliminary findings show that completion rates during the pilot phase ranged between 30% and 40% (Dahle et al., 2024). Although the pilot programme still being under evaluation, the government decided to adopt modularised VET as the primary model for adult vocational education, starting in 2024.

The third measure introduces various schemes aimed at supporting immigrants within the VET system. These initiatives specifically target adult immigrants who are entitled to upper secondary education but lack the Norwegian language proficiency needed to participate in regular programmes. In 2023, three counties began pilot testing different 4-year models for specialised VET classes for adult immigrants, with the goal of developing national models based on the results of these pilot programmes. So far, two classes for chefs and one for professional drivers have been established, reflecting the needs of the labour market. A key feature of these models is the combination of early workplace-based learning within a dual training framework, alongside additional support for learning Norwegian. An overview of both established and new VET pathways for adult learners is presented in Table 3.2.

Discussion

The aim of this chapter was to focus on the opportunity structures for adult VET students in Norway to gain a better understanding of how the opportunities for adults to complete a VET qualification have changed since Reform94. We have described implemented measures and in what ways the participation of adults in VET has changed. We were mainly interested in major shifts in educational policy during the last 10 years that affect the transition system between education and work for adults.

Table 3.2 Established and new pathways in Norwegian VET for adult learners

Established pathways	*Characteristics*
Ordinary VET organised for adults	At school (2 years) followed by an apprenticeship period (2 years), total 4 years. Academic subjects as well as vocational subjects
The experience-based trade certification (*Praksiskandidatordningen*)	For workers already holding a relevant job without formal VET qualifications. Requires at least 5 years of relevant full-time work experience. Candidates keep their wage/salary in the workplace. Exception from academic subjects (L1, L2, maths etc.). Passing a theoretical examination is required before the trade examination can be arranged, awarding candidates a trade certificate
New measures since 2017	*Characteristics*
Trade certificate at work, from 2018 (*Fagbrev på job*)	For workers already holding a relevant job without formal VET qualifications. Requires at least 1 year of full-time relevant work experience in the vocation. Duration up to 4 years. Otherwise, the same regulations apply as for the experience-based trade certification
Modularised VET education for adults Pilot 2017–2023, main model from 2024	Offered in 13 vocations. Organised in 4–7 modules, focusing on competencies needed for the vocation, with relevant competencies from general subjects integrated in vocational training. Possible (and desirable) to start company training (as an apprentice) already during the first year. Recognition of skills and qualifications mandatory, possibly reducing the time frame of the education. Ordinary trade examination for the trade certificate
Different targeted models for immigrants, either following the ordinary model or dual models Pilot financed by IMDi from 2022 to 2026	Pilot projects financed by the directorate of integration and diversity (IMDi) as well as local pathways developed according to regional needs. Emphasis on dual training models, with company training used earlier and more often than in ordinary models. Language support is integrated with workplace learning. Extensive collaboration with employment authorities (NAV) to recruit unemployed immigrants

The analysis was guided by policy documents and administrative statistical data of participants in VET.

The various policy measures introduced over the past decade clearly demonstrate that adult learners have been a central focus in Norway‘s latest educational reforms. Over the 30 years since Reform94, there has been a gradual expansion of rights for adult learners, significantly improving their formal rights, and thereby their opportunities to access VET. Reform94 marked a major shift by transforming a VET system that traditionally included many older youths and adults as apprentices, to one that prioritised the 16–19 age group (Høst et al., 2008). Today, access to VET in upper secondary education has been made equal across all age groups. The most recent reform grants lifelong access to VET for students without a qualification, ensuring they can continue their education until a qualification is completed (Ministry of Education and Research, 2016). Furthermore, measures implemented in 2017 extended access rights to young adults aged 19–24 and immigrants with foreign qualifications, thereby addressing gaps and eliminating inconsistencies in the statutory right to access VET. These reforms have the potential to simplify the often bureaucratic and time-consuming application processes within regional education authorities (Dæhlen et al., 2013). Moreover, they may contribute to the growing participation of adults in VET, which was a key focus in one of our research questions.

Until recently, the Norwegian transition system for adults consisted of two distinct structural features: a workplace-based route and the traditional dual route. In response to the need for better-targeted VET structures for adult learners, our analysis of the research question addressing measures introduced to support adults in completing a VET qualification reveals that two new pathways have been implemented. Additionally, a third pathway, specifically designed for adult immigrants, is currently in the pilot phase.

The new pathway outside the traditional school system, the *Trade Certificate on the Job*, expands the opportunities for adults to formalise skills and knowledge acquired through work-based learning in one's current job. This scheme bears some similarities to the well-established experience-based trade certificate, which has been recognised as a crucial second chance for adults without formal qualifications from upper secondary education (Bratsberg et al., 2017; Tønder & Aspøy, 2017). However, the new pathway may better serve individuals with low self-efficacy in theoretical subjects, particularly those who struggle with motivation for independent preparation for the trade examination. A key distinction is that in the new approach, the company assumes broader responsibilities for both guidance and training for their employee enrolled in the scheme (Lossius, 2021).

A completely new feature of the transition system for VET in Norway is being implemented with the modular programmes for 13 vocations, representing a significant transformation of the VET system. Similar modular approaches have been adopted in many European countries (CEDEFOP, 2015), with their advantages often cited in terms of flexibility and individualisation (Li & Pilz, 2017). The Norwegian government emphasises this approach's ability to cater to the individual needs of adult learners (Ministry of Education and Research, 2016). A modular structure allows for greater adaptation to a candidate's prior knowledge and personal circumstances, potentially leading to a shorter and more efficient pathway to the final trade examination. This flexibility can serve as a strong motivational factor for adults. Dahle et al. (2024) report that during the pilot phase of the modular program, the average time to earn a trade certificate was 2 years. However, there were significant variations between candidate groups. For example, candidates with a higher proficiency in Norwegian complete the programme more quickly.

The third pathway is targeted at immigrants, and the intention of the government is to develop national programmes for this group to increase participation and integration in working life. However, the modular programme already attracts a significant number of immigrants. More than 60% of participants are born outside of Norway (Dahle et al., 2024). Additionally, our analysis of the immigrant proportion of participants reveals a sharp increase in the proportion of adult immigrant students in VET over the last 25 years, from less than 10% to 50% of participants. This development signals that the VET system already attracts a considerable number of participants with an immigrant background. This development suggests that the necessity of introducing additional pathways for immigrants may be overstated.

In line with our findings, Aspøy et al. (2023) indicate that immigrants' rights to education have become more clearly defined in policy and certain barriers have been removed over time, making upper secondary education an increasingly more widely used tool for integration. A central measure that could explain the growth of immigrant participants, is that newly arrived immigrants could enrol full-time in upper secondary education as a part of the scheme called *the introduction programme* from 2016

(Aspøy et al., 2023). One important explanation of the total increase in adult participation in VET is therefore growing immigrant participation—probably due to the influx of immigrants in Norway, as well as in Europe, from 2015.

The white paper on adult learning and integration strategy documents emphasise upper secondary VET as a key tool for integrating vulnerable groups and immigrants with unstable connections to the labour market (Ministry of Education and Research, 2016, 2018). This is unsurprising, as the literature on VET highlights its strong potential for integration. Dual VET systems offer opportunities to gain work experience and become socialised into the written and unwritten workplace norms of the host country (Aerne & Bonoli, 2023; Jørgensen, 2022; Jørgensen et al., 2021).

However, the consequences of using the VET system in welfare and integration policies are rarely discussed or questioned in the Norwegian context. Aerne and Bonoli (2023) argue that collective skill formation systems, such as Norway's, are complex and depend on coordinated agreements between the government and trade unions. Governments attempting to enhance the system's integrative capacity, without considering employers' perspectives, may encounter resistance. Such initiatives may be risky, and in some cases, governments have had to backtrack. In Denmark, for example, increased access for disadvantaged youth harmed the reputation of the VET system, leading employers to withdraw (Aerne & Bonoli, 2023). Conversely, if employers perceive mutual beneficial outcomes—particularly in occupations facing labour shortages—the use of VET in integration policy may be considered to be more successful (Aerne & Bonoli, 2023; Stalder et al., 2023).

It is too early to assess the extent to which the new opportunities that surround adult learners in VET lead to increased participation and higher employment rates. However, the extension of formal rights within a transition system that offers multiple pathways to a trade certificate for adult learners lays a strong initial foundation for an inclusive system.

References

Aerne, A., & Bonoli, G. (2023). Integration through vocational training. Promoting refugees' access to apprenticeships in a collective skill formation system. *Journal of Vocational Education & Training*, *75*(3), 419–438. https://doi.org/10.1080/13636820.2021.1894219

Andersson Joona, P. (2019). Labour market policies: What works for newly arrived immigrants? In L. Calmfors & N. S. Gassen (Eds.), *Integrating immigrants into the Nordic labour markets* (pp. 85–110). Nordic Council of Ministers. Retrieved from https://norden.diva-portal.org/smash/get/diva2:1317928/FULLTEXT01.pdf

Arbo, P. (2022). Livslang læring—Fra slagord til reform [Lifelong learning—From slogan to reform]. *Uniped, 45*(2), 82–92. https://doi.org/10.18261/uniped.45.2.2

Arendt, J. N. (2022). Labor market effects of a work-first policy for refugees. *Journal of Population Economics, 35*(1), 169–196. https://doi.org/10.1007/s00148-020-00808-z

Arendt, J. N., & Bolvig, I. (2020). *Early labor market entry, language acquisition and labor market success of refugees*. VIVE: The Danish Center for Social Science Research.

Aspøy, T. M., Kavli, H. C., & Tønder, A. H. (2023). Da klokka klang … Utdanningens plass i integreringspolitikken [The role of education in integration policy]. In T. Fløtten, H. C. Kavli, & S. Trygstad (Eds.), *Ulikhetens drivere og dilemmaer* (pp. 237–257). Universitetsforlaget. https://doi.org/10.18261/9788215065403-23-14

Blommaert, L., Muja, A., Gesthuizen, M., & Wolbers, M. H. J. (2020). The vocational specificity of educational systems and youth labour market integration: A literature review and meta-analysis. *European Sociological Review, 36*(5), 720–740. https://doi.org/10.1093/esr/jcaa017

Bratsberg, B., Nyen, T., & Raaum, O. (2017). Fagbrev i voksen alder [Vocational qualifications in adulthood]. *Søkelys på Arbeidslivet, 34*(1–2), 24–43. https://doi.org/10.18261/issn.1504-7989-2017-01-02-02

Bratsberg, B., Nyen, T., & Raaum, O. (2020). Economic returns to adult vocational qualifications. *Journal of Education and Work, 33*(2), 99–114. https://doi.org/10.1080/13639080.2020.1722988

Bredgaard, T., & Thomsen, T. L. (2018). Integration of refugees on the Danish labour market. *Nordic Journal of Working Life Studies, 8*(S4), 7–26. https://doi.org/10.18291/NJWLS.V8IS4.111161

Busemeyer, M. R., & Trampusch, C. (2012). The comparative political economy of collective skill formation. In M. R. Busemeyer & C. Trampusch (Eds.), *The political economy of skill formation* (pp. 3–38). Oxford University Press.

CEDEFOP. (2015). *The role of modularisation and unitisation in vocational education and training* (Working paper no 26). Retrieved from https://www.cedefop.europa.eu/files/6126_en.pdf

Dæhlen, M., Danielsen, K., Strandbu, Å., & Seippel, Ø. (2013). *Voksne i grunnskole og videregående opplæring [Adults in primary, lower and upper secondary education]* (Report 7/13). Retrieved from https://oda.oslomet.no/oda-xmlui/bitstream/handle/20.500.12199/5056/6820_1.pdf?sequence=1

Dahle, M., Nordhagen, I. C., Jones, H., & Monsen, M. (2024). *Fag- og yrkesopplæring på voksnes premisser: Sluttrapport fra følgeevaluering av forsøk med Modulstrukturert fag-og yrkesopplæring* (Rapport 2/2024). Retrieved from https://ideas2evidence.com/nb/publications/sluttrapport-fra-evaluering-av-forsok-med-modulstrukturert-fag-og-yrkesopplaering-og

Geven, S., & Spörlein, C. (2023). Who benefits from school-to-work linkages in the labour market? A comparison between natives, migrants educated abroad, and those educated domestically. *European Sociological Review, 39*(3), 418–432. https://doi.org/10.1093/esr/jcac038

Høst, H. (2008). Inclusion of adults in VET. In H. Høst (Ed.), *Continuity and change in Norwegian vocational education and training* (pp. 53–72). NIFU. Retrieved from https://nifu.brage.unit.no/nifu-xmlui/handle/11250/283433

Høst, H., Gitlesen, J. P., & Michelsen, S. (2008). How the number of apprenticeships are influenced by policy and economic cycles. In H. Høst (Ed.), *Continuity and change in Norwegian vocational education and training* (pp. 17–29). NIFU. Retrieved from https://nifu.brage.unit.no/nifu-xmlui/handle/11250/283433

Jørgensen, C. H. (2015). Challenges for work-based learning in vocational education and training in the Nordic countries. In S. Bohlinger, U. Haake, C. H. Jørgensen, H. Toiviainen, & A. Wallo (Eds.), *Working and learning in times of uncertainty: Challenges to adult, professional and vocational education* (pp. 157–171). Sense Publishers.

Jørgensen, C. H. (2022). Are apprenticeships inclusive of refugees? Experiences from Denmark. In L. M. Herrera, M. Teräs, P. Gougoulakis, & J. Kontio (Eds.), *Migration and inclusion in work life: The role of VET* (pp. 342–372). Atlas Förlag.

Jørgensen, C. H., Järvinen, T., & Lundahl, L. (2019). A Nordic transition regime? Policies for school-to-work transitions in Sweden, Denmark and Finland. *European Educational Research Journal, 18*(3), 278–297. https://doi.org/10.1177/1474904119830037

Jørgensen, C. H., Hautz, H., & Li, J. M. (2021). The role of vocational education and training in the integration of refugees in Austria, Denmark and Germany. *International Journal for Research in Vocational Education and Training, 8*(3), 276–299. https://doi.org/10.13152/IJRVET.8.3.2

Kis, V., & Windisch, H. C. (2018). *Making skills transparent: Recognising vocational skills acquired through workbased learning* (OECD Education Working Papers No. 180).

Li, J., & Pilz, M. (2017). Modularisation in the German VET system: A study of policy implementation. *Journal of Education and Work, 30*(5), 471–485. https://doi.org/10.1080/13639080.2016.1243233

Lossius, R. (2021). *Fagbrev på jobb: En alternativ modell for opplæring og styrket rekruttering i telekommunikasjonsmontørfaget* [Master's degree, NTNU]. Retrieved from https://ntnuopen.ntnu.no/ntnu-xmlui/handle/11250/2976527

Manhica, H., Berg, L., Almquist, Y. B., Rostila, M., & Hjern, A. (2019). Labour market participation among young refugees in Sweden and the potential of education: A national cohort study. *Journal of Youth Studies, 22*(4), 533–550.

Ministry of Education and Research. (2004). *Kultur for læring* [White paper/St.meld. nr. 30 (2003–2004)]. Utdannings- og Forskningsdepartementet.

Ministry of Education and Research. (2006). *... og ingen sto igjen. Tidlig innsats for livslang læring* [White paper/St.meld. nr. 16 (2006–2007)]. Kunnskapsdepartementet.

Ministry of Education and Research. (2016). *Fra utenforskap til ny sjanse—Samordnet innsats for voksnes læring*. [White paper/Meld. St. 16 (2015–2016)]. Kunnskapdepartementet. Retrieved from https://www.regjeringen.no/no/dokumenter/meld.-st.-16-20152016/id2476199/

Ministry of Education and Research. (2018). *Integrering gjennom kunnskap: Regjeringens integreringsstrategi 2019–2022.* Kunnskapsdepartementet. Retrieved from https://www.regjeringen.no/contentassets/519f5492da984d1083e8047011a311bd/regjeringens-integreringsstrategi-2019-22.pdf

Ministry of Education and Research. (2021). *Fullføringsreformen—med åpne dører til verden og fremtiden* [White paper/Meld. St. 21 (2020–2021)]. Kunnskapdepartementet.

Ministry of Education, Research and Church Affairs. (1992). *Kunnskap og kyndighet: Om visse sider ved videregående opplæring* [White paper/St.meld. nr. 33 (1991–1992)]. Kirke, -Utdanning og Forskningsdepartementet.

Ministry of Education, Research and Church Affairs. (1998). *Kompetansereformen* [White paper/St.meld. nr. 42 (1997–1998)]. Kirke, -Utdanning og Forskningsdepartementet.

Ministry of Education, Research and Church Affairs. (1999). *Videregående opplæring* [White paper/St.meld. nr. 32 (1998–1999)]. Kirke, -Utdanning og Forskningsdepartementet.

Müller, W., & Gangl, M. (2003). *Transitions from education to work in Europe. The Integration of Youth into EU Labour Markets*. Oxford Univeristy Press.

NOU 1991:4. (1991). *Veien videre til studie- og yrkeskompetanse for alle*. Kirke,-Utdannings- og Forskningsdepartementet.

NOU 2003:16. (2003). I første rekke—Forsterket kvalitet i en grunnopplæring for alle.. Utdannings- og Forskningsdepartementet.

OECD. (2000). *From initial education to working life*. OECD. https://doi.org/10.1787/9789264181816-en

OECD. (2023). *Education at a glance 2023: OECD indicators*. OECD. https://doi.org/10.1787/e13bef63-en

Pilz, M. (2016). Typologies in comparative vocational education: Existing models and a new approach. *Vocations and Learning, 9*(3), 295–314. https://doi.org/10.1007/s12186-016-9154-7

Pilz, M., Li, J., Canning, R., & Minty, S. (2018). Modularisation approaches in Initial Vocational Education: Evidence for policy convergence in Europe? *Journal of Vocational Education & Training, 70*(1), 1–26. https://doi.org/10.1080/13636820.2017.1392994

Raffe, D. (2008). The concept of transition system. *Journal of Education and Work, 21*(4), 277–296. https://doi.org/10.1080/13639080802360952

Smyth, E., Gangl, M., Raffe, D., Hannan, D., & McCoy, S. (2001). *A comparative analysis of transitions from education to work in Europe (CATEWE): Final report*.

Stalder, B. E., Kammermann, M., Michel, I., & Schönbächler, M.-T. (2023). Successful integration of refugees in vocational education and training: Experiences from a new pre-vocational programme. In *Migration, education and employment: Pathways to successful integration* (pp. 133–154). Springer International Publishing.

Sterri, E. B., Andresen, S., Reegård, K., & Rogstad, J. (2015). *Grenser for utdanning: Rett til videregående opplæring for voksne [Barrieres for education: The right to upper secondary education for adult learners]* (Fafo-rapport 2015:27).

Støren, L. A., Skjersli, S., & Aamodt, P. O. (1998). *I mål? Evaluering av Reform 94: Sluttrapport fra NIFUs hovedprosjekt (NIFU-rapport 18/98)*. NIFU.

Tønder, A. H., & Aspøy, T. M. (2017). When work comes first: Young adults in vocational education and training in Norway. *International Journal for Research in Vocational Education and Training, 4*(3), 270–288. Retrieved from https://www.econstor.eu/handle/10419/172022

Vangen, T. (2007). *Nasjonal utdanningsdatabase. Dokumentasjonsrapport*. SSB. Retrieved from https://www.ssb.no/a/publikasjoner/pdf/notat_200754/notat_200754.pdf

Walther, A. (2006). Regimes of youth transitions: Choice, flexibility and security in young people's experiences across different European contexts. *Young, 14*(2), 119–139.

CHAPTER 4

100+ Years of Historical Opportunity: Vocational/Career and Technical Education in the State of Ohio

Chris Zirkle

In the United States, the country's educational systems at all levels are decentralized, and in accordance with the U.S. Constitution, each state is responsible for the education of its students. Due to its 50 states and differing systems, there can be variability in how grade levels are constructed, specifically at the elementary and secondary levels. This can also impact curriculum, teacher qualifications, school completion/graduation requirements, and many other education-related aspects. Needless to say, this decentralization also affects the delivery of vocational education (the commonly used term in the U.S.) courses and programs. If the significant diversity of economic, geographic and societal variables

C. Zirkle (✉)
Workforce Development and Education, College of Education and Human Ecology, The Ohio State University, Columbus, OH, USA
e-mail: zirkle.6@osu.edu

Norwegian University of Science and Technology, Trondheim, Norway

E. Smeplass and C. Zirkle (eds.), *Opportunity Structures in Vocational Education and Training*, Palgrave Studies in Technical and Vocational Education and Training,
https://doi.org/10.1007/978-3-032-04410-5_4

within the states is factored in, the result is substantial differences in the type, quantity and quality of vocational education programming across the country.

For years, vocational education had a negative connotation in educational circles. It was perceived as lower quality, "second class" education that offered limited economic and educational outcomes for students. While this perception existed for a significant period of time across the country, over time the importance of vocational education became apparent, and the term "career and technical education", or CTE, began to be used in the late 1990s. Thus, due to the change in terminology, the author will use the current term, career and technical education (CTE), when referring to the present-day, in lieu of vocational education.

There have always been pockets of merit within various American states with respect to vocational education. This chapter will focus on one such state, Ohio, which has a long history of excellence with vocational education. For more than 100 years, Ohio has held a national reputation for delivering one of the best state-developed systems in the United States of America. The chapter will highlight the many opportunity structures that exist within the present Ohio career and technical education system.

The State of Ohio

Ohio has long been an American state with a need for well-trained workers, dating back even before the first federal legislation in 1917 to fund vocational schooling in the state. Ohio's geography consists of mostly flat plains, and the state is comprised of approximately 29 million acres, of which 47% is dedicated to agricultural production in such areas as corn and soybean production, livestock, poultry and dairy products (United States Department of Agriculture, 2023). The state of Ohio ranks among the top producers of cheese, eggs and soybeans, and with more than 1500 food companies (Jobs Ohio, 2024a), there are numerous food manufacturing facilities within the state. International corporations including Campbell's, Dannon, General Mills, Kraft and Nestle all have a significant presence within the state. As a result, virtually every school district in Ohio has some type of career and technical education course or program focused on aspects of agriculture, and particularly in rural schools, agriculture is the primary career and technical focus.

Ohio also has a lengthy history related to industry and manufacturing. Indeed, manufacturing contributed significantly to the growth of major cities in Ohio, including rubber manufacturing in Akron, glassmaking in Toledo, the development of a "mechanical money drawer" (Ohio Federal Writers Project, 1930) otherwise known as a cash register, in Dayton and steel manufacturing in Youngstown. These products, and many others, have long required workers with the skills developed in vocational courses and programs. Today, Ohio is the third-largest manufacturing state in the country, with over 13,500 manufacturing companies employing over 800,000 workers (Industry Select, 2024). Some notable companies include General Electric, which manufactures turboprop and turboshaft engines for cargo, executive, and passenger jets, Honda, known worldwide for production of a range of passenger vehicles, and the Procter & Gamble Company, a multinational consumer goods corporation renowned for its wide range of household and personal care products. As with the agricultural economy, the needs of manufacturing industries have resulted in many career and technical courses and programs related to subjects such as computer-aided design and manufacturing, precision machining and welding.

Lastly, and most recently. Ohio has become a hub for healthcare and life sciences. More than 4200 bioscience companies are located in Ohio, employing approximately 84,000 life science-related employees (Jobs Ohio, 2024b). Healthcare research among universities and other research organizations exceeded 1 billion USD in fiscal year 2023 (Allchin, 2024) and the state has nationally ranked hospitals such as the Cleveland Clinic and The Ohio State University Wexner Medical Center. Healthcare focused on the welfare of children is significant, with Cincinnati Children's Hospital Medical Center and Nationwide Children's Hospital in Columbus, ranking #1 and #6 in the nation, respectively among hospitals specializing in the healthcare of younger children (Jobs Ohio, 2024b). Innovations in healthcare have impacted career and technical education in the state as well, with new courses and programs focused on exercise science, health informatics and bioscience.

The U.S. Education System: An Overview

The structure of the formal U.S. education system includes 12 years of regular schooling, potentially preceded by a year or two of pre-school education and followed by a four-stage higher education degree system (associate, bachelor's, master's, doctorate) plus various non-degree certificates and diplomas (United States Department of State, 2025). Within each state, particularly within the Kindergarten-12 system, there can be variations in how grade levels within schools can be structured. A general diagram of the system is depicted in Fig. 4.1.

In all states, services are provided for special education students and career and technical education. The diagram in Fig. 4.1 also provides information as to when career and technical education courses and programs are offered. In the United States, career and technical education is defined as education at below the baccalaureate level (Zirkle, 2016), so the shaded areas within the diagram illustrate this characteristic. Within this system, there may also be adult basic education programs for adults needing remediation, continuing education for professional education, and recreational learning programs, but how those are offered and by what educational entity may vary from state to state.

The Ohio CTE System: Built for Multiple Opportunities

As mentioned, career and technical education (CTE) is defined as education at below the baccalaureate level. In Ohio, courses and programs focused on CTE subjects can be found as part of the K–12 elementary and secondary system, 2-year colleges, universities, Ohio Technical Centers for adults, and through "second chance" and nonprofit providers. Career and technical education courses and programs at the K–12 level are administered through the Ohio Department of Education and Workforce, while the state's 2-year colleges, universities and Ohio's Technical Centers are overseen by the Ohio Department of Higher Education. Second chance and nonprofit providers have their own specific administrative structure(s). This wide variety of providers ensures that residents of Ohio have many opportunities to access career and technical courses and programs, while also considering their personal and professional education and economic needs. The administrative oversight provides quality verification of the courses and programs and ensures compliance with any curricular requirements, safety assurances and other regulatory obligations.

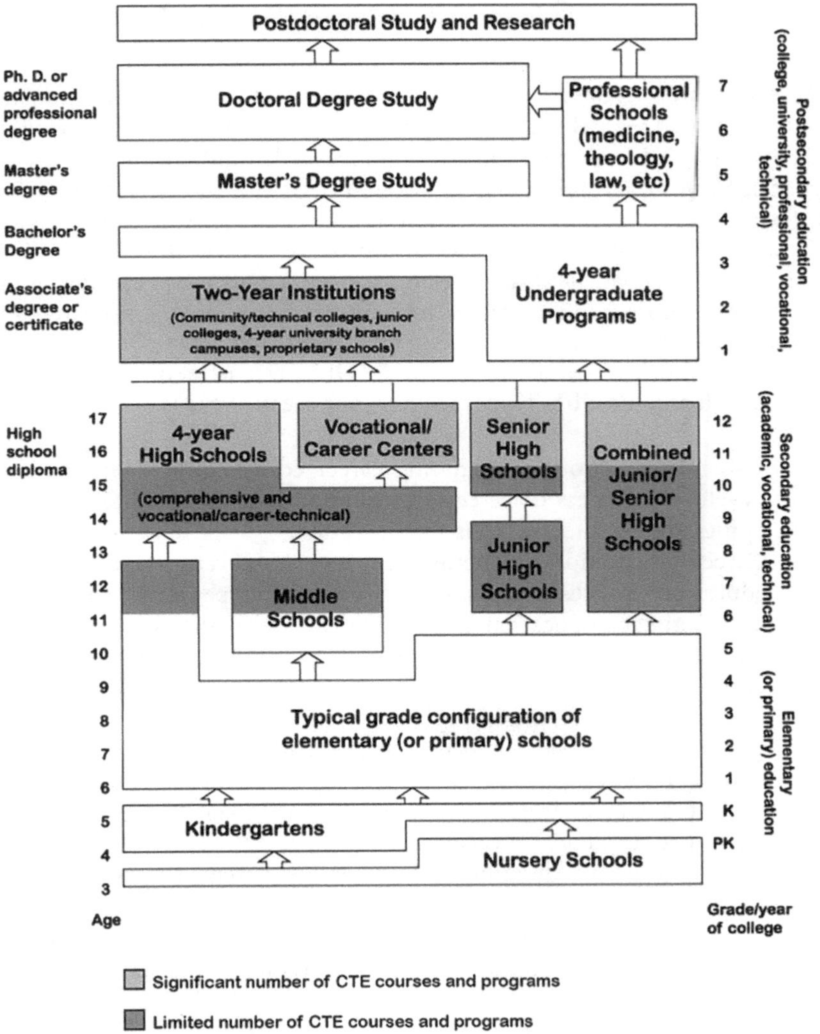

Fig. 4.1 The structure of education in the United States. Note: (Source: author's own compilation based on U.S. Department of Education, National Center for Education Statistics, Annual Reports Program)

K–12 Career and Technical Education in Ohio: The State's Largest Provider

According to the latest data from the Ohio Department of Education and Workforce (2025a), there are more than 1.7 million students enrolled in Ohio's K–12 education system with more than 136,000 of these students participating in a CTE course or program. Within the department, the Office of Career-Technical Education has direct supervisory responsibility for all of the state's CTE courses and programs at the K–12 level. The vision of the department as it relates to career and technical education is "Every Ohio secondary career tech program graduate is prepared for successful employment and ongoing education, and every Ohio business has the workforce it needs to prosper" (Ohio Department of Education and Workforce, 2025b). The office has four areas of focus:

- Career connections—integrating career connections (a program to develop awareness of careers) throughout the K–12 system, developing student success plans for all students
- Career preparation—increasing enrollment in career and technical education courses and programs, promoting pathways aligned to postsecondary education and in-demand occupations and ensuring the integration of high-level academic and technical knowledge and skills
- Career readiness—increasing the number of students obtaining credentials, certificates and licenses, preparing students for multiple options after high school, gainful employment and/or postsecondary study and collaborating with Ohio businesses to ensure they have the workforce needed to succeed
- Program quality—designing accountability systems so students have access to quality courses and programs, providing teachers with the professional development they need to meet high academic, technical and pedagogical standards and reducing barriers in credentialing career tech teachers through an alternative licensure program (Ohio Department of Education and Workforce, 2025b)

Course and Program Structure in K–12 Career and Technical Education

Ohio's K–12 school system is widely diverse in terms of students and also with respect to any variables related to the administrative structure of school buildings, grade levels, leadership and funding. Figure 4.1 provides some guidance on the basic grade level composition, but Ohio has 611 individual school districts (Ohio Department of Education and Workforce, 2025c), each with differing levels of capacity to arrange these variables in ways that work best for their specific situation. The following discussion of the course and program structure in K–12 career and technical education will capture the "big picture" but there are some singular and unique settings within the state that cannot be addressed in this chapter due to the sheer number and uniqueness of the numerous districts, schools and school buildings within the state.

Middle Schools: A New Area of Focus and Opportunity

Historically, the middle grades within the K–12 system (generally grades 6–8) have not been among the grade levels offering career and technical education. At most, short courses in such subjects as industrial arts and home economics were offered to introduce students to basic tools and homemaking skills. Many schools in Ohio were unable to offer even these short courses due to space or funding constraints. The focus at this grade level has generally been on primary academic subjects such as English, the natural sciences, mathematics and social studies, along with physical education and personal health.

Research has shown that middle school is a time when students can begin to benefit from career exploration, a process of building self-awareness, learning about potential careers, and developing a plan for reaching future goals (Bishop & Harrison, 2021: Callahan et al., 2019; Main & Whatman, 2023). Lacking these, as students exit middle school, they also can be unprepared for high school and grade nine is the grade level at which the greatest number of students drop out of high school.

In 2014, then-Ohio Governor John Kasich signed legislation which required schools to provide career and technical education courses in seventh and eighth grades by the 2015–2016 school year. Thus, Ohio became one of the few states to require the availability of career and technical education courses to middle school students. Districts that were unable or did not wish to provide these middle school courses had to submit a waiver to the Ohio Department of Education to avoid sanctions.

As a result of this legislation, Ohio has implemented a significant effort to offer more career and technical education courses to these grade levels with the goal of career exploration in order to discover work environments and understand the various aspects of the workplace (Ohio Department of Education, 2022). Most of the courses offered are exploratory in content but are aligned with CTE courses and programs at the high school level. For example, a middle school course titled *Agriculture, Food and Natural Resources* is designed to be the initial entry in a series of related courses such as Science and Technology of Food, Environmental Science for Agriculture and Natural Resources, and Business Management for Agricultural and Environmental Systems.

An additional option for schools wishing to offer CTE courses at the middle school level, but which may not have the specific facilities and required equipment, is to offer Career Connections. This course is designed to foster an understanding of the training and skills required for success in occupations within Ohio's 16 career clusters. Teachers and schools have some latitude in selecting classroom instruction and activities focused on specific career clusters relating to Ohio's in-demand jobs as well as skills and activities that align with local priorities (Ohio Office of Career-Technical Education, 2023).

High Schools: Career and Technical Education for Everyone

High schools in Ohio (most commonly grades 9–12) offer a wide slate of courses for all students (Zirkle, 2021) including academic (general) education, advanced placement and college preparatory courses, assistance for special needs learners, and career and technical education. In Ohio's high schools, the career and technical education courses and programs tend to reflect the local economy and workforce needs. For example, in rural areas, agriculture and family and consumer science courses and programs will be offered, while in urban areas, business and healthcare-related courses and programs are likely to top the list. One exception might be courses and programs in the technical trades, i.e., automotive technology, carpentry, cosmetology and welding. These are somewhat universal.

These schools have unique challenges in meeting the needs of all their students. Due to significant economic challenges such as persistent economic inequality, racial and gender wealth gaps, chronic unemployment and stagnant wages (The Center for American Progress, 2021), larger urban districts, such as Cleveland Municipal City Schools, in which

100% of students live in poverty, and rural districts like Western Local Schools, in which 82% of students are living in poverty (Ohio Department of Education and Workforce, 2025d), must recognize that these students struggle with many other issues in addition to completing their high school education. School personnel are challenged to provide additional services and assistance for these students and sadly, the relationship between poverty and dropping out of school is clearly supported by research (American Psychological Association, 2012; Bradley, 2022; Taylor, 2019). In addition to students living in poverty, schools have learners with special needs and English as a second-language learners.

In addition to meeting these challenges, many high schools face financial constraints in terms of the courses and programs they can offer, especially those outside the state-mandated academic (general) courses such as English, mathematics, the sciences and social studies. For decades high schools have been restricted in their CTE offerings due to space limitations, financial resources and difficulties in hiring career and technical teachers. In many instances, high schools have been limited to one or perhaps two courses and programs within their building, and as mentioned previously, these offerings will only reflect the local economy and workforce needs, and do not provide for a broadening of a student's outlook regarding potential job prospects or postsecondary education options.

Despite the challenges, Ohio's high schools within a school district have many different curricular options for offering CTE courses and programs. Historically, these courses and programs have been categorized into six broad fields of study (with sample course and program titles):

- Agriculture
  - Agribusiness, Animal Science, Horticulture, Production Agriculture
- Business
  - Accounting, Business Foundations, Business Information Systems, Finance
- Family and Consumer Science
  - Child Development, Food Science, Personal Finance, Textiles
- Health Occupations

  - Dental Assisting, Medical Assisting, Nurse Assisting
- Marketing
  - Entrepreneurship, Logistics, Marketing Management
- Trade and Industrial
  - Carpentry, Cosmetology, Drafting and Design, Electrical Trades, Welding

High schools could offer any of these courses and programs, pending approval from the Ohio Department of Education, and with appropriate space, equipment, funding and a qualified teacher. Most of these courses and programs were designed with one goal in mind: to give high school graduates sufficient knowledge and skills to enter the workplace into an entry-level job. Connections from these courses to postsecondary associate or bachelor's degrees were not well-developed.

Toward the end of the twentieth century, there was a realization that many vocational education programs taught outdated skills, were not responsive to the needs of the workplace, and left students with limited career options (Zirkle, 2016). In 1998, the term vocational education was replaced with career and technical education, and during this time, several organizations, including the U.S. Department of Education, the U.S. Office of Vocational and Adult Education and the National Association for State Directors of Career-Technical Education Consortium (now known as Advance CTE), began the development of "career clusters", a revised organization of career and technical education courses and programs. Sixteen career clusters were identified, as shown in Fig. 4.2.

These career clusters were developed to encourage the creation of inclusive, industry-responsive, and learner-centered career and technical education courses, programs and experiences (Advance CTE, 2025). Along with the development of these clusters, efforts to create articulated, or aligned, CTE coursework was initiated. When possible, these articulated courses begin with the exploratory courses in middle school, then in high school, the courses become more and more comprehensive and rigorous as the student moves through the "cluster" of courses. A possible middle/high school sequence might look like the following (using healthcare as the career cluster):

1. Career Connections (middle school grades 7–8)

16 Career Clusters

- Agriculture, Food, and Natural Resources
- Architecture and Construction
- Arts, A/V and Communications
- Business, Management, and Administration
- Education and Training
- Finance
- Government and Public Administration
- Health Science
- Hospitality and Tourism
- Human Services
- Information Technology
- Law, Public Safety, Corrections, and Security
- Manufacturing
- Marketing, Sales, and Service
- Science, Technology, Engineering, and Mathematics
- Transportation, Distribution and Logistics

Note. United States Department of Education, Office of Career, Technical and Adult Education.

Fig. 4.2 Sixteen career clusters. Note: United States Department of Education, Office of Career, Technical and Adult Education

2. Health Science and Technology (high school grade 9)
3. Principles of Allied Health (high school grade 10)
4. Lifespan Development (high school grade 11)
5. Human Anatomy and Physiology (high school grade 12)

This arrangement provides the student the opportunity to obtain a high level of knowledge and skills (there are other courses in the health-care cluster with significant technical skill development). The student is prepared to continue postsecondary education at a community college or university, or begin entry-level employment, dependent on course availability.

Developing sequences of courses requires significant resources on the part of a high school, which may not have the ability to offer such an in-depth curriculum within a particular cluster, especially for the 4 years of a student's high school experience. However, several decades ago, the state of Ohio had the foresight to see the challenges a stand-alone high school would have in offering high quality career and technical education courses and programs, and set about to develop a robust, collaborative effort to address this issue.

Career Centers: A Hallmark of Ohio Career and Technical Education

The limitations faced by high schools began to be addressed in 1963 when Ohio passed legislation clarifying the creation of joint vocational school districts. This legislation cited an imperative that all Ohio students would have access to vocational education courses and programs. The legislation permitted school districts with high schools to join together to create a collaborative school district focused on offering vocational education courses and programs to all students within the jointure, known at the time as a "joint vocational school district" or JVSD (Shoemaker & Parks, 2007). These newly designed districts were to be geographically "centered" within a set of school districts (and their respective high schools) in order to make transportation accessible. Each school district would contribute a small amount of their property tax income to support the operation of an area vocational school. These area vocational schools would offer a broad slate of vocational education courses and programs and students from each of the high schools could elect to attend the area vocational school in lieu of their "home" high school. Larger "comprehensive" districts, such as Columbus City Schools, Dayton Public schools and Cleveland Metropolitan School District, with large enrollments, were not part of these jointures, and some districts decided to create CTE "compacts", a partnership between two or more school districts to provide Career-Technical Education (CTE) services to their students. However, the majority of school districts were banded together in the joint vocational school district configuration. Figure 4.3 provides a visualization of this arrangement.

Area vocational schools had an immediate impact by making significantly more courses and programs available to greater numbers of

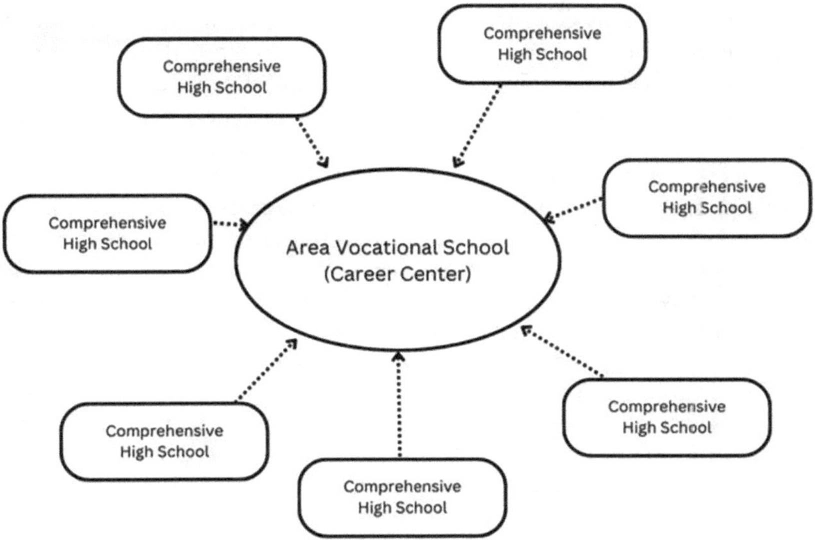

Fig. 4.3 Visualization of Ohio's career concept

students. It was also more cost-effective, as the expense of operating the schools was spread out among the participating school districts.

Today the original structural concept of the joint vocational school districts has been rearranged into 91 career-technical planning districts (CTPD's) (Ohio Department of Education and Workforce, 2025e), consisting of 25 comprehensive districts, 15 compact districts, two correctional districts (discussed later in this chapter), and 49 career center districts. The career center districts have one or more career centers (the new name for area vocational school) as the focal point of their efforts to provide career and technical education for the students in their geographic region.

In many respects, these career centers are very similar to their high school counterparts. They offer the academic (general) courses students need to fulfill high school graduation requirements, services and assistance for learners with special needs and English language learners, guidance and counseling and other ancillary services found in high schools. The major difference is in their ability offer a wide variety of career and technical education courses and programs.

Students generally elect to attend a career center in the last 2 years of high school (grades 11 and 12). They choose to leave their home high school and undertake a chosen career and technical program at the career center, and in most cases, spend the entire school day, taking academic courses for half of the day, and instruction and training in the chosen career-technical program for the other half (about 7 h in total). When all studies are completed, the student receives a certificate of completion from the career center (along with any technical, industry-based certifications) and a high school diploma from their home high school. The student is now prepared to move onto postsecondary education, enter the workforce, begin an apprenticeship program (see section "Apprenticeship", following) or enter the military, which is also an educational entity for further education and training.

The career center, originally developed as a way to make vocational education accessible to more Ohio students, now has become the primary educational choice for many Ohio students, Instead of the second-class status held in year's past, CTE is now at the forefront of Ohio's educational endeavors, as Ohio has seen CTE enrollment grow from 120,357 in 2012 to 136,691 in 2024, even as the number of secondary school students across the state has decreased over the same time period (Ohio Department of Education and Workforce, 2025f).

Ohio's 2-Year College System

A significant provider of career and technical education courses in Ohio is the state's well-established system of community colleges. These colleges are aptly named, as they are designed to generally serve students in their nearby communities. Community colleges were originally called junior colleges or 2-year colleges and date back to the Morrill Act of 1862, which expanded access into public higher education (Drury, 2003). However, prior to the development of community colleges in Ohio, the 2-year colleges in the state were technical colleges, with a strict focus on the skills needed to enter the workforce. Some of these technical colleges actually began as a technical institute and were an outgrowth of night classes offered by local high schools. Columbus State Community College began as the Columbus Technical Institute and met at Central High School, a high school that was part of the Columbus City School District. Clark State Community College began in 1962 as the Springfield and Clark County Technical Education Program. In 1966 the name was changed to Clark County Technical Institute and was chartered by the Ohio Board of Regents as Ohio's first technical college. In 1972 the institution's name was changed to Clark Technical College. In 1988 the Ohio Board of Regents approved the status change of Clark Technical College to Clark State Community College.

Over the years through the many name changes, the educational and curricular missions of these institutions have also significantly changed, enrollments have grown exponentially, and the state's community colleges now play a significant role with respect to both academic (general) education and career and technical education.

These institutions are supported financially in part by the state of Ohio. Currently there are 23 community colleges, geographically dispersed across the state (Ohio Department of Higher Education, 2025a). In the latest available data, more than 137,000 students were enrolled in Ohio's community colleges in the fall of 2024, with Columbus State Community College the largest with 26,781 students and Belmont Technical College the smallest with 839 students (Ohio Department of Higher Education, 2025b).

The Multiple Curricular Missions of Community Colleges in Ohio

Community colleges are known for their efficient response to the local workforce needs of their communities. As such, they offer not only associate degrees, but also certificate programs keenly focused on one aspect of workplace needs and customized in-company training specifically designed to meet the needs of a given company or organization.

The first type of associate degree offered by community colleges is an Associate of Arts degree. This degree, as the name suggests, is focused on subject areas related to the arts and humanities, and as such, does not have much connection to career and technical education or preparation for the workforce. An example of this degree would be an Associate of Arts degree in English or history. The degree itself is also designed to be the first 2 years of a Bachelor of Arts degree to be ultimately completed at an Ohio university. As part of the Ohio Guaranteed Transfer Pathways (Ohio Department of Higher Education, 2025c), courses required as part of an Associate of Arts degree are guaranteed to transfer and count toward meeting the same requirements of a bachelor's degree.

The second type of associate degree is an Associate of Science degree. This degree option functions much the same as the Associate of Arts with respect to guaranteed transferability but also may offer students with sufficient technical coursework to either enter the workforce or continue on to a Bachelor of Science degree. Examples of this degree include Associate of Science degrees in computer and information science, engineering technology and physics.

The third type of associate degree offered by community colleges in Ohio is the Associate of Applied Science degree. This degree has the closest connection to CTE, as the primary focus of the degree, once completed, is for the graduate to enter the workforce. The degree consists of a large proportion of technical coursework, with minimal academic (general) education. Courses completed as part of Associate of Applied Sciences do not generally transfer cleanly into a university degree program due to the heavy concentration of technical courses that will not meet university academic (general) education requirements.

The U.S. economy, along with employers in many technical fields, is facing significant shortages of technically skilled workers. Even though the associate degree pathway can be completed in 2 years, community colleges in Ohio have responded to this workforce need by creating numerous certificate programs that can be completed in 1 year of study or less. For example, Owens Community College in Toledo has certificate programs in computer programming, dental assisting and plumbing and pipefitting, all designed to be completed in 1 year or less. One additional benefit of many of these certificate programs is that they are articulated/aligned with an associate degree in a related field so if a student obtaining a certification wishes to continue their education, they have that opportunity.

Community colleges can also serve the community and create additional revenue by offering in-company training programs to area businesses and organizations. These programs, many times technical in nature, are customized to meet the specific needs of the client, which could be a local small business or a large corporation. The programs generally operate on a "fee-for-service" contract basis, with the client directly compensating the community college for services provided.

Ohio's University System: Still Technically a Provider of Career and Technical Education

Although career and technical education has been defined as education at below the baccalaureate level, Ohio's 4-year universities still play a significant role in several ways. First of all, despite primarily offering baccalaureate level and higher degrees, several universities offer associate degrees through their regional campuses. For example, Kent State University, through its eight regional campuses, offers associate degrees in such fields as cybersecurity, environment management, radiologic technology and respiratory care. Bowling Green State University offers associate degrees in digital content creation, diagnostic medical sonography and computer networking/cybersecurity. The Ohio State University, in addition to four regional campuses offering associate and bachelor's degrees, has one campus, the Agricultural Technical Institute, that specifically offers associate degrees in agricultural fields such as agricultural business management, crop management and soil conservation, livestock production and management and power equipment.

Another aspect of the university system is the ability to transfer credit from some of these associate degrees into a related bachelor's degree. While the primary focus of most associate degrees is for entry into employment, the transfer option is another opportunity for students to further their education. Many students take a job and carry on their studies on a part-time basis, while others may continue their education immediately after completion of the associate degree. Still others may enter employment, work for a few years, then return to university to complete the baccalaureate degree.

The largest universities in Ohio are financially supported by the state and are more likely to offer associate degrees in areas related to career and technical education. One exception is the University of Northwest Ohio, a private university which offers certificate, diploma, associate and bachelor's degrees in such fields as automotive technology and related fields, business and healthcare.

Ohio Technical Centers: A Special Focus on Adults

An outgrowth of Ohio's career centers, Ohio's Technical Centers are an association of independently operated career-technical centers offering educational programs specific to local workforce needs to adult students (Ohio Technical Centers, 2025). These centers operate cooperatively with the career centers that serve high school career and technical students. For example, during the day, a career center may offer a welding program to high school students, then in the evening, the technical center may offer the same, or similar program to adults. In this way, more opportunities are provided to potential students, and educational facilities and equipment are efficiently utilized.

Ohio's Technical Centers are focused on short-term, intensive education and training (significantly less time than an associate's degree from a community/technical college) and offer courses and programs designed to accommodate students who are changing jobs, have lost a job and/or are seeking to improve their current job situation by upskilling. Some courses and programs offered by Ohio's Technical Centers include the construction trades, healthcare, information technology, public safety and transportation.

Courses and programs at Ohio's Technical Centers are approved by the Ohio Department of Higher Education and in some cases, specific third-party national credentialing and accrediting agencies, such as The Council on Occupational Education, based in Atlanta, Georgia and The Accrediting Commission of Career Schools and Colleges, located in Arlington, Virginia. Both of these organizations are recognized by the U.S. Department of Education as the leading agencies for validating the quality of career and technical education courses and programs for adults.

Apprenticeship

The United States has had a system of apprenticeship for many years, although this method of training workers has not had the same impact or level of utilization found in many European countries. However, as more focus has been placed on the increasing costs of postsecondary education and accompanying student financial debt, apprenticeship has generated new interest within the U.S. population. In addition, the nationwide shortage of technically skilled workers has caused many individuals to consider apprenticeship as a training opportunity.

In Ohio, apprenticeship programs are overseen by the Ohio Department of Job and Family Services (ODJFS) through its ApprenticeOhio program (Ohio Department of Job and Family Services, 2025). As with many types of apprenticeship programs, apprentices are hired by employers and earn while they learn. Currently in Ohio, more than 22,000 individuals are engaged in a registered apprenticeship program, with approximately 170 employer partners (Ohio Bureau of Labor Market Information, 2025). More than half of the apprentices in Ohio are being trained in some aspect of the construction trades, with plumbers and pipefitters, electricians and sheet metal workers having the largest numbers of apprentices. Apprenticeships within manufacturing and those training nursing assistants represent the largest groups of apprentices not engaged in the construction trades.

Second Chance Career and Technical Education

In any society, certain individuals make decisions that ultimately are detrimental to their future. In Ohio's largest cities, there are significant numbers of students each year who do not complete their high school education, which not only limits employment opportunities but places these students into challenging economic conditions. Individuals who make further unwise decisions often end up in the court system, which may be followed by incarceration in a correctional institution.

Each of these situations merit second chances, and Ohio has provided these opportunities through different institutions. Career and technical education courses and programs play a prominent role in this effort. A first focus is on youth ages 16–24 who may have dropped out of high school, persons who have been adjudicated through the court system and those individuals who have few options to improve their personal situation (United States Department of Labor, 2025). For these individuals, the U.S. Department of Labor's Job Corps program provides general (academic) education for those who need a high school diploma or equivalent, and also provides CTE courses and programs in many areas within the construction trades as well as advanced manufacturing, hospitality and information technology. Ohio has three Job Corps centers, located in Cincinnati, Cleveland and Dayton.

A second focus occurs with individuals (both youth and adults) incarcerated in Ohio's correctional institutions. Education and training opportunities for them are provided by two school districts chartered by the Ohio Department of Education and Workforce. For adults, the Ohio Central School District is administered by the Ohio Department of Rehabilitation and Correction (2025) and provides programming. For youth, the Buckeye United School District (2025) is overseen by the Ohio Department of Youth Services. The educational programming in both districts is grounded in research that shows inmates who participate in correctional education programs, and specifically those with a focus on vocational training, have 43% lower odds of returning to prison than those who do not, and that every dollar spent on prison education saves four to five dollars on the costs of re-incarceration (Davis et al., 2013). Career-technical courses and programs offered by these institutions include automotive technology, barbering and cosmetology, construction trades, culinary arts, horticulture and welding. These courses and programs are developed using labor market data and consider potential barriers to entry upon release, i.e., only courses and programs where restored citizens can apply for employment are offered.

Nonprofit Organizations: Another Source of Career and Technical Education

Most states have citizens who are either unemployed or underemployed, and Ohio is no exception to this set of circumstances. Since most schooling past the K–12 system requires the payment of tuition, these individuals may find it difficult to access further education and training.

Nonprofit agencies and organizations assist in addressing this difficulty and offer these individuals opportunities to further their education and training at minimal to no cost, as well as providing “wrap-around” services such as bus/transportation assistance, guidance and counseling, and job placement assistance. Many of the education and training courses and programs are specific to the employment opportunities in the local area. For example, Goodwill Industries, a worldwide nonprofit which has a workforce development focus among its many other service initiatives, offers information technology and healthcare training at its center in Columbus, where many employment opportunities exist in these fields. Youth Opportunities Unlimited in Cleveland, in existence for over 40 years, has focused on preparing teens and young adults to become economically self-sufficient while also enhancing Northeast Ohio’s educated, skilled, and inclusive workforce (Youth Opportunities Unlimited, 2025). The Employers’ Workforce Coalition in Dayton, is a recently created nonprofit focused on working to fill the workforce gaps in science, technology, engineering, and mathematics (STEM) by ensuring that the future workforce has equitable access to STEM education, as well as resources and pathways to sustainable livable wages (The Dayton Foundation, 2025).

Linked Opportunities: Ohio's Career and Technical Education Is Connected

For many decades, vocational education and training in the United States (including Ohio to a degree) was seen as less than desirable, and academically capable students were steered away to focus on university preparatory studies, rather than consider vocational studies. However, over the past 30 years, there have been marked improvements in what is now called career and technical education (Zirkle, 2016). Perhaps the one significant event most responsible for this improvement in both the quality of the courses and programs as well as public perception, has been the dismantling of the two disparate paths: vocational vs. academic. There is a marked effort to blur the lines between the two disciplines by developing "career pathways". This is an outgrowth of the career clusters framework in order to create aligned K–12 and postsecondary education systems and, in particular, the career and technical education services provided within and across program providers (Perkins Collaborative Resource Network, 2025).

In the Columbus, Ohio, area, the healthcare industry is one of the largest employers, and the staffing of various personnel in the healthcare field is a major challenge for area hospitals and related businesses and organizations. Thus, a notable example to address this need is the collaboration among Columbus City Schools (the largest public school district in Ohio), Columbus State Community College (the largest community college) and The Ohio State University (the largest university) to develop a number of educational career pathways within the healthcare field.

As the shortage of nursing professionals is particularly acute, the three institutions partnered to create a nursing pathway by linking their respective curricula. Students entering the pre-nursing program at Columbus City Schools obtain academic and technical knowledge and skills. Through an Ohio program called College Credit Plus (Ohio Department of Higher Education, 2025d) via an agreement between the respective high schools within Columbus City Schools and Columbus State Community College, students can earn college credit for courses taken while in high school. This can significantly reduce the number of credit hours, as well as the accompanying tuition costs, needed to earn an associate degree. After high school, the student has the option to enter the workplace (where an employer may subsidize the cost of any additional education) or begin work on an associate degree at Columbus

State Community College. After completion of the associate degree, the student again has the option to enter the workplace, or transfer to The Ohio State University, where the student can complete the Bachelor of Science degree in nursing in approximately two years (and this option is available in an online format). This unique arrangement provides a student multiple opportunities to begin working or pursue further education. Table 4.1 provides a deeper understanding of this cooperation among the three institutions.

Examples of this aligned and articulated curriculum including the three levels (high school, community college, university) did not exist just a few years ago. Today, a student completing a career and technical education pathway may choose to enter the workforce, return to school, then reenter the workforce, and pause at different points, without many of the drawbacks inherent in earlier vocational programs. This example illustrates

Table 4.1 Nursing career pathway between three Ohio educational institutions

Institution	*Program*	*Description*	*Potential additional credentials*	*Options after completion*
Columbus City Schools	Pre-Nursing State-Tested Nursing Assistant	Two years in high school (grades 11–12). One college course earned to transfer to Columbus State Community College	State-Tested Nursing Assistant credential	1. Workplace 2. AAS degree at Columbus State Community College
Columbus State Community College	Associate of Applied Science in Nursing	Two years of postsecondary education (less transferred courses) from Columbus City Schools	Registered Nurse credential	1. Workplace 2. BSN degree at OSU
The Ohio State University	Bachelor of Science in Nursing	Four years of postsecondary education (or 2 years if AAS degree from Columbus State Community College)	Registered Nurse credential	1. Workplace 2. Master's Degree in Nursing

how the former vocational and academic paths are now merging, which provides students with more options to improve their personal and professional lives. This work is ongoing, in order to offer more opportunities for the citizens of Ohio.

Conclusion

Over time, the state of Ohio has created and developed a well-defined and articulated system of career and technical education. This system provides multiple opportunities for students to explore and obtain a variety of education and training credentials, as well as offering provisions for second chances or an interest in upskilling. This chapter provided valuable examples of how these opportunities can emerge from collaborations between various stakeholders in creating inclusive and relevant pathways, to the benefit of students, schools, business and industry and society in general.

References

Advance CTE. (2025, January 2). *About the national career clusters framework*. Retrieved from https://careertech.org/career-clusters/about-the-national-career-clusters-framework/

Allchin, T. (2024). *Ohio reaches $1 billion in NIH funding*. Retrieved from https://www.jobsohio.com/blog/ohio-reaches-1-billion-in-nih-funding

American Psychological Association. (2012). *Facing the school dropout dilemma*. Author. Retrieved from http://www.apa.org/pi/families/resources/school-dropout-prevention.aspx

Bishop, P., & Harrison, L. (2021). *The successful middle school: This we believe*. Association for Middle Level Education.

Bradley, K. (2022). *The socioeconomic achievement gap in the U.S. public schools*. Retrieved from https://ballardbrief.byu.edu/issue-briefs/the-socioeconomic-achievement-gap-in-the-us-public-schools

Callahan, J., Ito, M., Rea, C., Wortman, S., & Wortman, A. (2019). *Influences on occupational identity in adolescence: A review of research and programs*. Connected Learning Alliance.

Davis, L., Bozick, R., Steele, J., Saunders, J., & Miles, J. (2013). Evaluating the effectiveness of correctional education: A meta-analysis of programs that provide education to incarcerated adults. Rand Corporation. Retrieved from https://bja.ojp.gov/sites/g/files/xyckuh186/files/Publications/RAND_Correctional-Education-Meta-Analysis.pdf

Drury, R. (2003). Community colleges in America: A historical perspective. *Inquiry, 8*(1).

Industry Select. (2024, October 21). *Top 10 manufacturing companies in Ohio*. Retrieved from https://www.industryselect.com/blog/top-10-manufacturing-companies-in-ohio

Jobs Ohio. (2024a, December 28). *Food and agribusiness*. Retrieved from https://www.jobsohio.com/industries/food-and-agribusiness

Jobs Ohio. (2024b, December 28). *Healthcare and life sciences*. Retrieved from https://www.jobsohio.com/industries/healthcare

Main, K., & Whatman, S. (2023). *Health and well-being in the middle grades: Research for effective middle level education*. Information Age Publishing.

Ohio Bureau of Labor Market Information. (2025, January 5). *Statewide registered apprenticeship dashboard*. Retrieved from https://ohiolmi.com/Home/Dashboards/ApprenticeOhio_Dashboard

Ohio Department of Education. (2022). *Expanding career-technical education in Ohio*. Retrieved from https://education.ohio.gov/getattachment/Topics/Career-Tech/Planning-Funding-and-Accountability/Expanding-Career-Technical-Education-in-Ohio.pdf.aspx?lang=en-US

Ohio Department of Education and Workforce. (2025a, January 2). *About*. Retrieved from https://education.ohio.gov/About

Ohio Department of Education and Workforce. (2025b, January 6). *Career technical education vision and mission*. Retrieved from https://education.ohio.gov/Topics/Career-Tech/Contacts-and-Resources/Career-Technical-Education-Vision-and-Mission

Ohio Department of Education and Workforce. (2025c). *Facts and figures*. Retrieved from https://education.ohio.gov/Topics/Data/Frequently-Requested-Data/Facts-and-Figures

Ohio Department of Education and Workforce. (2025d). *Rural education*. Retrieved from https://education.ohio.gov/Topics/School-and-District-Improvement/Rural-Education

Ohio Department of Education and Workforce. (2025e). *Ohio Career-Technical Planning Districts (CTPDs)*. Retrieved from https://education.ohio.gov/getattachment/Topics/Finance-and-Funding/State-Funding-For-Schools/Career-Technical-Funding/Ohio-CTPDs.pdf.aspx

Ohio Department of Education and Workforce. (2025f, January 8). *Enrollment data*. Retrieved from https://education.ohio.gov/Topics/Data/Frequently-Requested-Data/Enrollment-Data

Ohio Department of Higher Education. (2025a, January 3). *Ohio's campuses*. Retrieved from https://highered.ohio.gov/about/ohios-campuses/map

Ohio Department of Higher Education. (2025b, January 4). *Preliminary headcount (fall 2024)*. Retrieved from https://highered.ohio.gov/data-reports/data-and-reports-sa/enrollment/preliminary-headcount/preliminary-headcount-2024

Ohio Department of Higher Education. (2025c, January 3). Ohio guaranteed transfer pathways. Retrieved from https://transfercredit.ohio.gov/students/student-programs/ogtp

Ohio Department of Higher Education. (2025d, January 4). *College credit plus*. Retrieved from https://highered.ohio.gov/initiatives/access-acceleration/college-credit-plus

Ohio Department of Job and Family Services. (2025, January 2). *ApprenticeOhio*. Retrieved from https://jfs.ohio.gov/job-services-and-unemployment/job-services/job-programs-and-services/apprentice-ohio/overview

Ohio Department of Rehabilitation and Corrrection. (2025, January 3). *Education and training*. Retrieved from https://drc.ohio.gov/systems-and-services/3-education-and-training/overview/education-and-training

Ohio Department of Youth Services. (2025, January 3). *Buckeye United School District*. Retrieved from https://dys.ohio.gov/facilities/buckeye-united-school-district

Ohio Federal Writers' Project. (1930). *National cash register toolmaking department*. Retrieved from http://www.ohiomemory.org/cdm/ref/collection/p267401coll34/id/7693

Ohio Office of Career-Technical Education. (2023). *Career connections course manual*. Retrieved from https://education.ohio.gov/getattachment/Topics/Career-Tech/CTE-Middle-Grade-Programming/Career-Connections-Course-Manual-2023-CC-PMG-2023-04-05.pdf.aspx?lang=en-US

Ohio Technical Centers. (2025). *About us*. Retrieved from https://ohiotechnicalcenters.com/about-us/

Perkins Collaborative Resource Network. (2025, January 7). *Career pathways systems*. Retrieved from https://cte.ed.gov/initiatives/career-pathways-systems

Shoemaker, B., & Parks, D. (2007). *A history of vocational and career education in Ohio: 1828–2000*. IUniverse, Inc..

Taylor, K. (2019, July 25). *Poverty's long-lasting effects on students' education and success*. Retrieved from https://www.insightintodiversity.com/povertys-long-lasting-effects-on-students-education-and-success/

The Center for American Progress. (2021, January 12). *The basic facts about children in poverty*. Retrieved from https://www.americanprogress.org/article/basic-facts-children-poverty/

The Dayton Foundation. (2025). *Employers' workforce coalition*. Retrieved from https://www.daytonfoundation.org/dayton-communities-center-for-talent-development

United States Department of Agriculture. (2023). *Ohio farm numbers*. Retrieved from https://www.nass.usda.gov/Statistics_by_State/Ohio/Publications/Current_News_Releases/2024/nr2410oh.pdf

United States Department of Labor. (2025, January 3). *Careers begin at Job Corps*. Retrieved from https://www.jobcorps.gov/

United States Department of State. (2025, January 4). *Structure: General information*. Retrieved from https://usinfo.org/enus/education/overview/edlite-structure-us1.html

Youth Opportunities Unlimited. (2025, January 2). *The support teens and young adults need to pursue a job or career in Northeast Ohio*. Retrieved from https://www.youcle.org/

Zirkle, C. (2016). A qualitative analysis of high school level vocational education in the United States: Three decades of positive change. In M. Pilz (Ed.), *Youth in transition: Vocational education and training in times of economic crisis* (pp. 321–337). Springer.

Zirkle, C. (2021). Trends and issues in career and technical education in the United States. In Y. Lee & L. S. Lee (Eds.), *Trends and issues in international technical and vocational education in the Indo-Pacific region*. Wu-Nan Books.

CHAPTER 5

Advancing Secondary Vocational Education System: A Case Study of Korea's Meister High School

Seong Ji Jeong

Introduction

Vocational education and training (VET) plays a critical role in addressing national demands for a skilled workforce while also supporting individual growth and development (Gray & Herr, 1998). Despite this importance, VET in Korea faces persistent challenges, largely due to credentialism, with 74% of high school graduates enrolling in college (Korea Education Statistics Service, 2025a). Vocational high schools are often stigmatized, perceived as pathways for academically or economically disadvantaged students and associated with poor working conditions (Kim & Ryu, 2022). As of 2015, 73% of vocational high school graduates opted for

S. J. Jeong (✉)
Workforce Education and Development Program, Learning and Performance Systems Department, College of Education, Pennsylvania State University, University Park, PA, USA
e-mail: sfj5484@psu.edu; mercyme1277@gmail.com

E. Smeplass and C. Zirkle (eds.), *Opportunity Structures in Vocational Education and Training*, Palgrave Studies in Technical and Vocational Education and Training,
https://doi.org/10.1007/978-3-032-04410-5_5

college enrollment, while only 19% pursued direct employment (Korea Education Statistics Service, 2015). However, such trends highlight a persistent mismatch between education and the labor market, imposing youth unemployment and economic inefficiency (Montt, 2017). This misalignment has drawn considerable attention from policymakers and researchers (Kim et al., 2023).

To address these challenges, the Korean government launched a new VET initiative at the upper secondary level: Meister High School (MHS), in 2010, among a select group of regular vocational high schools. This program aims to strengthen vocational education and develop a technically skilled workforce for the nation's industry. Unlike regular vocational high schools (i.e., Specialized High Schools), MHS operates under school-based management, granting autonomy over curriculum design, hiring, and industry partnerships. Additional features include small class sizes, extensive financial support, and industry-level facilities (Kim et al., 2023). A transformative complementary policy, the *Work First, Study Later* approach, allows students to secure immediate employment after graduation while providing expansive incentives for higher education later (Kim & Ryu, 2022). With these advantages, MHS attracts highly achieving students and prepares them for direct workforce entry through its industry-tailored curriculum.

Since its inception, MHS has consistently produced strong employment outcomes, maintaining a high employment rate even during economic downturns: 90.3% in 2013, 91.2% in 2015, 71.2% in 2020, and 73.7% in 2023 (Korea Education Statistics Service, 2024a). As of 2024, 54 Meister high schools operate nationwide, specializing in traditional industries as well as emerging fields such as robotics, semiconductors, smart manufacturing, and smart farming (Ministry of Education, 2024c). The MHS system has gained international recognition, with countries like Malaysia, Indonesia, and Saudi Arabia, exploring MHS as a model for secondary VET reform (Yu et al., 2021). Research on MHS effectiveness highlights its role in enhancing career readiness, self-efficacy (Maeng, 2014; Park, 2022), and employment outcomes, such as wages, job retention, and employment status (Ko et al., 2021; Park & Yu, 2023). The MHS opportunity structure provides students with technical certifications and industry experiences, offering a strong alternative to academic pathways while fostering the next generation of skilled technology workers.

Investigating how the MHS program achieves such positive outcomes and enhances social mobility for VET students provides valuable insights into the key factors behind the success of VET systems. Korea's MHS model offers important lessons for countries seeking to strengthen secondary vocational education and improving school-to-work transitions. Operating within a government-driven education framework, it establishes a strong foundation for sustaining high-quality programs, ensuring legal, fiscal, and institutional structures, as well as stakeholder accountability among students, teachers, schools, and industry partners. The government's efforts to enhance industry engagement and improve the social perception of vocational education offer key takeaways for enhancing the sustainability and quality of VET systems.

This chapter examines Korea's MHS system, focusing on how it creates an opportunity structure that enhances the quality of vocational education and nurtures a competitive, technically skilled workforce. It explores how MHS strengthens vocational education through an industry-aligned curriculum, school-employer-government partnerships, and rigorous performance management. The chapter concludes by discussing MHS's sustainability amid demographic shifts and technological advancements, offering insights into its potential adaptation in other national contexts.

Foundational Framework of Meister High Schools

Korea's Secondary Vocational Education System

Korea's secondary vocational education system highlights the distinct role of MHS compared to regular vocational high schools. Examining the secondary education system first, after completing 6 years of primary school and 3 years of middle school, students choose from four high school types: General High Schools (69.0%), Specialized High Schools (20.5%), Special Purpose High Schools (6.7%), and Autonomous High Schools (3.8%)—totaling 2,380 schools as of 2024 (Korea Statistics, 2024).

Among them, vocational education is primarily offered through three programs (see Fig. 5.1). First, specialized high schools (SHS) are regular vocational high schools that provide specialized vocational courses within a 3-year curriculum, accounting for the largest share of vocational high schools (477 schools). Second, MHS (54 schools) also follows a 3-year

Program Type	Distribution	Features
Specialized High Schools	• 477 schools • Around 173,000 enrolled students • Around 14% of entire high school population	• Regular vocational high schools • Offer vocational specialized courses in **three-year** curriculum
Meister High Schools	• 54 schools • Around 16,800 enrolled students • Around 1.3% of entire high school population	• Classified under *Special Purpose High Schools* that offer gifted education • Offer vocational specialized courses in **three-year** curriculum
Vocational Courses in General High Schools	• 49 schools • Around 5,200 enrolled students • Around 0.4% of entire high school population	• Offer **one-year** vocational curriculums by sending students to vocational agencies, industrial schools, or 2-year colleges

Fig. 5.1 Korea's Secondary Vocational Education System. (Source of Distribution: Korea Education Statistics Service (2024b, 2025b))

vocational curriculum but focuses specifically on direct employment after graduation. A notable feature is its classification as a special purpose high school that offers gifted education. By placing MHS within this category, the program enhances the perception of vocational education among students, parents, and industry stakeholders. Together, SHS and MHS constitute the majority of Korea's secondary vocational education system, and many researchers and policymakers compare MHS specifically with SHS as its counterpart.

Lastly, a small number of general high schools (49 out of 1,642) offer vocational courses for students transitioning from the general education track. Typically, seniors (12th graders) participate in vocational training through industrial schools or community colleges. This 1-year program is overseen by the Ministry of Employment and Labor (MOEL), while SHS and MHS are overseen by the Ministry of Education (MOE).

Background of Establishing MHS Programs

Korea's vocational education history provides context for the MHS program's introduction. The rapid economic development in the 1970s and 1980s expanded secondary vocational education to meet the growing technical workforce demand. However, during the 1990s, the shift toward computerization and globalization increased the need for highly skilled professionals, leading to the expansion of higher education (Chang, 2006). Consequently, vocational education focused on nurturing rudimentary workers rather than specialists (Kim & Park, 2020). Over time, vocational high schools faced significant challenges due to the social stigma associated with choosing vocational tracks over higher education. This led to a decline in the proportion of vocational high schools from 41.6% in 1995 to 17.2% in 2012 (Bae et al., 2013). Moreover, 73% of vocational high school students opted for college enrollment rather than employment (Korea Education Statistics Service, 2015). This overemphasis on higher education has led to considerable concern about the education-occupation mismatch and its socioeconomic costs (Kim et al., 2023).

The MHS initiative is rooted in this highly stratified education system (Kim et al., 2023). To address this challenge, the *High School Diversification 300* project was launched during the Lee Myung-bak administration in 2008 to reform high school programs (Kim & Park, 2020). Based on neoliberal principles, the initiative positioned education as a service provider for the labor market while increasing schools' accountability for workforce development (Park & Yu, 2023). To support this, the government provided extensive financial aid and regulatory autonomy, allowing schools to tailor curricula and operations. Named after the German word *Meister* (master), MHS aimed to produce technically skilled workers with a strong emphasis on direct employment (Primary and Secondary High School Law, Article 90:1). The program also responded to the 2008 economic crisis, which led to mass layoffs and increasing unemployment. Seeking to mitigate these challenges, Korea adapted elements of Germany's dual vocational education system while customizing it to fit the Korean context (Lee et al., 2016).

The first cohort of 21 MHSs were selected from high-quality vocational high schools (i.e., SHSs) and launched in 2010 (Yu et al., 2021). As of 2024, 54 schools are in operation, with five additional schools specializing in information technology and semiconductors set to open by 2026 (Ministry of Education, 2023, 2024b). Each MHS tailors its curriculum to a specific industry (see Table 5.1). The majority of schools focus on Mechanical and Metal and Electric and Electronics, reflecting South Korea's major industry.

Table 5.1 Meister high schools and their relevant industry fields

Category	*No. of MHS*	*No. of MHS entering students per year*	*Notable examples*
Mechanical & Metal	13	1,853	Busan Mechanical Technical High School, Daegu Il Meister High School, Hapduk Steel High School, etc.
Electric, Electronic, Semiconductors Nanotechnology, and Smart Manufacturing	9	1,148	Seoul Robot High School, Chungbuk Semiconductor high School, Asan Smart Factory Meister High School, etc.
Energy & Chemical	6	444	Ulsan Energy High School, Yeosu Petrochemical High School, Korea Nuclear Meister High School School, etc.
Agriculture, Food, Fisheries	6	472	Daegu Agriculture Meister High School, Korea Food Meister High School, Gimje Agricultural and Life Science Meister High School, etc.
Information Technology & Software	5	380	Mirim Meister High School, Busan Software Meister High School, Daedeok Software Meister High School, etc.
Aerospace & Shipbuilding	4	516	Hyundai Technical High School, Geoje Technology High School, Air Force Aviation Science High School, etc.
Medical & Bio Manufacturing	3	272	Korea Bio Meister High School, Wonju Medical High School, Gyeongbuk Bio Meister High School
Marine	3	300	Incheon Maritime High School, Korea, Korea Marine Meister High School
Harbor & Logistics	1	90	Korea Port Logistics High School
Overseas Construction	1	126	Seoul City Science Technical High School
Firefighting	1	64	Korea Firefighting Meister High School
International Trade	1	60	Korea Global Business Meister High School

(continued)

Table 5.1 (continued)

Category	*No. of MHS*	*No. of MHS entering students per year*	*Notable examples*
Gaming	1	80	Gyeonggi Game Meister High School
Total	54	5,909	–

Source: Ministry of Education (2024a)

Key Elements of MHS in Comparison with SHS

The distinguishing elements of MHS are as follows: (1) career pathways toward young Meisters, (2) regulatory exemptions for greater school autonomy, and (3) extensive fiscal support (Lee et al., 2016). First, career pathways offer incentives for graduates to pursue higher education while remaining employed, along with overseas employment support. Vocational high school graduates with 3 years of work experience qualify for alternative college admission procedures and government tuition assistance (Ministry of Education, 2024c). Additionally, alternative military service options allow male graduates to defer service for up to 4 years or serve as industrial technical personnel (Yu et al., 2021).

Second, regulatory reforms grant MHS full autonomy in curriculum and instructional material development, allowing them to align education with industry needs (Enforcement Rule of the Elementary and Secondary Education Act 73:2). Schools also recruit principals from industry professionals and ensure teacher quality through professional development programs. Finally, government fiscal support includes free tuition, dormitory accommodations, and student living stipends, alongside funding for facility upgrades (Kim et al., 2012).

For comparison, MHS aims to nurture professional technical workers to enter the workforce upon graduation, while SHS offers pathways to either employment or higher education. MHS operates with greater autonomy and stronger industry integration based on school-based management, while SHS follows standardized regulations set by the provincial Offices of Education. MHS maintains smaller class sizes (16 students or fewer per classroom) to enhance learning quality (MOE & KRIVET, 2024). Both MHS and SHS students receive tuition-free education, scholarships, and vocational training opportunities, while MHS

Table 5.2 MHS comparison with SHS

	(Regular) Specialized High Schools (SHS)	*Meister High Schools (MHS)*
Start	• Regular vocational high schools	• Began in 2010, enhanced vocational high schools
Objectives	• Nurturing technical forces pursuing *employment* and *college enrollments*	• Nurturing technical forces pursuing *employment* immediately upon graduation
Features	• Regulated by the provincial Office of Education in terms of curriculum and materials, personnel recruitment and appointments, etc. • Free tuition, fees, etc.	• School-based management, having complete control over curriculum and materials, hiring principals or staff, etc. • Free tuition, fees, dorms, live stipends, etc.
Numbers	• 477 schools, around 150,000 students	• 54 schools, around 15,000 students
Employment Rate	• 40.9% (2013), 46.6% (2015), 49.2% (2020), 53.3% (2023)	• 90.3% (2013), 91.2% (2015), 71.2% (2020), 73.7% (2023)

Note: The source of the employment rate comes from Korea Education Statistics Service (2015, 2024b).

provides free dormitory accommodations and living stipends (Yu et al., 2021). Table 5.2 illustrates the comparison between MHS and SHS.

Descriptive Trends of Student Enrollment

Figure 5.2 presents trends in SHS and MHS student enrollment over time. Notably, Korea's school-age population has declined significantly, reducing overall student enrollment in secondary and vocational education. The total number of high school students fell from 1,893,303 in 2013 to 1,278,269 in 2023, a 32.5% decrease (Korea Education Statistics Service, 2025b). This demographic shift has also led to a sharp decline in vocational education enrollments, with a 45.5% decrease over the same period.

Regarding MHS, enrollment has increased from 15,728 students in 2013 to 16,798 in 2023, despite a decrease in recent years. MHS maintains small student cohorts to uphold educational quality while expanding schools in line with the MOE's strategic goals, particularly in emerging industries. In line with this strategy, the MOE has announced plans to

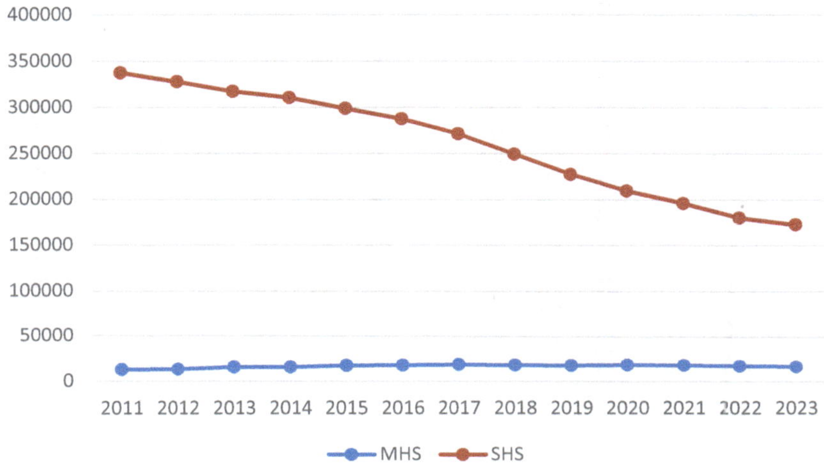

Fig. 5.2 Number of students enrolled in SHS and MHS. (Source: Korean Educational Statistics Service (2024a))

increase the number of MHS from 54 schools in 2024 to 65 by 2027 (Ministry of Education, 2023, 2024b). Amid declining youth populations, this shift reflects a focus on high-quality vocational education to meet evolving industry demands.

Figure 5.3 shows a detailed view of MHS student enrollment by gender. Gender distribution remains skewed, with 85.2% of MHS students being male in 2023, largely due to the concentration of MHS programs in traditionally male-dominated fields such as engineering and technology, mechanics, electronics, and information technology.

Figure 5.4 illustrates MHS student enrollment by program field over years. Engineering and Technology has been the most dominant field, particularly in the early years of the program's introduction. This reflects the MOE's strategic decision to prioritize MHS programs in Engineering and Technology, a key industry in Korea. Over time, MHS has diversified beyond this field, with additional enrollment in Agriculture, Marine, Fishery, Home Economics, and Business and Trade (Ministry of Education, 2024d).

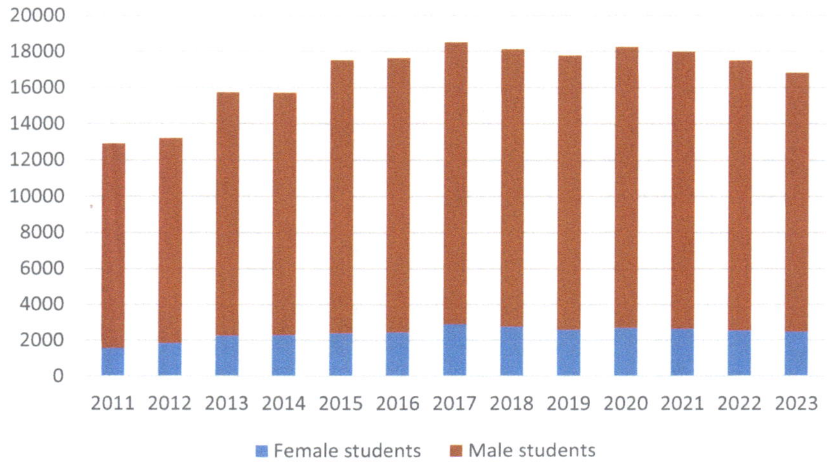

Fig. 5.3 Number of students enrolled in MHS by gender. (Source: Korean Educational Statistics Service (2024a))

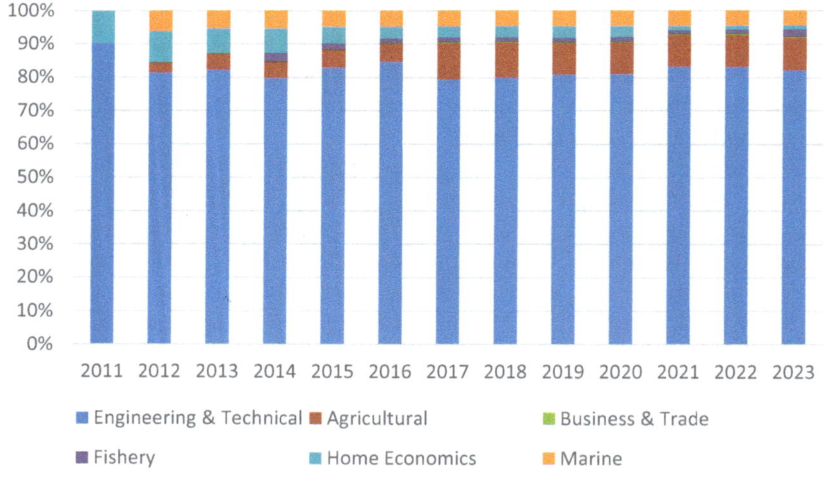

Fig. 5.4 Number of students enrolled in MHS by program fields. (Source: Korean Educational Statistics Service (2024a))

Strengths of Meister High School Program

Successful vocational education programs feature specialized training, strong industry ties, and direct employment opportunities (Raffe, 2003). This section investigates MHS's strengths, including a flexible curriculum, school-employer-government partnerships, performance management, and policy support. Particularly, it presents a case study of Busan National Mechanical Technical High School (BMT) to illustrate how these components function at the school level.

BMT, a public school established in 1967, aims to nurture skilled workers for the mechanical engineering industry. As of 2025, the school offers four programs: Machine (Precision Machine and Polymechanics), Industrial Equipment, Electronic Control (Robotics Tech and Mechatronics) and Electrical Engineering. It serves 710 students with 115 teachers and 68 staff members. BMT was among the first MHS cohorts designated in 2009, overseen by the Ministry of SMEs (Small and Medium Enterprises) and Startups, and has since maintained its competitiveness through high student employment rates, success in the SkillsKorea Competitions, and other achievements (Busan National Mechanical Technical High School, 2024).

Curriculum Tailored to Industry Demand

To keep up with technological advancements and evolving industry needs, MHS implements a flexible curriculum system through the industry-government-school cooperation steering committee. At BMT, 10th graders build foundational occupational skills before selecting a major in their second semester. In 11th grade, students focus on specialized skills and obtain course-based national technical qualification if they successfully complete competency-based courses aligned with the National Competency Standards (NCS). Finally, 12th graders take advanced specialized and elective courses within or across majors. By graduation, students complete 70 credits of general courses, 118 credits of specialization courses, and 18 credits of creative experiential activities, totaling 206 credits—14 credits more than the MOE's standard requirements of 192 credits (Ministry of Education, 2022).

Project-based learning (PBL) is integral to classroom instruction. At BMT, students take two or four credits of PBL courses per semester, applying critical thinking, decision-making, and problem-solving skills.

PBL involves hands-on experience in industrial design, machining, welding, plumbing, electrical work, and electronics with continuous teacher feedback. These activities help students develop both technical expertise and soft skills such as communication and teamwork. Some student projects have led to patent applications, intellectual property competitions, entrepreneurship proposals, and participation in the Korean Olympiad in Informatics (Busan National Mechanical Technical High School, 2017).

A significant curriculum reform was the introduction of *the elective credit system* in 2020. Previously, students followed a rigid 3-year curriculum with fixed course schedules, but the new system allows for individualized learning experiences and expands career pathways. For example, BMT students majoring in electronics can take elective courses from other majors, such as mold fabrications or robot tool development. BMT is also expanding elective courses through industry partnerships. Beyond the curriculum, BMT offers *after-school activities* in job search techniques, computer skills, foreign languages, writing, and book reading. Students prepare for industrial certifications in computer-applied milling, lathe operations, and mechanical assembly. They also dedicate after-school hours for the SkillsKorea competition, patent projects, entrepreneurship projects, and global training programs.

School-Employer-Government Relationships

MHS fosters strong industry-school-government partnerships to ensure graduate employment. Government agencies facilitate Memorandums of Understanding (MOUs) between schools and industries. The MOE has secured agreements with major corporations such as Samsung, LG, and Hyundai, as well as public enterprises, to expand job opportunities for MHS graduates (Kim et al., 2023). Additionally, the MOE has introduced the Intellectual Property Meister (IP Meister) program in collaboration with the Ministry of SMEs and Startups and the Korean Intellectual Property Office, enabling vocational high school students to develop and transfer their ideas and innovations to industry.

Individual schools establish industry ties through faculty, alumni, and employer networks. At BMT, 734 companies have signed MOUs, allowing for industry-aligned curricula. Schools also operate Industry-Government-School Cooperation Steering Committees (IGSCSC), comprising the principal, industry representatives, school staff, and public institutions. Industry partners advise on skill needs, curriculum development, work-based learning, employment, teacher training, and equipment upgrades (Busan National Mechanical Technical High School, 2025). Government and association representatives, such as those from the Busan Regional SMEs and Startups Office or the Korea Venture Business Association, support industry partnerships and student recruitment efforts.

Beyond the MOE, other government ministries collaborate with MHS to provide sector-specific support in planning, implementation, and performance management, offering consulting, financial aid, and employment opportunities (Ministry of Education, 2016). Since 2012, the Ministry of National Defense has trained Air Force aviation technicians at MHS (Air Force Aviation Science High School), with graduates serving 7 years as non-commissioned officers before transitioning to military or civilian aviation careers (Korea Policy Briefing, 2011). Similarly, BMT has directly partnered with and overseen by the Ministry of SMEs and Startups. Table 5.3 lists MHS schools and their respective government partnerships.

Table 5.3 A list of MHS partnering with Ministries

Government Ministries	*Number of MHS*	*Examples of MHS*
Ministry of Small and Medium Enterprises (SMEs) and Startups	5	Busan Mechanical Technical High School, Asan Smart Factory Meister High School
Ministry of National Defense	1	Air Force Aviation Science High School
Ministry of Agriculture, Food and Rural Affairs	6	Daegu Agricultural Meister High School, Korea Food Meister High School
Ministry of Culture, Sports and Tourism	1	Gyeonggi Game Meister High School
Ministry of Trade, Industry and Energy	3	Korea Nano Meister High School, Seoul Robot High School
Ministry of Science and ICT	4	Busan Software Meister High School
Ministry of Oceans and Fisheries	4	Busan National Maritime High School; Wando Fisheries High School

Source: Ministry of Education (2024a)

Enhancing Standards of Students, Teachers, and Schools

MHS ensures quality through continuous performance monitoring of students, teachers, and schools. It upholds high standards in student achievement, teacher development, and a 5-year school evaluation cycle.

Students

To ensure graduate competency, MHS awards the Young Meister Certification alongside high school diplomas, assessing occupational skills, foreign language proficiency, computer literacy, and professionalism (KRIVET, 2018). Students must master both technical expertise and soft skills, such as communication and teamwork, to succeed in the workplace (Yu et al., 2021).

At BMT, certification criteria include basic occupational (300), specialized vocational (300), foreign language (200), computer (200), and interpersonal (200) skills, totaling 1200 points. Assessments include course-based evaluations, the National Foundational Competency Assessment (mandatory for all vocational high school students), industrial qualifications, SkillsKorea competitions, foreign language proficiency tests, and other relevant evaluations. Graduates receive certification at five levels: A (1080+), B (960–1079), C (840–959), D (720–839), and E (below 720). Regularly updated, the certification helps place graduates in jobs and provide access global training to potential employers and global training programs, while promoting best practices among MHS institutions (Yu et al., 2021).

Research shows that MHS students report higher school satisfaction, career maturity, and self-efficacy compared to SHS and general high school students (Bae, 2012). Confidence in vocational curricula, problem-solving skills, and self-efficacy further contribute to a clearer perception of career pathways (Lee & Chung, 2019). Nearly 70% of MHS students intend to enter the workforce upon graduation, and half plan to pursue higher education while employed (Song et al., 2023). Their higher career maturity reflects greater confidence in career planning, stronger employment aspirations, and enhanced employability (Park, 2022).

Teachers

Teachers play a critical role in vocational education, particularly those with industry experience. MHS recruitment requires hiring a minimum percentage of teachers with industrial backgrounds and attracting highly qualified educators from other schools (MOE & KRIVET, 2024). However, the proportion of teachers with industry experience declined from 18.2% in 2012 to 15.9% in 2023 (KRIVET, 2023). Yu et al. (2021) highlight a growing shortage of industry-certified teachers, suggesting further investigation into the causes.

MHS teachers balance multiple duties, including curriculum planning, classroom management, academic advising, career counseling, student recruitment, event coordination, and safety management. They also oversee SkillsKorea competition training, industrial certification preparation, facility and equipment management, and cooperative education programs. A study found that specialized vocational subject teachers report higher workloads than general subject teachers (Ahn & Yang, 2023). Despite this, MHS teachers often report greater job satisfaction, citing professional growth, strong school culture, and supportive classroom environments (Lee et al., 2022).

Professional development is essential for aligning the MHS curriculum with evolving industry needs. Schools offer on-site training each semester in emerging fields, such as semiconductors, robotics, and artificial intelligence, through partnering firms, training institutes, or community colleges. In schools, teachers also engage in professional learning communities through monthly committee meetings on curriculum development, school safety, grading and evaluation, industry partnerships, and cooperative education. Senior teachers mentor early-career educators, providing guidance on student advising and instructional strategies.

Schools

MHS undergoes a rigorous 5-year performance management cycle, overseen by the MOE through the Korea Research Institute for Vocational Education and Training (KRIVET) (Kim et al., 2012). Schools are either re-designated or annulled based on evaluation results in the fifth year (Yu et al., 2021). This system ensures program quality through continuous feedback and assessment.

The application process starts when the MOE announces its plan for MHS designation in in-demand industries. Candidate specialized high schools (SHSs) must demonstrate clear educational objectives, curriculum feasibility, workforce targeting strategies, recruitment plans for principals and teachers, student admission and support frameworks, facility and equipment plans, support from the provincial Office of Education, and established partnerships with the provincial government, industry, or relevant ministries (MOE & KRIVET, 2024).

Once designated, schools receive approximately ₩5 billion ($3.4 million) over 3 years to prepare for reopening and ₩500 million ($0.34 million) annually for school operation (MOE & KRIVET, 2024). The MHS Support Center at KRIVET assists with selection, onboarding, implementation, and evaluation of MHS in accordance with MOE guidelines. Its roles include consulting on school operations, curriculum design, industry partnerships, and teaching strategies. It also oversees performance management, including field inspections, interim reviews, and preliminary and final evaluations (Kim et al., 2012). The final evaluation includes curriculum implementation, industry partnerships, employment outcomes, and stakeholder satisfaction from students, graduates, employers, parents, and teachers (Yu et al., 2021). These efforts ensure that MHS aligns with industry needs and remains a leader in secondary vocational education.

Advancing Career Pathways: "Work First, Study Later" Program

The *Work First, Study Later* initiative, introduced by the MOE, supports vocational high school graduates in securing immediate employment while ensuring pathways to higher education once they are employed. This approach offers an alternative to direct college enrollment, allowing students to gain work experience before pursuing further studies.

Vocational high school graduates with at least 3 years of work experience can apply to colleges and universities through an alternative admission process (Enforcement Decree of the Higher Education Act 29:2). Unlike traditional admissions, which heavily rely on standardized test scores (the College Scholastic Ability Test) or high school grades, this system allows universities to admit students beyond their quota through alternative criteria. As of 2024, 5,130 students have entered twenty-three 4-year colleges through this special admission track (Korean Council for University Education, 2025). Jang (2023) highlights that this initiative reflects a broader shift in higher education's role in lifelong learning, particularly in response to declining student populations. Vocational graduates are eligible to receive financial aid from government for their lifelong learning opportunities.

Beyond traditional universities, Korea National Open University (KNOU) offers PRIME College programs, providing online degrees tailored for working professionals, including MHS and SHS graduates, in fields such as business, engineering, and artificial intelligence (Korea National Open University, 2025). Another alternative is contract-based college programs, established through partnerships between industry and government to train workers for specific sectors. These programs offer specialized courses for both current employees and new hires, who enroll with guaranteed job placements upon graduation. Similarly, corporate colleges, where employers run MOE-accredited educational programs, provide another pathway to higher education. These institutions provide firm-specific training that ensures employees acquire up-to-date industry knowledge while earning degrees. For instance, Samsung Institute of Technology, part of Samsung Semiconductor, has offered bachelor's, master's, and doctoral degrees to nearly 1,100 working professionals (Samsung Semiconductor Newsroom, 2022).

While expanding higher education access for vocational graduates, the government has also promoted hiring incentives across both public and private sectors. This policy led to an increase in employment rates from 16.2% in 2009 to 53.6% in 2017. However, rates declined to 29.2% in 2021 (Ahn & Kim, 2022), due to demographic shifts, the COVID-19 pandemic, and ongoing concerns about job quality for high school graduates. Nonetheless, policymakers, researchers, and educators continue to advocate for expanding alternative career pathways for vocational students (Jang, 2023). Ensuring the long-term sustainability of this initiative requires a sufficient supply of quality jobs, strengthening public and corporate support, and further research into its impact on student career trajectories.

Opportunity Structure of MHS Programs

This section examines MHS graduates' employment trends, societal perceptions, and global influence. Research using the nationally-approved Vocational High School Graduates Employment Statistics (VHSGES) (Korea Education Statistics Service, 2024b) as well as the Korean Education and Employment Panel (KEEP) longitudinal survey, funded by KRIVET, has significantly contributed to understanding MHS graduates' employment transitions and the cost-effectiveness of the program.

Employment Rate Trend

Since the first MHS graduates in 2013, employment rates have consistently remained high, ranging between 80% and 90% in the first 6 years (90.3% in 2013, 91.2% in 2015), compared to significantly lower rates for SHS graduates (40.9% in 2013, 46.6% in 2015) (Korea Education Statistics Service, 2015). However, after peaking in 2017, employment rates for both groups declined. In 2020, the VHSGES system was introduced, integrating student enrollment data with employment and health insurance records, overseas work visas, and employment in agriculture, forestry, and fisheries. Recent data indicate employment rates have stabilized around 70% for MHS graduates and 50% for SHS graduates (Korea Education Statistics Service, 2024b) (See Fig. 5.5). For SHS graduates, the rate of students who went on to further education (advanced rate) was comparable to or higher than the employment rate.

By gender, employment rates in 2024 were 71.2% for male graduates and 80.4% for female graduates, with military service partially explaining

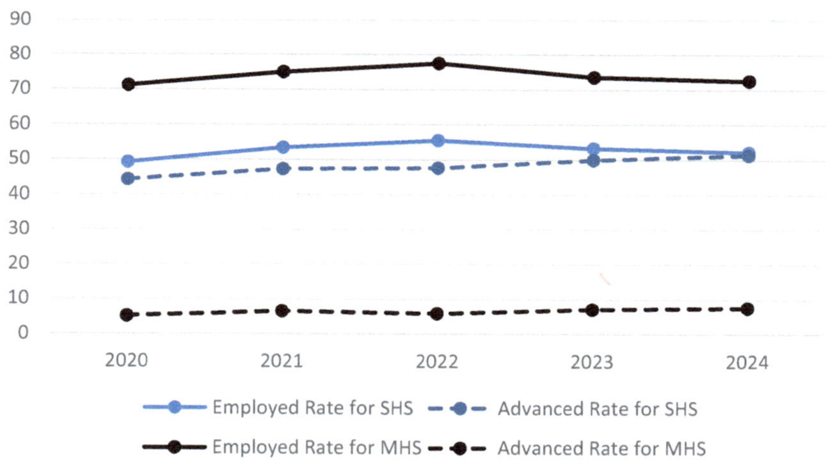

Fig. 5.5 Employed rate and advanced rate (college entrance) by school type (Source: Korea Education Service Statistics (2024b)). Note: The employment rate is calculated as the number of graduates employed at the time of the survey (April each year) divided by the total number of graduates, excluding those who are economically inactive (e.g., college enrollment, military service)

the lower male rate. By region, employment rates were 68.6% in urban areas and 76.1% in rural areas, indicating potential labor oversupply in cities. Retention rates highlight MHS graduates' employment stability. In 2024, the 6-month retention rate stood at 88.1%, while 69.7% remained employed after a year—higher than SHS retention rates (81.1% and 65.6%, respectively) (Korea Education Statistics Service, 2024a).

Research on School and Student Outcomes

Studies confirm MHS's effectiveness in improving employment outcomes, cost efficiency, and stakeholder perceptions. Im and Hwang (2015), analyzing responses from 1,221 teachers, students, and parents, found that MHS improved the perception of vocational education, increased career pathway flexibility, and enhanced employment prospects. In terms of cost-effectiveness, an early research based on three private MHS schools suggested MHS required higher government investment than SHS (Lee et al., 2016). However, a later study using a larger set of data from KEEP cycle 2 (91 MHS and SHS cases), found that MHS was more cost-efficient in terms of job placement, with lower per-student expenditure for job placement (Ko et al., 2021).

Longitudinal data from KEEP has significantly contributed to studies on MHS and SHS graduates' employment transitions. Park and Yu (2023) used propensity score matching to control for background characteristics when comparing MHS and SHS graduates. They found that MHS graduates had higher earnings, greater satisfaction, and more stable job status, compared to SHS graduates.

Since the introduction of the MHS program, vocational high school graduates have experienced more diverse career pathways. Among the studies that used the KEEP data, Ryu (2023) found that, compared to an earlier cohort of graduates from 2005, more recent graduates (2018) entered the workforce directly after graduation, demonstrating the intended outcomes of the MHS program. Ryu and Suh (2023) identified three primary career pathways: direct employment (50%), military service to employment (28%), and employment followed by higher education (7%). However, 18% were classified as Not in Education, Employment, or Training (NEET) category, indicating a need for targeted career support.

Social Outcomes

Looking at perceptions of employers toward MHS, a study surveying 620 HR professionals from firms employing MHS graduates reported high satisfaction with their competencies (Kim & Kim, 2017). Over 90% of respondents were satisfied with MHS graduates' subject knowledge, and 89% expressed satisfaction with their technical skills. Nearly 90% of firms planned to continue hiring MHS graduates, citing their fit with job requirements (29.6%), strong competencies (25.6%), and high learning capacity (16.2%).

MHS has also gained international recognition (Nahm, 2017; World Economic Forum, 2013). Countries such as Malaysia, China, Indonesia, and Saudi Arabia have studied Korea's MHS model for potential adoption (Yu et al., 2021). Individual schools engage in global training initiatives, such as Sudo Electric Technical MHS's nuclear internship for students from the UAE (Kim, 2010) and teaching strategy workshops for educators from Kazakhstan (Jo, 2024). The Korea International Cooperation Agency also launched a global internship program for vocational graduates to provide technical education in developing countries (KOICA, 2025).

Transforming VET: Implications from the MHS Model

Ensuring Quality Assurance and System Support

The success of MHS can be attributed to two key factors: quality assurance and system support. Performance management systems have been instrumental in maintaining high standards for students, teachers, and schools. The integration of MHS into Special Purpose High Schools, with extensive fiscal and regulatory support, has strengthened vocational education's social standing and industry alignment (Im & Hwang, 2015). A key element of quality assurance is the Young Meister Certification, which signals of graduates' competencies to industries (Yu et al., 2021). Positive labor market outcomes further reinforce MHS's ability to attract career-oriented students. Small class sizes and continuous teacher professional development also contribute to sustaining educational quality. The performance management framework ensures a balance between autonomy and accountability among stakeholders.

Furthermore, an evidence-based support system—comprising parallel policies, research institutions, data accumulation, and government entities—has positioned MHS as a compelling model that provides alternative career pathways. One of the most notable policies is the *Work First, Study Later* program, which has laid the foundation for MHS graduates to secure initial employment (Kim & Ryu, 2022). Through this initiative, they can enter the workforce, gain practical experience, and later pursue higher education through alternative admission procedures with government financial aid. Many MHS graduates who followed this pathway found it meaningful, especially when they recognized the need for further skill development aligned with their work experiences (Ryu et al., 2023).

In addition, the national research institute, KRIVET, plays a crucial role in bridging policy (MOE) and practice (MHS schools) by conducting policy research, providing school consultations, and overseeing performance management. The KRIVET-funded longitudinal panel study, KEEP, has been essential in analyzing MHS students' school experiences, career aspirations, and labor market performance (Ko et al., 2021; Park & Yu, 2023; Ryu & Suh, 2023). Furthermore, various government entities, including the Ministry of Startups and SMEs and the Ministry of Agriculture, Fisheries, and Rural Affairs, oversee sector-specific MHS, leveraging industry expertise and workforce policies.

Looking Ahead Toward Job Quality and Sustainability

Despite its strengths, challenges persist in this government-led system. Korea's vocational training has historically been government-driven since the industrialization period of the 1960s–1980s (Fleckenstein et al., 2024). While inspired by the industry-driven training models of Germany and Switzerland, MHS emerged primarily as a government response to excessive higher education enrollment and education-job mismatches. This framework has positioned MOE as a key player in school-industry relations and program implementation. To ensure long-term sustainability, the Korean government has taken steps to encourage greater industry participation, introducing incentives such as tax credits, preferential interest rates, and subsidies for trainers (Yu et al., 2021).

Still, employer participation remains limited, with many companies reluctant to take the lead in training initiatives (Kim et al., 2023). This hesitancy has contributed to persistent issues in workplace safety, workers' rights, and fair treatment for vocational students in cooperative education programs or graduates in workplaces. One example is the MOE's revision of cooperative education guidelines, shifting from on-site industry training to school-based training (Ministry of Education, 2017). This shift, along with a declining school-age population and the impact of the COVID-pandemic, has negatively impacted graduates' employment rates since 2017 (Jang, 2023). To address this issue, industry should play a more active role by ensuring safer working conditions, clearer career growth opportunities, and fair treatment for vocational high school graduates (Kim et al., 2023).

Another critical issue is job quality for high school graduates. While MHS has successfully introduced an alternative career pathway for graduates' direct employment, its evaluation system remains heavily focused on employment rates, often overlooking job quality (Kim & Ryu, 2022)–including remuneration, benefits, career growth, work-life balance, and retention (Im & Hwang, 2015; Park & Yu, 2023). These factors contribute to Korea's dualized labor market, where college graduates receive a wage premium while high school graduates have limited access to career mobility (Jeong et al., 2023). Improving employment quality is essential for enhancing the social recognition of vocational education and ensuring that MHS graduates can build sustainable careers.

The availability of quality jobs also affects regional workforce retention, a growing concern amid Korea's declining youth population in suburban or rural areas. While MHS graduates tend to earn higher wages and experience better retention rates than SHS graduates, many relocate to the Seoul metropolitan area for better job opportunities rather than remaining in their home regions (Park & Yu, 2023). This trend underscores the need for stronger collaboration between industries and local communities to expand quality employment opportunities outside metropolitan areas. Retaining graduates in local communities can help alleviate labor shortages and prevent rural depopulation (Choi et al., 2023).

Key Takeaways

These success factors and challenges offer key lessons for vocational education worldwide. A government-driven model provides a strong foundation for sustaining high-quality vocational education by ensuring legal, fiscal, and institutional support and enhancing stakeholder accountability. However, for long-term sustainability, strong industry and community involvement are key to expanding job opportunities for vocational graduates and ensuring that vocational education remains relevant to labor market demands.

Korea's MHS system is transitioning beyond its initial establishment toward a focus on job quality and sustainability. Prioritizing policies that improve job quality—ensuring clear career progression and family-sustaining wages—will make vocational education more appealing and socially recognized by graduates. Korea's MHS model is paving the way forward in this movement, serving as a reference for countries seeking to develop or refine vocational education systems tailored to their labor market needs.

Conclusion

This chapter examined Korea's MHS model, which has played a key role in advancing secondary vocational education and preparing a new generation of technically skilled workers to sustain national competitiveness. Despite Korea's credentialism and dual labor market structure, MHS has created an opportunity structure and sustained a compelling model for vocational education by aligning autonomous curricula with industry needs and implementing a continuous performance management system. This framework provides students with foundational workplace skills, technical competencies, and industry-recognized credentials, sending a strong signal to employers who continue to hire MHS graduates for their better job fit and higher learning capacity.

Additionally, the opportunity structure of MHS serves as a competitive alternative career pathway to traditional academic tracks through the *Work First, Study Later* initiative. This program allows vocational students to enter the workforce immediately after graduation while maintaining access to lifelong learning through alternative admission procedures and financial incentives. By enabling students to transition between work and study, this approach reinforces career mobility in an evolving labor market.

Looking ahead, ensuring the long-term sustainability of MHS requires continued collaboration among government, industry, and local communities to enhance job quality and employment stability for vocational graduates. Korea's MHS model offers valuable insights into how vocational education can promote upward mobility, job readiness, and career adaptability amid technological advancements and demographic shifts.

References

Ahn, J.-Y., & Kim, S. (2022). *Trends in career paths after graduating from vocational high schools and the economic indicators* (KRIVET issue brief 236). KRIVET.

Ahn, J.-Y., & Yang, Y.-J. (2023). A study on the perception of vocational high school teachers' work and the needs for reorganization of teacher organization. *Journal of the Korean Institute of Industrial Educators, 48*(1), 1–29. https://doi.org/10.35140/kiiedu.2023.48.1.1

Bae, S. H. (2012). The effects of school type on career maturity and career decision making self-efficacy of second-year high School students from Meister, vocational, and general high schools. *The Journal of Educational Administration, 30*(1), 525–548.

Bae, S. H., Jang, H. Y., Oh, S. H., Jang, C. S., & Lee, T. H. (2013). Vocational education research in Korea: A trend analysis of the Journal of Vocational Education Research. *The Journal of Vocational Education Research, 32*(2), 45–71.

Busan National Mechanical Technical High School. (2017). *A study on development strategies for Meister high schools to establish a leading model for secondary vocational education*. Busan National Mechanical Technical High School.

Busan National Mechanical Technical High School. (2024). *Education plan 2024*. Busan National Mechanical Technical High School.

Busan National Mechanical Technical High School. (2025). *Industry-government-school cooperation steering committee*. Retrieved from https://school.busanedu.net/bmt-h/main.do

Chang, W. (2006). *Pedagogy of work*. Hakjisa.

Choi, S., Ryu, J., & Yoon, J. (2023). The local settlement of vocational high school graduates (KRIVET issue brief 249).

Fleckenstein, T., Lee, S. C., & Park, J. (2024). Skills and training in hierarchical capitalism: The rise and fall of vocational training in South Korea. *Journal of Contemporary Asia, 54*(2), 260–281. https://doi.org/10.1080/00472336.2023.2168207

Gray, K. C., & Herr, E. L. (1998). *Workforce education: The basics*. Allyn and Bacon.

Im, S., & Hwang, J. (2015). Analysis of achievement of policy for Meister high schools through recognition of School constituents. *The Journal of Vocational Education Research, 34*(3), 45–66.

Jang, M. (2023). For the success of "Work First Study Later" program. *Korean Education News*. Retrieved from http://www.hangyo.com/news/article.html?no=99062

Jeong, S. J., Choi, S. J., & Hawley, J. (2023). Middle-aged adults' career trajectories and later-life financial security: Evidence from Korea. *Longitudinal and Life Course Studies*. https://doi.org/10.1332/175795921X16843342371110

Jo, J. (2024, December 15). Visit of the delegation of key university officials in the field of technology and industry from Kazakhstan to the capital electric power company. *Electric News Paper*. Retrieved from https://www.electimes.com/news/articleView.html?idxno=347719

Kim, D. (2010, March 2). Sudo Electric Technical Meister High School. *Asian Economy*. Retrieved from https://www.asiae.co.kr/article/2010030210324227906

Kim, J., & Kim, S. (2017). *Trends in corporate satisfaction with Meister high school graduates* (KRIVET issue brief 115). Korean Research Institute of Vocational Education and Training.

Kim, J., & Park, H. (2020). An activity theory analysis of Korean secondary vocational education curriculum: A case study of Electronics and Media Meister high school. *KEDI Journal of Educational Policy, 17*(1), 3–22. https://doi.org/10.22804/KJEP.2020.17.1.001

Kim, H., & Ryu, K. (2022). An analysis of research trends on the work first-then to college system in specialized vocational high school and Meister high school. *Korean Education Inquiry, 35*(4), 59–83.

Kim, J., Jang, M., Choi, S. J., & Hur, Y. (2012). Meister high school policy performance evaluation and challenges (2012–8). KRIVET.

Kim, K., Kim, N., Park, D.-H., Kim, H., & Chang, W. (2023). Institutional changes in the Meister high school policy: Perspectives of historical institutionalism. *The Journal of Vocational Education Research, 42*(6), 131–165. https://doi.org/10.37210/JVER.2023.42.6.131

Ko, G. Y., Min, J., & Park, J. (2021). Cost-effectiveness analysis of employment performance by Meister High School. *Journal of Vocational Education and Training, 24*(2), 37–65.

KOICA. (2025). *Oversea volunteer official development assistance programs*. The Korea International Cooperation Agency (KOICA). Retrieved from https://www.koica.go.kr:443/koica_kr/947/subview.do?

Korea Education Statistics Service. (2015). *Vocational high school graduates' employment rates in 2015*. Ministry of Education.

Korea Education Statistics Service. (2024a). *Enrollment and postgraduation trends of Meister and specialized high school students* [Dataset]. Retrieved from https://kess.kedi.re.kr/post/6673785?code=&words=&since=&until=&page=1&itemCode=04&menuId=m_02_04_03_01

Korea Education Statistics Service. (2024b). *Vocational high school graduates' employment statistics in 2024*. Korea Education Statistics Service.

Korea Education Statistics Service. (2025a). *Enrollment rate and advanced rate in education*. Retrieved from https://www.index.go.kr/unity/potal/main/EachDtlPageDetail.do?idx_cd=1520

Korea Education Statistics Service. (2025b). High school population by School type. Retrieved from https://www.index.go.kr/unity/potal/main/EachDtlPageDetail.do?idx_cd=1541

Korea National Open University. (2025). *Prime College Program requirement*. The Prime College Undergraduate Program. Retrieved from https://smart.knou.ac.kr:443/smart/5861/subview.do?epTicket=LOG

Korea Policy Briefing. (2011). *Air force aviation science high school became MHS*. Retrieved from https://www.korea.kr/news/policyNewsView.do?newsId=148709696

Korea Statistics. (2024). *Distributions of high schools and students by school type*. Retrieved from https://www.index.go.kr/unity/potal/main/EachDtlPageDetail.do?idx_cd=1541

Korean Council for University Education. (2025). *2025 Colleges and universities admissions implementation plan*. Korean Council for University Education.

KRIVET. (2018). *Meister school*. Korean Research Institute of Vocational Education and Training.

KRIVET. (2023). *Human resources development indicators*. KRIVET.

Lee, J., & Chung, H. (2019). An analysis of student and school factors on the career maturity of specialized high school and Meister high school students. *Studies on Korean Youth, 30*(1), 279–311.

Lee, K. W., Kim, D. H., & Lee, H. K. (2016). Is the Meister Vocational High School more cost-effective? *International Journal of Educational Development, 51*, 84–95. https://doi.org/10.1016/j.ijedudev.2016.07.003

Lee, C.-J., Song, N.-H., Kang, C., & Ryu, B. (2022). A comparative analysis study on the happiness index of teacher in vocational high school, Meister high school, and general high schools. *Journal of the Korean Institute of Industrial Educators, 47*(1), 1–21. https://doi.org/10.35140/kiiedu.2022.47.1.1

Maeng, H. J. (2014). Comparative analysis on work values and the career maturity of Meister school students and specialized high school students. *Journal of Research in Curriculum Instruction, 18*(3), 579–599.

Ministry of Education. (2016). *Meeting with government ministries and Meister high school teachers gathered to discuss ways to promote high school employment*. Ministry of Education.

Ministry of Education. (2017). Guidelines for improving cooperative education in vocational high schools. Ministry of Education.

Ministry of Education. (2022). *Ministry of Education notice 2022–3: Special Education Curriculum Amendment*. Retrieved from http://117.110.143.100/upload/sse/na/bbs_377/ntt_5032374/doc_09c089f3-638c-498d-bfd4-0b2e97265e2c1642a6332b17213.pdf

Ministry of Education. (2023). *New designation of three Meister high schools tailored to industry demand*. Ministry of Education. Retrieved from https://www.moe.go.kr/boardCnts/viewRenew.do?boardID=294&lev=0&statusYN=W&s=moe&m=020402&opType=N&boardSeq=95567

Ministry of Education. (2024a). *Designation and industry partnership of Meister high schools*. Retrieved from https://www.moe.go.kr/boardCnts/viewRenew.do?boardID=316&lev=0&statusYN=W&s=moe&m=0302&opType=N&boardSeq=98391

Ministry of Education. (2024b). *New designation of two Meister high schools tailored to industry demand*. Retrieved from https://www.moe.go.kr/boardCnts/viewRenew.do?boardID=294&lev=0&statusYN=W&s=moe&m=020402&opType=N&boardSeq=101367

Ministry of Education. (2024c). *Specialized high school and Meister high school portal: Hi-five*. Retrieved from https://www.hifive.go.kr

Ministry of Education. (2024d). *Statistics of specialized high schools and Meister high schools 2024*. Retrieved from https://www.moe.go.kr/boardCnts/viewRenew.do?boardID=316&lev=0&statusYN=W&s=moe&m=0302&opType=N&boardSeq=101534

MOE & KRIVET. (2024). *2024 Information session on new designation of Meister high school*. MOE, KRIVET.

Montt, G. (2017). Field-of-study mismatch and overqualification: Labour market correlates and their wage penalty. *IZA Journal of Labor Economics, 6*(1), 2. https://doi.org/10.1186/s40172-016-0052-x

Nahm, M. M. (2017). Korea's vocational education training sector in a globalized world: Current practices and future plans. In L. T. Tran & K. Dempsey (Eds.), *Internationalization in vocational education and training* (Vol. 25, pp. 167–185). Springer International Publishing. https://doi.org/10.1007/978-3-319-47859-3_10

Park, J. (2022). *The mediating effect of self-efficacy and the moderating effect of subjective job demand in the multi-level path model of employability and student and school-level variables among employment-seeking Meister high school students*. Doctoral Dissertation, Seoul National University.

Park, K. H., & Yu, J. (2023). Labor market performance gaps and the role of secondary vocational education at Meister high school in Korea. *Cogent Education, 10*(1), 2168407. https://doi.org/10.1080/2331186X.2023.2168407

Raffe, D. (2003). Bringing academic education and vocational training closer together. *Futures of Education, 1*, 49–65.

Ryu, J. (2023). *Changes in the labor market transition status of vocational high school graduates: A comparison between 2005 and 2018 graduates* (KRIVET issue brief 253). KRIVET.

Ryu, J., & Suh, Y. (2023). School-to-work transition patterns of vocational high school graduates by school type. *Journal of Vocational Education & Training, 26*(2), 93–125.

Ryu, J., Kim, S., Ahn, J.-Y., & Kang, S. (2023). The experience and characteristics of young vocational high school graduates' career paths formation. *Journal of Agricultural Education and Human Resource Development, 55*(1), 27–54. https://doi.org/10.23840/agehrd.2023.55.1.27

Samsung Semiconductor Newsroom. (2022). *A visit to Samsung Electronics Engineering University (SSIT) that helps employees grow consistently*. Retrieved from https://tinyurl.com/4ermcus9

Song, N., Lee, & Choi, H. J. (2023). A study on vocational high school students' perception of school changes and department restructure. *Journal of the Korean Institute of Industrial Educators, 48*(1), 78–103. https://doi.org/10.35140/kiiedu.2023.48.1.78

World Economic Forum. (2013). *Education reform in Korea* (OECD Economics Department working papers 1067). OECD. https://doi.org/10.1787/5k43nxs1t9vh-en

Yu, J., Kim, S.-N., Oh, M., & Yoo, E. (2021). *Meister high school system in Korea*. Ministry of Education, Korea Research Institute for Vocational Education and Training.

CHAPTER 6

Opportunity Structures for Who and What? The Discursive Struggle Between Educational Equality and Economization in Europeanized Finnish Vocational Education and Training

Antti Seitamaa and Petteri Hansen

A Brief Socio-Historical Introduction to the Changing Opportunity Structure in Finnish Vocational Education and Training

There has been growing interest among researchers studying vocational education and training (VET) toward the concept of opportunity structure, originally introduced by American sociologist Robert Merton in the 1930s to explain how in modern industrial society, crime and delinquency are the result of contradictions between structural and cultural goals. When economic and political factors prevent individuals from accruing

A. Seitamaa (✉)
Department of Education, University of Helsinki, Helsinki, Finland
e-mail: antti.seitamaa@helsinki.fi

E. Smeplass and C. Zirkle (eds.), *Opportunity Structures in Vocational Education and Training*, Palgrave Studies in Technical and Vocational Education and Training,
https://doi.org/10.1007/978-3-032-04410-5_6

material wealth and social recognition through legitimate opportunity structures, they tend to pursue these goals illegally through crime or revolution (Cloward & Ohlin, 1960). Just as there are many different ways to theorize the social dimension of vocational education (McGrath, 2012), there are also many ways to understand the concept of opportunity structure (Smeplass & Zirkle, 2025). In this chapter we adopt a Mertonian perspective to Finnish VET, viewing it as a particular social mechanism that creates conditions for social stability, democracy and economic growth. However, we also understand VET as a complex domain, which constantly seeks a balance within and between structural expectations of education, economy and politics.

Different institutional arrangements shape the opportunity structure. For example, in Nordic welfare states VET is both part of the education system and part of the broader institutional safety-net (Di Stasio & Solga, 2017). In recent years there has been intense public debate about the end of the welfare state, as well as anxiety over the nature and prevalence of crime, particularly among stigmatized groups. More and more frequently, VET is portrayed as a potential social policy solution, a ladder or, if you will, an opportunity structure, for 'at-risk' populations to climb out of their predicament (Åslund et al., 2018; Di Stasio & Solga, 2017).

When considering VET as an opportunity structure in the Nordic context, one should also consider how the VET systems in Nordic countries vary. At the end of ninth grade, Finnish comprehensive school students choose to continue in either general upper secondary school

---

P. Hansen
Department of Pedagogy, Norwegian University of Science and Technology, Trondheim, Norway
e-mail: petteri.h.hansen@ntnu.no

(GUS), VET,[1] or preparatory education.[2] General upper secondary school is the traditional academic track leading to higher education, particularly at the university level, while vocational upper secondary education is the traditional track to employment and, increasingly, entrepreneurship (Virolainen & Stenström, 2014; Välimaa, 2019). From an international perspective, Finnish VET is a relatively popular choice, with approximately 40% of the post-comprehensive youth cohort continuing in one of ten vocational fields (Finnish National Agency for Education, 2018; Rintala & Nokelainen, 2022). To compare, the enrollment rate for vocational upper secondary programs is 53% in Norway, while in Sweden enrollment has steadily declined during the last 15 years despite various government initiatives and is around 35% (Norwegian Directorate for Education and Training, 2024; ReferNet Sweden Cedefop, 2022). Apprenticeships, often suggested as an alternative to school-based VET, are hardly a panacea; enrollment rates for VET in Denmark, where there is a strong dual-VET model, have also been steadily declining in recent years (Aarkrog, 2020).

Although there is no universally accepted explanation for what makes VET more attractive in some countries than others, many researchers have noted growing tensions between VET's traditional goal of preparing young people for employment and mounting pressure to prepare young people for higher education (Billett et al., 2022; Esmond & Atkins, 2022; Jørgensen et al., 2018). These tensions can also be seen in Finnish

[1] The vocational qualification structure in Finland consists of three different types of vocational qualifications; vocational upper secondary qualifications and further vocational qualifications are equivalent to National and European Qualification Framework level 4, while specialist vocational qualifications are equivalent to National and European Qualification Framework level 5. Further and specialist vocational qualifications are intended only for adult learners with extensive prior work experience in the field, while vocational upper secondary qualifications are for people of all ages. All three are integral parts of the Finnish VET system, along with 'preparatory education for vocational training' (Act on Vocational Education and Training 531/ 2017).

[2] The expansion of compulsory education from 16 to 18 in 2021 resulted in the institutional re-organization of different forms of preparatory education into a single entity. Of the approximately 5.4% of students that currently continue in preparatory education, a clear majority eventually pursue a vocational upper secondary qualification (Statistics Finland, 2023). It is also important to note that 5–8% of VET students choose to complete a so-called 'dual qualification', which combines vocational upper secondary qualification with the matriculation examination (Lietzén, 2023). In practice, this is a demanding route that usually requires an additional 1–2 years of study.

education policy, which emphasizes the centrality of higher education for maintaining national competitiveness in a globalized knowledge economy, while at the same time serenading VET as high-quality preparation for employment, entrepreneurship and life-long learning (Virolainen & Tønder, 2018). While the importance of maintaining both vocational and academic pathways is well-established in Finland, there are enduring historical and political differences regarding how much and what kind of education is desirable for which individuals and social groups (Kettunen, 2013; Lietzén, 2023).

The history of Finland in the twentieth century can be read as the story of a newly independent nation trying desperately to create conditions of political stability and social consensus through the gradual expansion of public education (Kauko, 2019; Kettunen, 2013; Simola, 2015). The role of VET in the Finnish opportunity structure has changed in response to economic and political factors. In the early twentieth century, VET helped create the foundation for future export-based economic growth by training technically skilled workers to work in the forestry industry (Stenström & Virolainen, 2018). In addition to its economic role, VET also supported conservative and nationalist goals of maintaining social order and fostering disciplined citizenship through the medium of work, as well as providing a structure for positioning people strategically in times of national crisis (Kettunen, 2013; Laukia, 2013). After the Second World War, for example, soldiers returning from the front, many of whom were disabled, received vocational training so they could be employed in industrial work to help pay back the war compensations Finland owed to the Soviet Union.

When Finnish industrialization started in full force in the 1950s, it uprooted the political and economic foundations of the conservative agricultural society (Heikkinen, 2001). Vocational institutes that had originally been established and controlled by private industrial interests came under municipal jurisdiction (Stenström & Virolainen, 2018). The majority of Finns still earned their livelihood from agricultural work until the 1960s. VET assumed a new role in the opportunity structure as it provided a pathway for social change in the context of rapid urbanization and economic transformation. Since folk schools were the highest achievable level for the vast majority of rural and working-class people, they wanted their children to have the same chances as children from the upper strata to matriculate in gymnasiums and enter university (Kauko, 2019; Simola, 2015).

The comprehensive school reform in the 1960s and 1970s played a critical role in transforming Finland into a Nordic welfare state with redistributive and generous welfare policies (Antikainen, 2006; Esping-Andersen, 1990). This reform changed the opportunity structure both institutionally and discursively; the former by creating a new legislative foundation for the organization and governance of basic education, and the latter by re-defining what was seen as 'sensible, realistic and legitimate' education policy (Koopmans & Statham, 1999; Dale & Parreira do Amaral, 2015). Inspired by the 'youth schools' in Sweden, the architects of the comprehensive school reform wanted to expand comprehensiveness to the upper secondary education level. They felt it was no longer enough for vocational schools to respond passively to the demands of the economy and train obedient workers. Instead, VET was supposed to train the next generation of well-educated workers that could participate as equals in democratic decision-making (Aho et al., 2006; Kettunen, 2013; Stenström & Virolainen, 2018).

The role of VET in the opportunity structure changed again after Finland joined the European Union in 1995 in the midst of a severe economic recession. This article examines the ensuing discursive struggle between educational equality and economization in the 2000s and 2010s, when the popularity of VET rose to new heights and European influences began shaping national VET policy. The emergence of transnational political entities such as the European Union provide the 'context of contexts' for our research, as more and more often the opportunity structure extends across borders (Dale et al., 2017). The transnational dimension is particularly important in VET, with the creation of a common European labor market and educational area exerting growing influence on how Member States develop national systems (Cort, 2010; Li & Pilz, 2021).

Viewed from the present day, the history of Finnish VET appears as a development story in which VET structures have successfully adapted to serve ever new societal tasks. On the other hand, behind the scenes of this development story, there has always been a discursive struggle over the goals and tasks of VET as a particular social mechanism. Next, we present the discursive approach that serves as the methodological framework of our study.

Research Material and Methodology: Expert Interviews as Windows into the Discourses Around Opportunity Structures

Over the past decades, there has been a diverse methodological discussion about how the idea of opportunity structures could be analyzed empirically. Some of the studies have highlighted especially the discursive dimension of opportunity structures in the institutional and cultural contexts of politics and education (Koopmans & Statham, 1999; McCammon, 2013; Smeplass et al., 2023). However, from the discursive perspective it is equally interesting how discourses and opportunity structures change over time as well as how opportunity structures are the object of constant discursive struggle (Smeyers & Depaepe, 2016).

In this study, we examine the discursive shaping of opportunity structures of vocational education in the light of Finnish educational policy expert interviews. Expert interviews provide a window into the change of discourses in two different ways. On the one hand, the expert interviews can provide an insight into the backstage happenings during the frontstage policy processes, which are often missing from the documentary material (Kosunen & Hansen, 2018). On the other hand, interviews are also opportunities for interviewees to return to the stage and justify their own actions at a certain point in time (Al'Abri et al., 2024).

Considering both frontstage and backstage perspectives, expert interviews can thus be thought of as narratives in which both past events and assumed futures are applied to support as well as challenge hegemonic ways of thinking about the goals and means of education (Hansen et al., 2020). This type of thinking relates also to the idea of communicative externalization (Schriewer, 1990), where different actors in and around education try to legitimize or resist changes by appealing, for example, to the practices of other countries and their own cultural values.

Another methodological question related to expert interviews is who can be considered experts in education policy and how the choice of interviewees affects the shaping of the discursive space. Halperin and Heath (2020) have conceptually separated elite interviews and expert interviews. The elite interviews are interested in those actors who have had a significant role and influence on decision-making, while expert interviews are mainly interested in the persons who have special knowledge needed for the decisions. As Hörmann and Sivesind (2024) argue, this classification is partly problematic because in expert-led and network-based policy space, these two actor-categories are often overlapping.

The data used in this study represent certain key organizations, but in a broader sense also the organizational field in which VET policy is made. The first author carried out 32 semi-structured interviews with leading Finnish experts in VET between 2020 and 2021, including senior government officials, representatives from stakeholder organizations, high-level management from VET institutes, experienced VET researchers, as well as political actors. Many of the participants had occupied several positions in the VET policy field during their long careers. Experts were assured of their complete anonymity and were asked at the end of each interview to recommend other highly respected individuals from the Finnish VET policy field for interviews.

The COVID-19 pandemic made it necessary to conduct half of the interviews in person and half online. The first author personally transcribed all of the interview data, which amounts to 600 pages after anonymization. Interviews lasted on average slightly less than 2 hours. The first author read through all the interview data several times. Based on this preliminary familiarization, the first author then began coding the data in ATLAS.ti, identifying sections where participants described the change in VET's role, with particular attention to the impact of the European Union and internationalization. A total of 430 quotations, averaging 4–5 sentences each were coded with 100 different codes. After merging the codes, we explored the data for descriptions of institutional discourses of change in VET policy from the upper secondary vocational education reform in the 1980s to the 2018 VET reform. The excerpts were selected based on their representativeness of the direct and indirect changes that have taken place in Finnish and European Union VET policy in the last two decades.

Results

In the results section, we explore discourses of change regarding the role of VET in the opportunity structure. We have divided the results into three sections. In the first section, experts explain what it meant that VET became part of the school form during the upper secondary vocational education reform in the 1980s and how the School Acts of the 1990s paved the way for a more work-based approach to VET in the following decades.[3] The second section examines VET policy during the 2000s when a new European opportunity structure emerged, and Finnish and European Union VET policy began to intertwine. In the third section, we focus on the organizational and pedagogical fragmentation of VET in response to changes in the national and international operational environment and how these were concretized in the historical VET reform that took place between 2015 and 2019.

Setting the Stage: Defending the School Form and Educational Equality Against the Demands of Business Life

Experts' perspectives on change often depend on how long they have been involved in education policy. This does not exclude the possibility that even younger education policy actors might have a strong opinion of historical changes through, for example, research literature or passed down from their colleagues.

In this study, some of the experts had witnessed firsthand the growth and development of the Finnish VET system over four decades, from the 1970s until the 2010s, and they spent a considerable amount of time explaining the continuity and tensions between different phases in this long historical arc. Many experts emphasized the importance of the upper secondary vocational education reform that took place between 1982 and 1988, planned as the follow-up to the comprehensive schooling reform in the 1970s (Aho et al., 2006; Stenström & Virolainen, 2018). The upper secondary vocational education reform was aimed at expanding comprehensiveness to upper secondary education by introducing significantly more educational elements into vocational schools for youths:

> *When they wanted to start building VET, it was seen kind of as a school—in a positive way—and that it should be on par with general upper secondary schools. But since that strong discourse—which certainly came*

[3] The School Acts were a comprehensive legislative package that profoundly re-shaped the Finnish education system. They introduced far-reaching changes to the organization of basic education, general upper secondary school, vocational education and training for youths and vocational education for adults, among others (for a critical perspective, see Simola, 2015). The Acts came into effect in 1998 and were implemented in the following years, with one of their stated goals to increase the attractiveness of VET.

from the Confederation of Finnish Industries and the Federation of Finnish Enterprises—about 'preparing [students] for working life' and 'working-life oriented', 'working life' and 'going to work quickly' (...) VET is no longer so strongly—in a positive sense—a school on par with general upper secondary school. Those positive features characteristic of a school institution are disappearing from there all the time. Expert interview 2 (stakeholder representative)

This expert condenses the modern history of Finnish VET from the 1980s to 2010s into an overarching struggle between powerful economic and political actors with different interests and goals (for historical equivalents from other Nordic countries, see Thunqvist et al., 2019). While progressive reformers in the 1980s succeeded in increasing institutional parity between vocational schools and general upper secondary schools by making VET more school-like, this changed when new political and economic winds started blowing in the mid-1990s and a more economistic approach emphasizing employers' needs took hold:

In '98 there came three-year vocational upper secondary education and training that gave general eligibility for higher education, with work-based learning and common units. Even though the common units are a bit of nonsense, at least when people want to compare them with the learning produced in the general upper secondary school syllabus. But it's better than nothing. Expert interview 15 (senior government official)

Participants tend to treat the School Acts of 1998 and the development of mandatory work-based learning in 2001 as essentially the same thing, although the two were a few years apart (for a more comprehensive timeline see Stenström & Virolainen, 2018; Virolainen, 2022). While the School Acts significantly improved the formal position of vocational students by giving them access to higher education, the expert sarcastically questions the credibility of the common units meant to guarantee eligibility (Isopahkala-Bouret, 2015; Lappalainen et al., 2019). Whereas the upper secondary vocational education reform sought explicitly to improve the parity between vocational schools and general upper secondary schools by including more educational elements in vocational upper secondary qualifications, the School Acts granted vocational students the theoretical possibility to continue in higher education, regardless of whether the common units were actually enough to make this possible:

> *[unifying vocational qualifications] was very important because it was only at that stage that you could say to parents that "your Ville (boy) or Tiina (girl), they can apply for university later if they want" because that is what is appreciated most in Finland. There are no dead-ends.* Expert interview 22 (senior government official)

The newly established universities of applied science were thought to provide a more natural pathway for vocational students to continue their studies than the traditional academic university (Isopahkala-Bouret, 2015; Välimaa, 2019). However, as they were still a recent invention, from a public relations point of view it was important to formally link VET with the prestige of academic university, even though in practice this was not reflected in the quality and duration of the common units. Access to university appears to be more of a selling point for students and parents than a genuine concern in VET policy:

> *Extending vocational qualifications to three years was about quality and attractiveness. But from the beginning, it contained a lot of elements that gave [students] full freedom to de-select anything that made qualifications broader.* Expert interview 4 (senior government official)

Whereas the upper secondary vocational education reform was based on strengthening the role of broader, non-technical education, particularly social studies (Aho et al., 2006), the new common units emphasized life-long learning and individual choice. Increasing flexibility meant students could opt out of subjects they felt were difficult or uninteresting. The School Acts simultaneously increased the attractiveness of vocational qualifications but failed to strengthen their educational credibility. Although the inclusion of formal eligibility for higher education in vocational qualifications was an important step forward for VET, the slimmed-down, increasingly optional common units on which this eligibility depended was a far cry from the previous goal of improving VET's parity with general upper secondary school. Many researchers consider the common units inadequate not only in terms of providing access to higher education, but also for providing a solid foundation for life-long learning and democratic citizenship for the entire youth cohort (Isopahkala-Bouret et al., 2014; Lappalainen et al., 2019; Lappalainen & Lahelma, 2015).

Introducing New Actors and Bigger Stages: Europeanization and New Economic Opportunity Structures

The School Acts were passed at the end of the 1990s when Finland was in the process of integrating itself economically and politically into the European Union and globalizing world order. One of the most consequential changes during this time period was the radical decentralization of power from the Ministry of Education to education providers across the country (Simola, 2015). The various funding arrangements from EU projects enabled local actors to engage in new types of network-based development activities, but also made VET providers responsible for developing and implementing the newly legislated system of mandatory work-based learning, which required building connections with local employers:

> *VET policy started changing EXTREMELY powerfully and new funding opportunities emerged with the EU. Project fundings. It opened our eyes very much because the point of view changed from the local level towards the national. And in Finland VET started to be developed in massive network projects with EU support (…) we started to realize that "my goodness, we can influence things.* Expert interview 26 (senior VET researcher)

Not only did EU funding help VET providers develop links with local employers, it also empowered VET providers to collectively advocate on behalf of their interests. Providers seized on the new possibilities to collaborate with each other within and beyond national boundaries. Interacting with VET providers from other countries provided a mirror for critical self-reflection, stretching the boundaries of what Finnish politicians and stakeholders thought possible. No longer would the purpose and structure of VET be solely decided by a small cabal of politicians, corporatist stakeholders and government officials at the Ministry of Education and the Vocational Education Directorate, but those responsible for implementing it in practice would have a far stronger voice. Against the backdrop of dynamic economic expansion and transformation, the old world of highly regulated and closed-off vocational schools with their ideological fixation on improving parity with general upper secondary schools appeared increasingly outdated and quaint.

The European Union launched multiple initiatives in the 2000s to pave the way for a common European labor market and education area

(Ante, 2016; Zaunstöck et al., 2021). In addition to laying the foundation for a common European identity through international mobility for students and staff, there was also a push for uniformity and standardization. These goals were first formulated in the Copenhagen Declaration in 2001, which have since been followed up in numerous other communications and recommendations. The drive for standardization resulted in common metrics like the European Qualification Framework, which Finland was actively involved in developing. Many participants described the first decade of the new millennium as a time of excitement and optimism, when there was also great forward momentum for developing VET as part of Europe's economic and political integration (Bohlinger, 2019; Cort, 2010).

National funding for VET grew steadily during the 2000s as it attracted more and more students, both youths and adults. The mood changed dramatically during the Global Financial Crisis and the Euro Crisis, when many Member States initiated austerity programs to improve their fiscal situations. The pressure on government institutions to produce better outcomes also found its way into Finnish VET policy debate. The idea of austerity programs and apprenticeship-based VET is associated both with the Europeanization of education, as well as VET policy promoted by centre-right political parties emphasizing the changing needs of employers and work:

> *A big political question is related to apprenticeships. For the right-wing, taken broadly as in the center-right, apprenticeship has been seen as something that needs to be increased because we have this school-based system. They've wanted to increase it to improve employment. And one particular problem is that for us apprenticeships were for a long time adult VET, and still are to a large degree. And they've wanted to increase youth apprenticeships. And THIS is one of those ideological dividers that exists. And the reasoning comes from employment, the model has been brought from Central Europe where, although I think they have slightly different approaches to those issues in Germany, Austria and Switzerland, the big picture appears the same from here.* Expert interview 15 (senior government official)

By the end of the 2000s, there were growing calls from the right-wing to develop a Finnish model of youth apprenticeship. When it came out that Finland had the highest proportion of NEET youths in the Nordic countries, economic and political pressure to improve school-to-work transitions in VET increased (Nordic Statistics Database, 2023).

Many national politicians thought that despite all the changes, youth VET was still too school-based and lacked vocational relevance (for a Nordic comparison see Thunqvist et al., 2019). As the economic situation in Finland grew progressively worse in the first years of the 2010s, VET's ability to produce employable workers rose to the forefront of educational policy debate, both on the national and European level (see Martínez-Izquierdo & Torres Sánchez, 2022). Until this point, although many Member States viewed VET primarily through a lens of economic utility, it was grouped together with the rest of the education sector in the Directorate General for Education, Youth, Sport, and Culture (so-called DG-EAC).

When the new European Commission started its term in 2014, it decided to use VET to help dig European economies out of the hole and transferred jurisdiction over VET to the Directorate General of Employment, Social Affairs and Inclusion. For many Finnish experts, this decision was not merely symbolic or technical but instead confirmed that the dominant way of thinking about VET in the EU differs strongly compared to Finland. Many experts described how this directorate overwhelmingly tended to view VET through the lens of employment and social policy and tried to keep it separate from education, although on paper the two directorates were supposed to collaborate closely. Although the transfer of VET on the European Union level was bureaucratic and did not change anything on the national level, in practice they felt the transfer weakened the relationship between VET and higher education. Multiple experts would later refer to this decision as indicative of the European Union's indirect, long-term influence on national VET policy. For them, it demonstrated that despite all of the EU's rhetoric about the value and importance of VET, the dominant view is that it is foremost an instrument for producing workers for the labor force:

> *The European Union's education policy is awfully fine on the clausal level: equality and getting the right amount and right kinds of people into working life. And securing mobility, that might be the leading idea: securing mobility in EU countries. And raising the skill level in all the EU countries. But in most EU countries they think of skill level only in terms of lifting the practical vocational skill level, and not the same way we do, of living as a human being*. Expert interview 21 (senior government official)

Some participants emphasized that Europeanization is not one-sided and that the Finnish approach to VET, with its emphasis on lifelong learning has also influenced the development of VET policy in the European Union. The interplay between individual Member States and EU VET policy is highly complex, taking place at multiple levels simultaneously:

> *In my opinion, we have been bringing this idea that [vocational education and training] is more than technical work skills. That's one thing that we have been bringing strongly into the conversation (...) as the policy conversation has been taking place, the language has become polished, and some countries have brought their own ideas into it and then all the countries have adopted them. And then of course the question is how do policy and practice meet.* Expert interview 15 (senior government official)

Member States have shared lessons and learned from each other, and brought their qualification frameworks into alignment (Ante, 2016; Cort, 2010; Maurer & Gonon, 2014). Although control over education policy remains firmly under national jurisdiction, the evolving transnational policy discourse has changed the way countries think about their own systems (Cort, 2010). In the Finnish case, national VET policy has developed in response to changes in both the national and international operational environment. Like the participant above, many others also believed that the 'Finnish way' of thinking about VET had found its way into the EU policy level. The Finnish experts involved in EU policymaking considered themselves particularly proactive and conscientious, both in promoting national ideas as well as implementing VET policy that has been negotiated and developed at the European Union level.

Epilogue: From Pedagogical Opportunities to Responsiveness and Organizational Fragmentation in VET

The role of VET changed dramatically in the mid-2010s in response to political and economic changes in the national and international operational environment (Seitamaa & Hakokongäs, 2022; Seitamaa & Rosenblad, forthcoming). Prior to this, the institutional separation of youth and adult VET had long been the subject of political debate, with the political right-wing wanting to cut costs and increase efficiency by combining the two (Stenström & Virolainen, 2018). The separate adult VET system was the darling of the left-wing parties and their historical allies in the Finnish labor movement, which viewed it as an integral part of the solidaristic, redistributive employment and welfare arrangements that were made in the 1980s (Stenström & Virolainen, 2018). Furthermore, separating the two was strategically important because it allowed for the possibility to someday again develop upper secondary education as a whole as had been the plan since the social democratic heyday in the 1970s and 1980s (Aho et al., 2006).

The coalition government that came to power in 2011 in the middle of the Euro Crisis originally planned to tackle youth drop-out rates by raising the age of compulsory education from 16 to 17 (Stenström & Virolainen, 2018). However, when the economic basis of the government program fell apart mid-term, the government decided to abandon this reform and announced that it would carry out major cuts to the VET budget demanded by the Ministry of Finance through a structural reform of upper secondary education from 2014 to 2015. This reform, spearheaded by the Social Democratic Party, was aimed at combining several education sectors, most importantly youth VET and GUS, into a single organizational framework and funding system. However, this reform also failed at the last-minute in spring 2015, creating a unique opportunity for the next government to carry out its own VET reform during the 2015–2019 government term (Stenström & Virolainen, 2018). This new right-wing-led VET reform abolished the separate adult VET subsystem and created a single legislation, funding system and individualized competence-based system of VET for all learners (Niemi & Jahnukainen, 2019; Rosenblad et al., 2022; Seitamaa & Rosenblad, forthcoming). Many experts attributed the dramatic change of direction in VET policy to the gradual permeation of a 'European' outlook among Finnish politicians and stakeholder:

> *In the same way as on the EU-level, VET in Finland has been thought of as 'skills' and 'training', not so much about 'education' (…) All of these international trends have been quite easy to see in the development of vocational education over the last few years, notoriously reaching their peak in the [2015–2019] VET reform. I still wonder how it passed because it was such a huge break with the way that at least some of us thought, that vocational education should be educational and provide real opportunities for continuous learning.* Expert interview 4 (senior government official)

The emphasis on skills and competences originated in European Union policy discourse. The European Qualification Framework and European Credit System for Vocational Education and Training that were developed during the 2000s required Member States to modify the vocational qualification structure to correspond with what had been agreed on a European Union-level (Bohlinger, 2019). In Finland, competency-based training started to be expanded in the vocational qualification structure in the early 2010s (Stenström & Virolainen, 2018; Rosenblad, 2023). Prior to this, only the adult VET subsystem was competency-based, as it was specifically geared for recognizing prior learning gained in practical work experience. The 2018 VET reform took this to a completely new level by turning 'competence units' into the new metric for performance-based funding (Seitamaa & Rosenblad, forthcoming). Measurable output purporting to represent real-life 'skills' overshadowed vocational knowledge and in-depth socialization into professional practice (Rosenblad, 2023).

The emphasis on measurable outcomes based on skills and competences is in sharp contrast with vocational Bildung and professional knowledge that were at the center of the upper secondary vocational education reform. The focus on skills and competences also extends to the common units included in vocational qualifications, which are completely different from the general education school syllabus (Lappalainen et al., 2019). These new goals reflect economic and political pressure to provide measurable improvements in taxpayer funded outcomes for a far more heterogeneous mix of students:

> *There is this constant pain because we educate youths that have finished comprehensive school for vocations and adulthood. Then there is the need to train people who have left something unstudied at some point in their lives or that have studied something else and need vocational education and training alongside it. And then there are different needs that are born along the career path or as a result of immigration or whatever else. And VET tries to respond to all of this at the moment and to be fair and of equal quality for all of these target groups.* Expert interview 5 (stakeholder representative)

The ever-expanding role of VET has changed the meaning of educational equality. The old goal of maintaining parity of VET with GUS is simply unrealistic when VET students come from such heterogeneous backgrounds. There are profound differences between 16-year-olds that have just finished comprehensive school and 40-year-olds that are changing careers or seeking further training to improve their position on the labor market. Both groups also include immigrants, many of whom did not have the chance to attend school regularly before arriving in Finland and struggle with learning the Finnish language. Even if they have participated in education, this is not always recognized or considered adequate. At the same time, immigrants are generally considered a 'problematic' group in terms of their disproportionately high unemployment rate and integrating them into the labor force is a top economic and political priority (Masoud et al., 2021; Seitamaa & Hakoköngäs, 2022; Zacheus et al., 2017). VET providers are expected to not only respond to the collective needs of each target group, but to each individual student and to do this in a fair and equal way.

Changes in VET reflect changes in the opportunity structure in Finland and the EU. The 2018 VET reform was based on the idea that VET must respond more effectively to the new operational environment where the need for adult education is growing constantly and the survival of the welfare state depends on improving the employment rate. The main goal of VET is to maximize the possibility for each individual to integrate into the labor market (Masoud et al., 2021; Seitamaa & Rosenblad, forthcoming). VET's educational objectives have been tailored to support this, and if some of the students end up continuing their studies in the universities of applied science, this is considered a nice additional success. Work is the main opportunity structure for the majority of students. All EU Member States are struggling to keep up with the pace of change and economic turbulence. Many experts took pride in the fact that Finland has created such a flexible, responsive system of VET:

> *If you look at preparations being done in the Commission, they are giving a recommendation for VET in which there are a lot of the same things that were in our reform and that we've discussed here, as in recognizing the growing significance of VET, the need for increasing flexibility, taking individual needs into consideration more than before, serving working life as well as strengthening resilience and including everyone.* Expert interview 3 (senior government official)

The fact that elements of the 2018 VET reform in Finland have found their way into the European Commission's official recommendations indicates growing consensus in the EU regarding the centrality of VET in addressing economic challenges (Martínez-Izquierdo & Torres Sánchez, 2022). Although the common units are still part of vocational qualifications in Finland, their form and content have changed so much over time that few would seriously claim that they are actually meant to increase parity between VET and GUS. Many of the stakeholders did not see this as a problem. Already in the School Acts, the goal of achieving parity with GUS was re-interpreted to mean parity of formal recognition vis-a-vis higher education rather than actually providing a comparably solid educational foundation. With each student individually responsible for building their own 'personalized' pathway, VET policy is no longer even nominally interested in reducing differences in educational outcomes related to social background.

Discussion

The role of VET in the opportunity structure changes depending on the economic and political context. While Finnish VET policy in the 1980s was unapologetically ideological in its pursuit of parity between vocational and general upper secondary schools, the focus on schooling was replaced with vocational relevance supported by life-long learning in the late 1990s. During the 2000s and early 2010s, the separation into youth and adult VET subsystems helped maintain a formal semblance of parity between VET and GUS, despite incessant criticism from various actors; employers constantly berated VET for failing to meet their needs, educational researchers focused on how education equality had failed, and VET providers demanded less regulation. When the 2018 VET reform finally succeeded in combining the youth and adult subsystems into a single legislation and funding system, the final threads of the already-flimsy pretext of educational equality finally came apart. Skills, competence and training became the cornerstones of a pragmatic, economistic approach that provides individualized solutions for a very heterogeneous body of students.

Participants' opinions of these changes range from the deeply cynical to carefully optimistic. Depending on one's perspective, budget cuts[4] and reform have either made vocational education and training more lean

[4] Between 2013, when initial budget cuts to VET were announced, and 2019, when the reform had largely been implemented, VET funding decreased by over 400 million euros, representing roughly 15–20% of the total budget (National Audit Office, 2021).

and fit-for-purpose or left it conceptually malnourished. There is more than a little sarcasm in the way experts describe the inherent ambiguity of VET, which is both chronically short of funding while overflowing with expectations and new responsibilities from politicians, employers and the public. Aside from disappointment and anger, many participants also communicated genuine excitement about the possibilities of the reformed VET system, and the vital role VET plays. Few pedagogical contexts are as dynamic and demanding as VET; or play as important a role in society. VET harbors great potential and shoulders huge responsibility; despite constant fluctuation in the institutional framework, hope persists even if often thwarted. VET has never played a more important role in the opportunity structure than now, nor has it been more subject to contradicting interests resulting from immigration and economic pressure (Masoud et al., 2021; Seitamaa & Hakokongäs, 2022; Seitamaa & Rosenblad, forthcoming).

The European Union is under growing pressure both internally and externally. The European social contract is in the process of being rewritten, with powerful voices demanding that Member States prioritize military spending over social security and welfare. The potential consequence of a new wave of aggressive austerity programs is greater political instability and growing anti-immigration sentiment across the continent, skillfully exploited both by populists, online demagogues and the EU's geopolitical rivals. In these unpredictable times, the goal of improving economic output by strengthening the link between education and work can be seen as a double-edged sword. On one hand, the opportunity structure must allow those Europeans left behind economically, as well as those that have recently arrived, to find their way forward. At the same time, the economistic emphasis on VET as straightforward skills and competence output threatens to eclipse its deeper democratic functions.

We would like to conclude by answering the question posed in our title about who and what VET serves as an opportunity structure for in Finland. As our analysis illustrates, VET is expected to provide something, anything or everything, for anyone that walks through its doors or logs in. Facilitated by deregulation and decentralization, VET has over time become the main institutional opportunity structure for people with few choices, in some ways returning to its starting point in the 1970s, except that charting a career path is far more complex for young people nowadays (Ågren, 2021; Kalalahti & Varjo, 2022; Masoud et al., 2021). Actors in the policy field find themselves between a rock and a hard place; while they understand that it has been necessary to expand access to VET for the purpose of social inclusion, they are well aware that this inevitably undermines VET's broader educational mission, a trend that is only intensified by performance-based funding which rewards VET providers for completed qualifications and competence units. It is perplexing how Finnish VET has managed to retain its popularity, despite more than a decade of very severe budget cuts and constant public criticism. VET somehow manages to offer opportunities for first timers and last-chancers alike, even if its new, outcome-based individual pedagogy is stunted in the eyes of the skeptical educational researcher. This endlessly dynamic show must go on.

References

Aarkrog, V. (2020). The standing and status of vocational education and training in Denmark. *Journal of Vocational Education & Training, 72*(2), 170–188. https://doi.org/10.1080/13636820.2020.1717586

Act on Vocational Education and Training 531/2017.

Ågren, S. (2021). Exploring vocational education students' visions of a successful transition to working life from the perspective of societal belonging. *Journal of Applied Youth Studies, 4*, 67–81. https://doi.org/10.1007/s43151-021-00037-5

Aho, E., Pitkänen, K., & Sahlberg, P. (2006). *Policy development and reform principles of basic and secondary education in Finland since 1968* (p. 2). The World Bank.

Al'Abri, K., Hogan, A., Lingard, B., & Sellar, S. (2024). Researching policy elites in education. In M. Stacey & N. Mockler (Eds.), *Analysing education policy: Theory and method* (pp. 235–249). Routledge.

Ante, C. (2016). Introduction. In C. Ante (Ed.), *Europeanisation of vocational education and training* (pp. 1–33). Springer International Publishing. https://doi.org/10.1007/978-3-319-41570-3

Antikainen, A. (2006). In search of the Nordic model in education. *Scandinavian Journal of Educational Research, 50*(3), 229–243. https://doi.org/10.1080/00313830600743258

Åslund, O., Grönqvist, H., Hall, C., & Vlachos, J. (2018). Education and criminal behavior: Insights from an expansion of upper secondary school. *Labour Economics, 52*, 178–192. https://doi.org/10.1016/j.labeco.2017.11.007

Billett, S., Dymock, D., Hodge, S., Choy, S., & Le, A. H. (2022). Shaping young people's decision-making about post-school pathways: Institutional and personal factors. In S. Billett, B. E. Stalder, V. Aarkrog, S. Choy, S. Hodge, & A. H. Le (Eds.), *The standing of vocational education and the occupations it serves. Professional and practice-based learning* (pp. 103–136). Springer Nature.

Bohlinger, S. (2019). Ten years after: the 'success story' of the European qualifications framework. *Journal of Education and Work, 32*(4), 393–406. https://doi.org/10.1080/13639080.2019.1646413

Cloward, R. A., & Ohlin, L. E. (1960). *Delinquency and opportunity: A study of delinquent gangs*. Free Press.

Cort, P. (2010). Stating the obvious: The European Qualifications Framework is not a neutral evidence-based policy tool. *European Educational Research Journal, 9*(3), 304–316. https://doi.org/10.2304/eerj.2010.9.3.304

Dale, R., & Parreira do Amaral, M. (2015). Discursive and institutional opportunity structures in the governance of educational trajectories. In M. Parreira do Amaral, R. Dale, & P. Loncle (Eds.), *Shaping the futures of young Europeans: Education governance in eight European countries* (pp. 23–41). Symposium Books.

Dale, R., Kazepov, Y., Rinne, R., & Robertson, S. (2017). Scales, discourse and institutions in the governance of educational trajectories in Europe. In A. Walther, M. P. D. Amaral, M. Cuconato, & R. Dale (Eds.), *Governance of educational trajectories in Europe: Pathways, policy and practice* (pp. 55–74). Bloomsbury Publishing Plc.

Di Stasio, V., & Solga, H. (2017). Education as social policy: An introduction. *Journal of European Social Policy, 27*(4), 313–319. https://doi.org/10.1177/0958928717728712

Esmond, B., & Atkins, L. (2022). *Education, skills and social justice in a polarising world: Between technical elites and welfare vocationalism*. Routledge.

Esping-Andersen, G. (1990). *The three worlds of welfare capitalism*. Polity Press.

Finnish National Agency for Education. (2018). *Cedefop opinion survey on vocational education and training in Europe: Finland*. Cedefop ReferNet thematic perspectives series. Retrieved November 1, 2024, from https://cumulus.cedefop.europa.eu/files/vetelib/2018/opinion_survey_VET_Finland_Cedefop_ReferNet.pdf

Halperin, S., & Heath, O. (2020). *Political research: Methods and practical skills*. Oxford University Press.

Hansen, P., Wallenius, T., Juvonen, S., & Varjo, J. (2020). Moving landscapes of Nordic basic education: Approaching shifting international influences through the narratives of educational experts. *Compare: A Journal of Comparative and International Education, 50*(6), 773–791. https://doi.org/10.1080/03057925.2018.1557509

Heikkinen, A. (2001). The transforming peripheries of vocational education: Reflections from the case of Finland. *Journal of Education and Work, 14*(2), 227–250.

Hörmann, B., & Sivesind, L. (2024). Ekspertintervjuer i det utdanningspolitiske feltet. In K. Sivesind (Ed.), *Hvordan studere utdanningspolitikk? Tekster, teorier, tilnærminger* (pp. 213–231). Cappelen Damm Akademisk.

Isopahkala-Bouret, U. (2015). 'It's considered a second class thing.' The differences in status between traditional and newly established higher education credentials. *Studies in Higher Education, 40*(7), 1291–1306. https://doi.org/10.1080/03075079.2014.881339

Isopahkala-Bouret, U., Lappalainen, S., & Lahelma, E. (2014). Educating worker-citizens: Visions and divisions in curriculum texts. *Journal of Education and Work, 27*(1), 92–109. https://doi.org/10.1080/13639080.2012.718745

Jørgensen, C. H., Olsen, O. J., & Thunqvist, D. P. (2018). *Vocational education in the Nordic countries: Learning from diversity*. Routledge.

Kalalahti, M., & Varjo, J. (2022). On the pathway to an unforeseeable future: An actantial analysis of career designs of young people. *Young, 30*(4), 327–343. https://doi.org/10.1177/11033088211057369

Kauko, J. (2019). The Finnish comprehensive school: Conflicts, compromises, and institutional robustness. In P. 't Hart & M. Compton (Eds.), *Great policy successes* (pp. 122–142). Oxford University Press.

Kettunen, P. (2013). Vocational education and the tensions of modernity in a Nordic periphery. In M. Buchardt, P. Markkola, & H. Valtonen (Eds.), *Education, state and citizenship* (NordWel studies in historical welfare state research) (pp. 31–55). Nordic Centre of Excellence Nordwel.

Koopmans, R., & Statham, P. (1999). Political claims analysis: Integrating protest event and political discourse approaches. *Mobilization: An International Quarterly, 4*(2), 203–221. https://doi.org/10.17813/maiq.4.2.d759337060716756

Kosunen, S., & Hansen, P. (2018). Discursive narratives of comprehensive education politics in Finland. *European Educational Research Journal, 17*(5), 714–732. https://doi.org/10.1177/1474904118764938

Lappalainen, S., & Lahelma, E. (2015). Subtle discourses on equality in the Finnish curricula of upper secondary education: Reflections of the imagined society. *Journal of Curriculum Studies, 48*(5), 650–670. https://doi.org/10.1080/00220272.2015.1069399

Lappalainen, S., Nylund, M., & Rosvall, P.-Å. (2019). Imagining societies through discourses on educational equality: A cross-cultural analysis of Finnish and Swedish upper secondary curricula from 1970 to the 2010s. *European Educational Research Journal, 18*(3), 335–354.

Laukia, J. (2013). Education of skilled workers and citizens—Vocational education in Finland. In K. Aaltonen, A. Isacsson, J. Laukia, & L. Vanhanen-Nuutinen (Eds.), *Practical skills, education and development—Vocational education and training in Finland* (pp. 8–18). Haaga-Helia University of Applied Sciences.

Li, J., & Pilz, M. (2021). International transfer of vocational education and training: A literature review. *Journal of Vocational Education & Training, 75*(2), 185–218. https://doi.org/10.1080/13636820.2020.1847566

Lietzén, O. (2023). The positioning of dual qualification studies in Finnish upper secondary education and government policy since the 1980s. *European Educational Research Journal, 22*(3), 368–393. https://doi.org/10.1177/14749041211065328

Martínez-Izquierdo, L., & Torres Sánchez, M. (2022). Dual vocational education and training and policy transfer in the European Union policy: The case of work-based learning and apprenticeships. *Cogent Education, 9*(1). https://doi.org/10.1080/2331186X.2022.2154496

Masoud, A., Holm, G., & Brunila, K. (2021). Becoming integrateable: Hidden realities of integration policies and training in Finland. *International Journal of Inclusive Education, 25*(1), 52–65. https://doi.org/10.1080/13603116.2019.1678725

Maurer, M., & Gonon, P. (2014). The challenges in policy transfer in vocational skills development—An introduction. In P. Gonon & A. Heikkinen (Eds.), *The challenges in policy transfer in vocational skills development* (pp. 15–34). Peter Lang AG. https://doi.org/10.3726/978-3-0351-0758-6

McCammon, H. (2013). *Discursive opportunity structure*. https://doi.org/10.1002/9780470674871.wbespm073

McGrath, S. (2012). Vocational education and training for development: A policy in need of a theory? *International Journal of Educational Development, 32*(5), 623–631. https://doi.org/10.1016/j.ijedudev.2011.12.001

National Audit Office of Finland. (2021). *Ammatillisen koulutuksen reformi [The reform of vocational education and training]*. Valtiontalouden Tarkastuskertomukset 2/2021 [in Finnish].

Niemi, A. M., & Jahnukainen, M. (2019). Educating self-governing learners and employees: Studying, learning and pedagogical practices in the context of vocational education and its reform. *Journal of Youth Studies, 23*(9), 1143–1160. https://doi.org/10.1080/13676261.2019.1656329

Nordic Statistics Database. (2023). *NEET01: Young people neither in employment nor in education and training (NEET rates) by reporting country, sex, age, educational attainment level and time*. Retrieved November 1, 2024, from https://www.nordicstatistics.org/news/nordic-neet-rates-on-eu-target-levels/#:~:text=Among%20the%20five%20Nordic%20countries,NEET%20rate%20below%209%20percent

Norwegian Directorate for Education and Training. (2024). *Flere søker yrkesfag [More are applying for vocational subjects]*. Retrieved November 21, 2024, from https://www.udir.no/tall-og-forskning/statistikk/statistikk-videregaende-skole/analyser/2024/sokertall-vgo-2024/#:~:text=53%20prosent%20av%20søkerne%20til,og%20studiespesialisering%20går%20mest%20ned

ReferNet Sweden Cedefop. (2022). Sweden: Increasing attractiveness of secondary VET through access to higher education. *National news on VET*. Retrieved November 21, 2024, from https://www.cedefop.europa.eu/en/news/sweden-increasing-attractiveness-secondary-vet-through-access-higher-education

Rintala, H., & Nokelainen, P. (2022). Promoting the standing of VET in Finland: Balancing amongst the different aims of VET. In S. Billett, B. E. Stalder, A. Vibe, S. Choy, S. Hodge, & A. H. Le (Eds.), *The standing of vocational education and the occupations it serves: Current concerns and strategies for enhancing that standing* (pp. 159–176). Springer Nature.

Rosenblad, N. (2023). *Knowledge and agency in vocational education: An analysis of competency-based training and student counselling utilising a lens of critical realism to cultural historical activity theory*. PhD dissertation, University of Helsinki.

Rosenblad, N., Schaffar, B., & Löfström, E. (2022). VET and the "Competency-Tetris": Inclusion of whom, to what, and where? *Nordic Journal of Comparative and International Education (NJCIE), 6*(3–4). https://doi.org/10.7577/njcie.4835

Schriewer, J. (1990). The method of comparison and the need for externalization: Methodological criteria and sociological concepts. In J. Schriewer & B. Holmes (Eds.), *Theories and methods in comparative education* (pp. 25–83). Peter Lang.

Seitamaa, A., & Hakokönggäs, E. (2022). Finnish vocational education and training experts' reflections on multiculturalism in the aftermath of a major reform. *Journal of Vocational Education & Training, 76*(3), 644–663. https://doi.org/10.1080/13636820.2022.2066559

Seitamaa, A., & Rosenblad, N. (forthcoming). Freedom to compete: Autonomy and control in Finnish vocational education and training reform. In M. Pereyra & M. Torres Sánchez (Eds.), *Challenges of the reform of vocational training from the German dual modality and the Spanish today's educational system: Analysis of cases and from a comparative perspective*. Brill.

Simola, H. (2015). *The Finnish Education Mystery: Historical and sociological essays on schooling in Finland*. Routledge.

Smeplass, E., & Zirkle, C. (2025). Introduction. In E. Smeplass & C. Zirkle (Eds.), *Opportunity structures in vocational education and training: National Studies and International Discussions* (pp. 1–19). Palgrave Macmillan.

Smeplass, E., Rapp, A. C., & Corral-Granados, A. (2023). Understanding how institutional dynamics can contribute to educational inequality in Nordic cities. *Oxford Review of Education, 50*(2), 290–307. https://doi.org/10.1080/03054985.2023.2274027

Smeyers, P., & Depaepe, M. (2016). Mutual dependencies: "Change" and "discourse.". In P. Smeyers & M. Depaepe (Eds.), *Educational research: Discourses of change and changes of discourse* (pp. 1–8). Springer.

Statistics Finland. (2023). *Entrance to education* [online publication]. Reference period: 2022. ISSN:1799–4527. Statistics Finland. Retrieved from https://stat.fi/en/publication/cl8cvm9et90ba09w27xqclksi

Stenström, M.-L., & Virolainen, M. (2018). The modern evolution of vocational education and training in Finland (1945–2015). In S. Michelsen & M.-L. Stenström (Eds.), *Vocational education in the Nordic countries* (Routledge research in international and comparative education) (pp. 102–123). Routledge.

Thunqvist, P. D., Tønder, A. H., & Reegård, K. (2019). A tale of two reforms: Institutional change in vocational education and training in Norway and Sweden in the 1990s. *European Educational Research Journal, 18*(3), 298–313. https://doi.org/10.1177/1474904118823104

Välimaa, J. (2019). *A history of Finnish higher education from the middle ages to the 21st century* (Higher education dynamics) (Vol. 52). Springer. https://doi.org/10.1007/978-3-030-20808-0

Virolainen, M. (2022). *Case study Finland. The future of vocational education and training in Europe. Volume 2. Delivering IVET: Institutional diversification and/or expansion.* Cedefop.

Virolainen, M., & Stenström, M. (2014). Finnish vocational education and training in comparison: Strengths and weaknesses. *International Journal for Research in Vocational Education and Training, 1*(2), 81–106.

Virolainen, M., & Tønder, A. H. (2018). Progression to higher education from VET in Nordic countries: Mixed policies and pathways. In C. H. Jørgensen, O. J. Olsen, & D. P. Thunqvist (Eds.), *Vocational education in the Nordic countries: Learning from diversity* (Routledge research in international and comparative education) (pp. 51–73). Routledge.

Zacheus, T., Kalalahti, M., & Varjo, J. (2017). Cultural minorities in Finnish educational opportunity structures. *Finnish Journal of Social Research, 10*(2), 132–144. https://doi.org/10.51815/fjsr.110772

Zaunstöck, T., Marhuenda-Fluixá, F., Ros-Garrido, A., & Fischer, M. (2021). Europeanisation of VET—The Spanish Vocational Education and Training system and the influence of European education policy. *Journal of Vocational Education & Training, 73*(2), 316–335. https://doi.org/10.1080/13636820.2021.1889645

CHAPTER 7

Quality Indicators for the Vocational Education and Training of Refugees: Acceptance of German Vocational School Principals

Katrin J. Posch, Junmin Li, and Matthias Pilz

Introduction

Vocational education and training (VET) is considered vital for integrating immigrants and refugees into both education and the workforce (Jeon, 2019). Since refugees often have basic or informal vocational education, VET schools play an important role in their education and to open pathways, and support the integration both into the labour market

K. J. Posch
WiSo Faculty at the University of Cologne, Cologne, Germany

J. Li (✉)
UNESCO Chair on TVET and Competence Development for the Future of Work, TU Dortmund University, Dortmund, Germany
e-mail: junmin.li@tu-dortmund.de

E. Smeplass and C. Zirkle (eds.), *Opportunity Structures in Vocational Education and Training*, Palgrave Studies in Technical and Vocational Education and Training,
https://doi.org/10.1007/978-3-032-04410-5_7

and into the society (Winther et al., 2018). Refugees are seeking vocational education mainly to get employment in the German labour market (Granato et al., 2011). Refugees have their own specific needs, which may differ greatly from standard schooling, and they may be confronted by different structural barriers in the educational system (Jørgensen et al., 2021). As a result, there are specific requirements for good quality schooling for refugees at vocational schools to build an opportunity structure according to their specific needs. This paper focuses on the quality aspects associated with refugee migration within VET schools in Germany.

The concept of quality in the course of school development is widely discussed and promoted (Hopkins, 1994; Fullan, 2007), however, most political approaches are general in nature and do not focus sufficiently on the level of individual training programmes within VET schools, and their specific target groups.

The Erasmus+ project "Indicators of good VET practice for refugees"developed quality indicators for the integration of refugees in VET in a European context. These quality indicators were systematically generated from experiences at the practical level of VET in Austria, Denmark, Germany and Italy and are dedicated to the quality development of a specific training programme. This type of quality development, which differs from generic quality development across all training programmes, leads to a specific type of quality management and school opportunity development (GoodVET, 2019).

Against this background, this paper examines the acceptance by school principals at German VET schools regarding the specific quality indicators for the specific target group of refugees in VET.

The subject of this research work relates to the context of the development of the individual school. Special attention is paid to the school principals' level and their acceptance and personal attitude towards the indicators. The level of school principals is used as these are people who,

---

M. Pilz
Business Education and International VET Research, University of Cologne, Cologne, Germany

German Research Center for Comparative Vocational Education and Training, Cologne, Germany

due to their function, have a general overview and the necessary information to be able to assess the quality indicators. At the same time, their function gives them a special role in relation to school developments (Heinrichs et al., 2016). Since a European quality framework for the integration of refugees in VET is analysed from the view of school principals in German VET schools, the insights of this study could also be relevant for the implementation of quality indicators for other vulnerable groups in other European VET schools.

The Quality Indicators for the Integration of Refugees in VET

In the GoodVET project, the project team developed an extended version of the input-process-output quality model (IPO model) according to Dubs (2003) and Draft (2008). It comprises quality dimensions that map the organisational structure and responsibilities in VET institutions. These dimensions and quality indicators were identified through 20 best practice examples in the field of the integration of refugees through VET in Austria, Denmark, Germany and Italy. The analysis was conducted by document analysis, interviews and expert workshops (GoodVET, 2019). These quality dimensions are institutions, teachers and trainers, learners and interactions with the environment. The dimensions of the IPO model and the supplementary four dimensions form a framework for the 27 quality indicators (see Table 7.1). These indicators have been identified as particularly important in the context of VET for refugees (GoodVET, 2019). The indicators served as the basis for developing the quality analysis concept.

The institutional dimension here encompasses governance, leadership and decision-making in the school and its environment. Through continuous and joint activities of these actors, the framework conditions can be created here to influence certain areas in the school (Altrichter, 2010).

The dimensions of teachers, trainers and learners are considered in a differentiated way. Teachers and training staff instruct learners or teach them new skills or knowledge regardless of the place of learning, while learners are taught in VET programmes. Regarding the indicators, it is crucial that these two groups interact on several levels (de Abreu & Hale, 2014).

Interactions with the environment are also an important dimension for refugees. They are often subject to post-traumatic experiences and

Table 7.1 Overview of the quality indicators from the GoodVET project (from GoodVET (2019), licensed under CC-BY-SA)

	Institution	*Teacher and trainer*	*Learner*	*Interaction with the environment*
Input	Learning goals and content Support system and coordination	Special training for teachers Joint preparation of all teachers	Individual demands Access and entrance requirenments Class composition Contact with learners	Sustainable demand for training programmes Cooperation between schools and workplaces Validation and recognition of learning and certifications
Process	Finance and funding	Selecting teaching methods Transfer orientation Formative and summative feedback	Classroom situation Learning language and culture	Target group orientated towards the mass media and social media Matching of governmental goals and refugees' goals
Output		Evaluation of teaching methods Assessment of learning progress	Absence and dropout rate	Measurement of the attainment of learning objectives Certification and their recognition in the labour market Employability Bildung, empowerment and the feeling of being an active member of society Follow-up courses

Source: GoodVET (2019)

face new challenges with regard to their first employment or recognition in a society that is new to them. These influences should be included and taken into account in the training programmes for refugees. The indicators are located at different levels of abstraction and are both qualitative and quantitative in nature. Their analysis is subject to very different approaches, which will not be discussed further here (Pilz, 2021).

Theoretical Foundation of Acceptance Research

The aim of this study is to analyse the extent to which the quality indicators are applicable at the level of an individual school in the context of refugee integration. To this end, the observations and expertise of selected school principals are used. The applicability of the indicators to the VET school system is purely hypothetical. Consequently, this is explicitly not implementation research. Implicitly, however, individual excerpts from implementation research serve as useful approaches, and selected elements from innovation management and change management can also be used as a supplement.

Since the potential implementation of individual indicators entails a development of the school and thus an increase in school quality, the study can be categorised in the area of school development and school quality research.

For the hypothetical applicability of the quality indicators, it is therefore first necessary for the school principals interviewed to confirm (potential) adoption by accepting and supporting each individual indicator. To be able to depict such approval, the focus is placed accordingly on the acceptance of the indicators.

In the academic literature, the term "acceptance" is often used in many different ways, which can be explained by the multidisciplinary contexts, among other things. Existing definitions are usually influenced by the respective research direction or by the object being analysed. In application, a definition was originally pursued by Pressmar (1982), in which acceptance is regarded as a consenting acceptance or an affirmative acceptance of a situation, an object or a person.

Research into acceptance can be seen as running in parallel with research in psychology and social sciences, which serves to analyse the acceptance (or rejection) of innovation subjects (Simon, 2001; Pressmar, 1982; Stiller & Wager, 2023). Acceptance approaches and theories are also examined in many other areas of research; they often even refer to each other. In recent years, there has been a strong development in technology acceptance research, in which a distinction can be made between an origin in the social sciences or in economics. Business management approaches, such as human resource management, are often used (Schnell, 2009). Theories from psychological approaches, norm activation theory (e.g., Huijts et al., 2012) or sociological implementation, innovation and diffusion research (e.g., Rogers, 2003) are also frequently referred to.

For the purposes of this study, however, acceptance research must be clearly distinguished from diffusion research—and here primarily from adaptation theory. Within adaptation theory, the focus is more on the object-related macro level to implement an innovation in an organisation. However, acceptance is increasingly subjective, which is why it is always linked to attitudes and behaviour (Liu et al., 2005). Acceptance research can therefore be located more on a micro level, in which the question of the acceptance of an innovation by an individual is investigated. Acceptance can therefore be seen as a positive decision in favour of the adoption of an innovation by a potential user (Simon, 2001).

As the school principals were asked about their attitude towards, and support for, the individual quality indicators, this part can be assigned to acceptance research on the basis of these considerations. However, due to the different and interdisciplinary approaches, it is still necessary to specify the understanding of acceptance applicable for this work in more detail.

In the social and economic literature, a categorisation is made into behavioural or attitudinal concepts, and sometimes a combination of both is used (Simon, 2001). At present, the approaches of a two-dimensional construct by Müller-Böling and Müller (1986) have become established, which differentiate between attitudinal and behavioural acceptance. Attitude is a key dimension of acceptance in that it appears in almost every definition. Müller-Böling and Müller separate three components within acceptance: an affective (emotional), a cognitive (intellectual) and a conative (action-orientated) component.

The affective component of acceptance addresses the motivational-emotional aspects and establishes a connection between a situation and a subjective inner attitude towards it (Müller-Böling & Müller, 1986; Meyer, 1997). Thus, the emotional attitude is presented, for example whether the head teachers are emotionally opposed to the quality indicators in order to help refugees achieve an improved learning environment, or whether they accept them as a welcome aid to improve school quality.

The second component of the cognitive aspect considers a comparison of the costs and benefits of an innovation, taking into account the respective personal context. In the present case, this cognitive aspect can refer not only to the financial resources but can also be adapted to the content of the quality indicators to be analysed. The cognitive aspect refers to a rational weighing up of the respective advantages and disadvantages, taking into account the personal context. The conative component means that although a tendency to act should be recognisable, actual action does not necessarily have to occur (Müller-Böling & Müller, 1986; Meyer, 1997).

According to Müller-Böling and Müller (1986), acceptance not only includes the attitude components just described, but also a behavioural component in this context: It is examined whether the underlying facts are actually implemented in reality, for example, whether certain quality indicators are actually used in the VET school context by the school principals surveyed.

Acceptance can usually be seen as a prerequisite for behavioural acceptance. In other words, it is assumed that the individual attitude will influence the resulting behaviour, provided that the behaviour is based on a voluntary acceptance (e.g., Taylor & Todd, 1995).

For the present research context, acceptance is considered to be a positive attitude towards a specific innovation object. It is also assumed that although there may be a positive willingness to potentially apply an innovation, it does not necessarily have to be actually implemented (Müller-Böling & Müller, 1986; Meyer, 1997).

Based on this understanding, it can be concluded that acceptance is a personal and therefore purely subjective factor that is usually the result of the experiences and personality of the individual.

Method

The selection of the federal state of North Rhine-Westphalia (NRW) as the research area is because NRW, as the most populous state in Germany, has the highest number of refugees; thus, more empirical values are to be expected.

When selecting the VET schools, care was taken to include different districts and different subject areas in the study.

To ensure the highest possible level of objectivity and validity for the study area, and to obtain a detailed overview of the conditions in NRW,

VET schools from three different administrative districts were selected for the study: Arnsberg, Düsseldorf and Cologne. In terms of population, the administrative district of Düsseldorf has the highest proportion of foreign students in NRW in 2020. According to section 3 of the Refugee Reception Act, Cologne takes in the most refugees in absolute terms (Bezirksregierung Arnsberg, 2020). The administrative district of Arnsberg is included alongside the two administrative districts with the largest cities in NRW because it tends to be less urban and more rural in character. In turn, schools from three different subject areas from each of the three selected administrative districts are included in the survey: (1) agriculture and nutrition, (2) health care and social and (3) business and administration.

A qualitative approach was chosen as the research method, as the subjective views of the interviewees are relevant, whereby these are expressed in the form of acceptance in this research work (Patton, 1990).

Due to their leadership function, school principals also have the opportunity to further develop the respective individual schools and implement the corresponding measures. School principals in Germany are experienced teachers, who have undergone an aptitude assessment procedure as a school principal. Because of the importance of individual experience, no person from outside of the school system would possess the essential knowledge and expertise to fill this role. Various research results confirm the special function of school principals through a significant connection between school processes and the actions of the school principals (Harris et al., 2003; Klein & Schwanenberg, 2022).

For a better understanding, it should be emphasised that the term "school management" covers all persons who hold a managerial position. On the one hand, these are the actual school principals and their deputies. As each school can be regarded as an individual organisational unit, other managers are also regarded as experts depending on the structure of the respective school, such as level heads or, in the specific context of VET schools, course leaders, area managers, course coordinators or department heads.

The thematic focus of the interviews is the applicability of the quality indicators at the level of the individual school, for which the acceptance of the school principals from NRW is surveyed. For this reason, the structure of the guidelines is based on the structure of the indicators from the GoodVET project and divided into input, process and output according to Dubs (2003).

The interviews were conducted shortly after contacting the respective VET schools and all took place in summer 2020. Of the ten interviews carried out, eight were conducted with one interviewee each; it was possible to motivate the school principals for five of the interviews and the extended school management boards (*erweiterte Schulleitung*) for three. The remaining two interviews were conducted with two interview partners each, whereby in one case the head of the educational programme will be interviewed together with the person responsible for school social work, and in the other case the school principal was interviewed together with the extended management board of the school. All interviews are conducted on site at the respective school. The interviews lasted just under 2 hours on average, with the longest interview lasting 177 minutes and the shortest 82 minutes. The transcripts and protocols are analysed using a qualitative content analysis. For this purpose, categories are worked out on the basis of the material and analysed systematically. Qualitative content analysis is described as an open, systematic and theory-led procedure (e.g., Mayring, 2014).

Results

It should be mentioned at the outset that the school principals do recognise that each indicator has an influence on the quality of schooling and, in particular, on the schooling of refugees. However, it can also be seen that the perception of the importance and necessity of the indicators differs greatly. Thus, the indicators are first grouped according to their importance as assessed by the school principals, and thus according to the level of acceptance, detached from the previous structure (input-process-output). For this purpose, the indicators can be divided into three categories: (1) high acceptance, (2) medium acceptance and (3) low acceptance.

Indicators with High Acceptance

The indicators "support system and coordination" from the dimension "institution" and "learning language and culture" from the dimension "learner" (see Table 7.1) are considered indispensable by the school principals without any restrictions. Even if language learning is part of the curriculum, the VET school principals consistently recognise the further need of the students, which in most cases the schools are unable, or insufficiently able, to meet with the available resources (personnel and financial). Learning about culture is also part of the curriculum at VET schools, but both aspects go beyond the formal form. As one school principal described:

> *Getting to know social life is a very important aspect. Getting to know cultural life, getting to know political life and then taking part. Only then does language acquisition work properly.* (B6, item 99).

Language and culture are also taught informally, for example during a work placement or through the composition of classes to promote interculturality.

Offering additional support for refugees within the schools is considered to be of inestimable value; however, the scope of possibilities depends heavily on the individual school and its experience, but also on its existing network. Central contact persons for a wide range of areas would be a possible option. Comprehensive counselling for refugees regarding VET should be considered an important aspect of support within schools because, a complex range of counselling services exists (Hilkert et al., 2020). Needs are primarily described in the optimisation of support, for which further resources are required. For example, a school principal mentioned:

> *There are so many individual cases that are then also picked up by the school or with the school, which then lead to structures forming, networks forming. And that's where the school social workers are absolutely, well, important, irreplaceable, so to speak. Then it's about making applications for education and participation or also accompanying them on visits, visits to the authorities or also in the case of parents, even if you can't always do much* (B8, item 19).

Further needs are recognised regarding external school support, for example in the insufficient number of trauma therapists, over which the school has no influence, which is why this facet is not part of school development.

In addition to these two indicators, further indicators can be added that are also classified as indispensable: "evaluation of teaching methods" from the dimension "teacher and trainer," and "finance and funding" from dimension "institution" (see Table 7.1). The existing vocational training measures for refugees can be assessed as financially secure, whereby the budget is regularly adjusted accordingly. In the current scenario, the head teachers describe the relevance of the longer-term predictability of the budget. In general, this indicator has a special significance insofar as the available monetary resources in turn have an influence on the feasibility of other indicators, as one school principal emphasised:

> *I need security in order to be able to plan for the long term, both financially and in terms of personnel, right? If the integration centres were to be taken away from us now in terms of the budget, then we would have a big problem. And that's also the case with the finances* (B10, item 130).

The indicators "joint preparation of all teachers," and "special training for teachers" from the dimension "teacher and trainer," and "certification, and their recognition on the labour market" from the dimension "interaction with the environment" (see Table 7.1) also have high acceptance. Under certain conditions, the indicators "class composition" and "classroom situation" from the dimension "learner" and the indicator "cooperation between schools and workplaces" from the dimension "interaction with the environment" (see Table 7.1) can also be placed in this upper priority group.

The acceptance of the indicator "evaluation of teaching methods" can be seen as unavoidable.

One frequently cited obstacle to implementation is the time factor. In terms of a holistic view of the school and corresponding development, an evaluation should not be limited to teaching and learning methods but should also be applied to other areas (e.g., Capaul et al., 2020). Such an expansion of the indicator is repeatedly recommended by school management, with reference being made to the external evaluation of the state of NRW. In direct contact between teachers and refugee students, the focus is placed on the "evaluation of teaching and learning methods".

The indicators "special training for teachers" and "joint preparation of all teachers" are aimed at the planned preparation of teachers for multicultural learning environments by means of further and advanced training. The management levels are in favour of these instruments and use them for the most part. Such knowledge is particularly important for all teachers who have direct contact with refugees. However, the school principals have also identified a greater need for teachers to take part in training courses. Due to the financial restrictions in this area, however, these resources are carefully allocated and planned over a longer period.

The results have shown that from the perspective of principals, "certifications and their recognition on the labour market" are of considerable importance to employers, especially in the context of refugee students, as this group often has insufficient evidence. The awarding of certificates in a school is a matter of course; for those involved in practice, such as potential employers, further meaningful evidence documenting the relevant competences, skills and knowledge is a key indicator in a recruitment decision. The school principals are in favour of more comprehensive certificates, primarily for practical relevance, and hope for a revision at the political level so that VET schools can act in the interests of the students.

Close thematic link between the indicators "class composition" and "classroom situation" has been recognised. It has been conveyed that a criteria-orientated composition of the respective classes can have an influence on the subsequent learning atmosphere within a class. As far as this point of initial contact is concerned, elements of the "contact with learners" indicator can also be recognised, as the composition of the classes is already designed here. During the interviews, the need for a course system with different teachers for different courses was identified several times, which would allow individual needs to be better catered for. However, this wish contradicts, at least to some extent, the fact that the school administrators within the so-called "International supporting classes" [*Internationale Förderklasse*][1] have also spoken out in favour of the class teacher to be able to guarantee the new students a certain degree of stability. Regarding the implementation of a course system, differentiation according to educational programme, or individual consideration within the individual school, would certainly be considered necessary.

The school principals confirm the importance of the indicator "cooperation between schools and workplaces" but see an extensive network with various stakeholders as insufficiently taken into account in the quality indicators presented. Such missing characteristics, such as networking between VET schools, are discussed in detail in the presentation of the quality characteristic "network with relevant stakeholders." Cooperation with workplaces in the sense of the dual system is an integral

[1] The schooling of young refugees who have not yet reached the age of 18 takes place at the vocational school within the framework of compulsory schooling in accordance with § 34 Para. 6 SchulG in conjunction with § 38 SchulG (compulsory schooling in upper secondary level) in the international support class in the 1-year training preparation programme.

part of VET in Germany anyway. Nevertheless, this characteristic should also continue to be considered, as the practical experience, especially for refugees, also represents learning about the (work) culture. An extensive network is considered particularly valuable, which can also be explained by the indicator "support system and coordination" from the dimension "institution."

Indicators with Medium Acceptance

In the following section, those indicators that have been accepted at school management level but are prioritised at a middle level are discussed. The following indicators can be counted in this category: "learning goals and content," "individual demands," "contact with learners," "access and entry requirements," "validation and recognition of learning and certification," "selecting teaching methods," "transfer orientation," "formative and summative feedback," "assessment of learning progress," "absence and dropout rate," "measurement of attainment of learning objectives," and "*Bildung*, empowerment and the feeling of being an active member of society." As one school principal expressed his point of view for "*Bildung*, empowerment and the feeling of being an active member of society" from the dimension "interaction with the environment" as follows:

> *I think the better the education, the better the chances of integration. That's the way it is. Well, and also for long-term integration, right? […]. But specifically, for someone to have an independent life here in the long term, because they earn their own money, that's exactly what I think they need education for. And, above all, language skills* (B7, item. 230).

Referring to the indicator of "absence and dropout rate" from the dimension "learner," one school principal describes the reasons for absence here:

> *Well, in the international classes, pupils who don't turn up go to work. And we officially know very little about it. Sometimes a bit under the table. […] And a second point for absenteeism is that they are here with their families and have to translate. They then have to go to the doctor with their mum, to the doctor with their dad, to every authority and they always have to act as translators and it seems that for the parents, school doesn't have the relevance that it has for us, right?* (B10, pos. 190).

The "learning goals and content" from the dimension "institution" can largely be regarded as being specified by the Ministry of Education and Cultural Affairs, even if the individual school is given a certain amount of room for manoeuvre. However, the ideal vision of the school principals requires more human resources and a change in the legal requirements.

Although the "individual demands" indicator from the dimension "learner" is closely linked to "contact with learners" in the same dimension, the two can be separated in terms of time. In terms of content, the indicator deals de facto with the expectations, ideas and wishes of the refugees. In practice, this process takes place at an external organisation and not directly at a VET school.

The "contact with the learners," in other words the students, then refers to the contact within an assigned individual school and is classified as extremely important by the experts. As far as the type of implementation and organisation is concerned, the schools can act individually. It would be desirable to have a personal interview with each individual student immediately after allocation, during which an initial assessment can be made for allocation to a class and possible misjudgements can be easily revised. In the interviews, however, the most important contact with the students is considered to be that which unfolds continuously during the school years. The focus is on building trust between the class teacher and students.

The admission and entry requirements for a VET school are determined by the training examination regulations for VEY schools in the respective federal state. Regardless of the educational programme, VET schools only have a minor influence on the requirements. There is a great need to change the entry requirements for dual training. There are no checks by school authorities regarding the language skills of young people. An existing training contract already legitimises participation in school lessons, which in most situations leads to considerable problems in schooling, as the refugees are usually unable to follow the lessons adequately due to the language barrier. A frequent consequence is that they fail the final examinations of the vocational programmes and therefore do not obtain a recognised vocational qualification. To eliminate this imbalance, however, it is not only necessary to readjust the entry requirements, there is also a need to amend the final examinations organised by the chambers.

The need to address the special circumstances of refugees is taken into account in the indicator "measurement of the attainment of learning objectives" from the dimension "interaction with the environment" (see Table 7.1). The experts agree that the aim should not be to simplify the examinations at the expense of the quality of vocational training, but primarily to rethink and revise certain requirements. A related consideration arises from the indicator "validation and recognition of learning and certifications" from the same dimension. The organisation of the recognition and certification of existing knowledge and prior experience can vary in scope. As the results indicate, validation procedures and the possibility of obtaining qualification modules already exist. The increased modularisation of training courses has also been under discussion for several years.

The results are evidence of an existing acceptance of the "selecting of teaching methods" from the dimension "teacher and trainer," whereby it must be taken into account that each class has different prerequisites and therefore no clear statement can be made as to which method will lead to the desired result. Once again, the possibility of establishing a course system would be desirable. The method of feedback (indicator "formative and summative feedback" from the same dimension) is seen as a core element in this context, as the target group is not sufficiently familiar with the local school culture. Regular feedback helps students to optimise their own learning, to assess themselves better and to make the system of awarding school grades more accessible to them. The summative feedback part of this indicator with the "assessment of learning progress" from the same dimension can hardly be differentiated.

A transfer orientation can certainly be identified through the reviews, but also through feedback and regular contact with the students. This is particularly evident in the context of language skills. Transferring what has been learnt is seen as an essential goal of a VET school in order to transfer the competences, skills and abilities acquired into practice. For this reason, this goal is taken into account in all learning fields and therefore in the curriculum (indicator "learning goals and content" from the dimension "institution" see Table 7.1).

In this context, the need for a central organisational unit to support the schools, especially relating to the absenteeism of students who are required to attend school, is evident. It would also be good if the scope of school social work could be extended to include home visits or if a school psychologist could be assigned to this task. In reality, the necessary personnel resources are lacking for implementation. References to the sub-aspects of the "support system and coordination" indicator from the same dimension can be recognised. However, the reduction of absenteeism and dropouts demands a special type of support, for which a number of authorities are responsible, which is why special attention is required.

The individual schools endeavour to create the best possible conditions for this within their framework of action. Even if vocational training is seen as an essential element of integration, there are other factors that cannot be directly influenced and that come into play at the macro and meso levels.

Indicators with Low Acceptance

The third category includes the indicators "sustainable demand for the training programmes," "target group orientated towards the mass media and social media," "matching of governmental goals and refugees' goals," "employability" and "follow-up courses" which are all in the dimension "interaction with the environment."

The educational programmes offered at a VET school are influenced by several factors external to the school and are to be understood as given. The schools themselves have no influence on the design and are responsible for the best possible implementation. The presentation of results has shown that the indicator "sustainable demand for the project" has a low level of acceptance, which can be explained precisely by the fact that the feasibility of this indicator cannot be localised at school level. Ultimately, it can be substantiated that there is no positive attitude towards this indicator.

The "target group orientated towards the mass media and social media" is already part of the curriculum and is therefore included in the learning objectives and content. The same applies to the use of various end devices, which are also used as part of the teaching and learning methods as soon as this is deemed appropriate. Although teaching the conscious and responsible use of these media is considered sensible, this part of the lesson is not especially prioritised in the case of refugee learners.

In addition to this topic, however, the results often highlight the importance of digitalising lessons.

The complexity of "matching government goals with refugees' goals" prevented the school principals from explicitly commenting on this. Nevertheless, an acceptance can be recognised that there should be a clear and consistent migration policy. In contrast, there were more extensive statements on education policy that extend to a direct influence on school life. Rather, the indicator has so far reflected issues of migration policy and the associated provisions that are to be localised at federal level in Germany.

From the perspective of school principals, the refugees see it as an essential task to achieve long-term "employability" on the labour market with suitable "follow-up courses." As a result, information and counselling are seen as an obligation of the VET school in order to convey the need for this subject area to the students. During such a consultation, however, great importance is also attached to providing individual and situation-specific advice.

Discussion

The categorisation of the quality indicators according to the three categories, (1) high acceptance, (2) medium acceptance and (3) low acceptance, does not indicate a high or low rejection of the quality indicators. By linking acceptance, this result indicates that the school principals interviewed categorise the indicators more or less in the functional and action area of their respective schools. This argument can also be supported by the theoretical foundation, as the aim of the study was to analyse the extent to which the quality indicators are applicable within an individual school. Furthermore, in the context of the understanding of this study, an individual school is seen as an independent organisational unit that independently drives its own school development forward within the legally possible framework.

It cannot be generalised that all those indicators which the school has no influence have a low level of acceptance. The indicators "learning goals and content" and "finance and funding" are also not within the direct sphere of influence of school management. However, these are aspects where the school is given room for manoeuvre, which is why it has a certain amount of control. For example, although the "learning goals and content" are prescribed by school law, the schools adapt these specifications individually to the needs and possibilities on site with the help of the annual didactic plan. The factors relating to the financing of a school must also be accepted as given under school law, and it has been demonstrated that some needs exist, such as the independent utilisation of the budget granted. Nevertheless, the school principals can prepare a viable calculation in accordance with the specified rules and the sustainability of the vocational training programmes is a matter of course. Within the framework of school legislation, the internal school budget can also be increased, or existing collaborations can be used to cover existing needs in other ways.

To summarise, it can therefore be seen that a differentiation of acceptance must be made regarding the application of the indicators in the context of school development. Among the indicators, there are those that are recognised independently by an individual school as part of a bottom-up strategy and developed according to individual needs. There are also indicators that require centralised control and development due to education policy. However, it must also be noted that, although a certain scope for action can be recognised and acceptance is given, a possible (further) development cannot be localised in an individual school.

Conclusion

In conclusion, this study shows that acceptance is significantly influenced by both the perceived demands and the feasibility of implementing these indicators. Consistent with findings from change management literature, it is evident that conducting a thorough demand assessment prior to implementation can enhance acceptance among stakeholders, particularly school principals (Lauer, 2010; Mandl & Burg, 2008; Venkatesh & Davis, 2000). For example, the school principals described the considerable demands of "support system and coordination" and "learning language and culture" for the successful integration of the refugee students during their interviews. However, while a clear understanding of specific demands can increase subjective acceptance, it does not guarantee it, highlighting the complexity of human factors in organisational change.

Furthermore, the analysis reveals that feasibility plays an indirect yet crucial role in shaping acceptance levels. Factors such as legal and organisational constraints, resource availability and inherent uncertainties can significantly affect how school principals perceive and accept new quality indicators (Seitter et al., 2015). For example, the indicator "sustainable demand for the project" was evaluated as a low level of acceptance, which was argued by the school principals by the fact that the feasibility of this indicator cannot be localised at school level.

The multi-level governance structure within the education system further complicates this dynamic, as principals must navigate various external requirements while adapting them to their unique local contexts (Bonsen, 2016).

Overall, this study contributes valuable insights into the factors influencing acceptance of quality indicators in schools. It emphasises the necessity for educational leaders to consider both the contextual demands and the feasibility of innovations to foster a more supportive environment of opportunity structure for the training of refugees. The study shows that VET for refugees needs different kind of quality indicators, since refugees have in some cases different contextual demands than regular VET students. Currently, the quality management systems for VET schools in Germany don't set special quality indicators for the integration of refugees. This study shows both a high demand on these quality indicators and the degrees of acceptance of the school principals on these quality indicators.

Future studies should continue to explore these dimensions to develop strategies that enhance acceptance and facilitate successful implementation of quality indicators in diverse educational contexts.

This study has several limitations that should be considered when interpreting the results. Firstly, the analysis is based on a relatively small number of interviews, which restricts the generalisability of the findings. Furthermore, the results pertain exclusively to the context of North Rhine-Westphalia and may not be applicable to other federal states or to the whole of Germany. Additionally, the specific conditions and challenges present in other EU countries can differ significantly from those examined here, further limiting the applicability of the findings in a European context. Future research should therefore aim to incorporate a broader data set and diverse geographical contexts to yield more comprehensive and generalisable insights.

References

Altrichter, H. (2010). Theory and evidence on governance: Conceptual and empirical strategies of research on governance in education. *European Educational Research Journal, 9*(2), 147–158. https://doi.org/10.2304/eerj.2010.9.2.147

Bezirksregierung Arnsberg. (2020). *Zuweisung nach dem Flüchtlingsaufnahmegesetz [Allocation in accordance with the Refugee Reception Act]*. Retrieved

Ocobter 24, 2024, from https://www.bezreg-arnsberg.nrw.de/themen/xyz/zuweisung_flueag/index.php

Bonsen, M. (2016). Schulleitung und Führung in der Schule [School management and leadership in the school]. In H. Altrichter & K. Maag Merki (Eds.), *Handbuch Neue Steuerung im Schulsystem [Handbook on new governance in the school system]* (pp. 301–324). Springer VS.

Capaul, R., Seitz, H., Keller, M., & Binneweg, J. (2020). *School management, leadership and improvement*. Haupt.

de Abreu, G., & Hale, H. (2014). Conceptualising teachers' understanding of the immigrant learner. *International Journal of Educational Research, 63*, 26–37.

Draft, R. L. (2008). *New era of management*. Thompson Corporation.

Dubs, R. (2003). *Qualitätsmanagement für Schulen, Studien und Berichte des IWP, Band 13 [Quality management for schools, studies and reports of the IWP, Volume 13]*. Institut für Wirtschaftspädagogik.

Fullan, M. G. (2007). *The new meaning of educational change*. Teacher College Press, Columbia University.

GoodVET. (2019). *Supranational manual. Erasmus+ project: Indicators of Good VET practice for refugees.* Retrieved from https://goodvet.uni-koeln.de/sites/goodvet/pdf/GoodVET_Guideline_ENGL.pdf

Granato, M., Münk, D., & Weiß, R. (Eds.). (2011). *Migration als Chance. Ein Beitrag der beruflichen Bildung [Migration as an opportunity. A contribution from vocational training]*. W. Bertelsmann Verlag.

Harris, A., Day, C., Hopkins, D., Hadfield, M., Hargreaves, A., & Chapman, C. (2003). *Effective leadership for school improvement*. Routledge/Falmer.

Heinrichs, K., Kärner, T., Ziegler, S., Feldmann, A., Reinke, H., & Neubauer, J. (2016). Die Implementierung neuer Konzepte zur Beschulung von Flüchtlingen und Asylsuchenden—Herausforderungen und Chancen aus organisationstheoretischer Perspektive. Gruppe. Interaktion [The implementation of new concepts for the schooling of refugees and asylum seekers—Challenges and opportunities from an organisational theory perspective. Group. Interaction]. Organisation. *Zeitschrift für Angewandte Organisationspsychologie (GIO), 47*(3), 231–241.

Hilkert, B., Pötter, N., & Diettrich, A. (2020). *Chancen des Zugangs zur beruflichen Bildung für bleibeberechtigte junge Geflüchtete: Möglichkeiten und Hindernisse in der Beratung und Unterstützung, Forschungsprojekt: Abschlussbericht [Opportunities for access to vocational education and training for young refugees entitled to stay: Opportunities and obstacles in counselling and support, research project: Final report]*. BIBB.

Hopkins, D. (1994). *School improvement in an era of change* (School development series). Books International, Inc.

Huijts, N. M. A., Molin, E. J. E., & Steg, L. (2012). Psychological factors influencing sustainable energy technology acceptance: A review-based comprehensive framework. *Renewable and Sustainable Energy Reviews, 16*(1), 525–531.

Jeon, S. (2019). *Unlocking the potential of migrants: Cross-country analysis*. OECD Publishing. Retrieved from https://www.oecd-ilibrary.org/education/unlocking-the-potential-of-migrants_045be9b0-en

Jørgensen, C. H., Hautz, H., & Li, J. (2021). The role of vocational education and training in the integration of refugees in Austria, Denmark and Germany. *International Journal for Research in Vocational Education and Training, 8*(3), 276–299. https://doi.org/10.13152/IJRVET.8.3.2

Klein, E. D., & Schwanenberg, J. (2022). Ready to lead school improvement? Perceived professional development needs of principals in Germany. *Educational Management Administration & Leadership, 50*(3), 371–391. https://doi.org/10.1177/1741143220933901

Lauer, T. (2010). *Change management*. Springer.

Liu, S. H., Liao, H. L., & Peng, C. J. (2005). Applying the technology acceptance model and flow theory to online e-learning users' acceptance behavior. *E-Learning, 4*(6), H8.

Mandl, H., & Burg, O. (2008). Acceptance of e-learning at the workplace: The influence of institutional framework. In W. J. Nijhof (Ed.), *The learning potential of the workplace* (pp. 213–224). Sense Publishers.

Mayring, P. (2014). Qualitative content analysis: Theoretical foundation, basic procedures and software solution. Beltz. Retrieved from http://nbn-resolving.de/urn:nbn:de:0168-ssoar-395173

Meyer, J.-A. (1997). The acceptance of visual information in management. *Information & Management, 32*(6), 275–287.

Müller-Böling, D., & Müller, M. (1986). Akzeptanzfaktoren der Bürokommunikation [Acceptance factors for office communication]. Oldenbourg Verlag.

Patton, M. Q. (1990). *Qualitative evaluation and research methods*. SAGE. https://doi.org/10.1002/nur.4770140111

Pilz, M. (2021). Qualität in der europäischen Berufsbildung—Ein Beispiel zur Analyse von Projekten für Flüchtlinge [Quality in European vocational education and training—An example of analysing projects for refugees]. In S. Dernbach-Stolz, P. Eigenmann, C. Kamm & S. Kessler (Eds.), *Transformation von Arbeit, Beruf und Bildung in internationaler Betrachtung, Festschrift für Philipp Gonon, Internationale Bildungsforschung [Transformation of labour, occupation and education from an international perspective, Festschrift for Philipp Gonon, Internationale Bildungsforschung]* (pp. 239–255). Springer Fachmedien.

Pressmar, D. (1982). Zur Akzeptanz von Computergestützten Planungssystemen [On the acceptance of computer-aided planning systems]. In H. Krall-mann (Ed.), *Unternehmensplanung und -steuerung in den 80er Jahren. Betriebs- und*

Wirtschaftsinformatik, Volume 3 [Corporate planning and management in the 1980s. Business and Information Systems, Volume 3] (pp. 324–359). Springer.

Rogers, E. M. (2003). *Diffusion of innovations* (5th ed.). Free Press.

Schnell, M. (2009). Einführung in die Akzeptanzforschung am Beispiel von Web-TV [Introduction to acceptance research using the example of web TV]. *WissenHeute, 62*(1), 4–12.

Seitter, W., Schemmann, M., & Vossebein, U. (2015). Bedarf—Potential—Akzeptanz. Integrierte Zusammenschau [Demand—potential—acceptance. Integrated overview]. In W. Seitter, M. Schemmann, & U. Vossebein (Eds.), *Zielgruppen in der wissenschaftlichen Weiterbildung—Empirische Studien zu Bedarf, Potential und Akzeptanz [Target groups in academic continuing education—Empirical studies on demand, potential and acceptance]* (pp. 23–60). Springer.

Simon, B. (2001). *E-Learning an Hochschulen—Gestaltungsräume und Erfolgsfaktoren von Wissensmedien [E-learning at universities—Scope for design and success factors of knowledge media].* Josef Eul Verlag.

Stiller, K., & Wager, S. (2023). Modelling e-learning acceptance of clinical staff using the General Extended Technology Acceptance Model for E-Learning. *The Online Journal of Distance Education and E-Learning, 11*(4), 3084–3106. Retrieved from https://tojdel.net/journals/tojdel/articles/v11i04/v11i04-11.pdf

Taylor, S., & Todd, P. A. (1995). Understanding information technology usage: A test of competing models. *Information Systems Research, 6*(2), 144–176.

Venkatesh, V., & Davis, F. D. (2000). A theoretical extension of the technology acceptance model: Four longitudinal field studies. *Management Science, 46*(2), 186–204.

Winther, E., Jordanoski, G., & Deutscher, V. K. (2018). Zum Umgang mit fehlenden formalen Qualifikationen [On dealing with a lack of formal qualifications]. *Unterrichtswissenschaft, 46*(1), 7–20. https://doi.org/10.1007/s42010-018-0012-1

CHAPTER 8

A Review of Career and Technical Education in the United States: Why CTE Is Essential

Filiz Oskay and Jay Plasman

Introduction

Vocational education (now referred to as career and technical education, or "CTE") has a long history in the United States, with formal recognition as part of the public education system stemming back to the Smith-Hughes Act of 1917 (Fletcher, 2006). This initial policy tasked vocational education with preparing students with the knowledge and skills needed for entry-level jobs, though it also created a separation of academic and vocational programming in secondary schools (Fletcher et al., 2013; Gordon & Schultz, 2020). Over time, however, vocational education has become increasingly re-integrated with academic education to meet evolving labor market needs (Castellano et al., 2003). The shifting focus of vocational education in the United States reflects broader global trends

F. Oskay · J. Plasman (✉)
Department of Educational Studies, The Ohio State University, Columbus, OH, USA
e-mail: plasman.2@osu.edu

E. Smeplass and C. Zirkle (eds.), *Opportunity Structures in Vocational Education and Training*, Palgrave Studies in Technical and Vocational Education and Training,
https://doi.org/10.1007/978-3-032-04410-5_8

in vocational education and training (VET), where systems have increasingly integrated academic learning with career preparation, aiming to equip students with both general knowledge and specific technical skills. Internationally, many VET systems use a dual approach, blending classroom learning with hands-on workplace training, which contrasts with the traditional U.S. model of separated academic and vocational tracks (Cedefop, 2017; Imperatore & Hyslop, 2017; Moodie, 2008; Plasman et al., 2020). However, recent iterations of federal policies governing CTE in the United States, collectively referred to as Perkins legislation, emphasize the importance of expanding opportunity structures by integrating academic/theoretical learning with technical/practical skills (Brand et al., 2013; Plasman et al., 2021). Perkins IV (2006) introduced rebranded vocational education as "career and technical education" to reduce stigma and highlight its role in both college and career readiness. In the 2018 of the Perkins Act (Perkins V), legislators emphasized incorporating high-quality work-based learning opportunities into CTE programming—thereby aligning the U.S. vocational system more closely with global VET models (Congressional Research Service, 2018). These changes in U.S. policy reflect growing international recognition of the importance of providing students with educational opportunities that are both academically rigorous and aligned with labor market needs (Cedefop, 2017; Moodie, 2008). Subsequently, the emphasis on vocational education in the United States shifted from providing narrow, entry-level workforce preparation to offering academic, technical, and employability knowledge and skills to succeed regardless of postsecondary pursuits (Gordon & Schultz, 2020). This shift aligns with global discussions on VET, which increasingly emphasize flexibility, pathways to higher education, and career readiness.

As such, contemporary CTE programming is more diverse in offerings and focuses on providing an array of opportunities in high-demand, high-skill fields for a diverse set of learners, including those most traditionally marginalized in education (Congressional Research Service, 2018). Importantly, there is growing evidence of the benefits of CTE participation, including positive labor market outcomes (Dougherty, 2018), improved school engagement (Plasman et al., 2021), and increased chances of high school completion (Bozick & Dalton, 2013; Gottfried, 2015; Plasman & Gottfried, 2018). Although CTE has expanded significantly over the last century, there is still work to be done to meet the needs of individuals, the nation, and the global economy.

In this chapter, we examine the history and current state of vocational education (i.e., CTE) in the United States and how this form of education aligns with and contributes to international discussions on vocational education and training (VET). We then discuss how CTE fits into the broader purposes of education in the United States and how perceptions toward and offerings within career-focused education have evolved. Through this exploration, we discuss how CTE has become an essential element within the opportunity structures that education provides, shaping equitable and sustainable pathways into career for individuals across diverse backgrounds and aspirations. By broadening access to both immediate employment prospects and lifelong personal and professional growth, CTE fosters inclusive educational ecosystems that support meaningful, adaptable trajectories in society (Gordon & Schultz, 2020).

History of CTE in the United States

Federal policy first recognized the importance of vocational education through the Morrill Act in 1862, which allocated federal land to establish institutions focused on careers in agriculture and the mechanical arts, highlighting the role of practical education in national development (Grubb & Lazerson, 2011). These land-grant institutions, as they are known, still exist across the country to serve the needs of their communities with respect to job preparation.

In the early twentieth century, the Smith-Hughes Act formalized the role of vocational education in secondary education (Fletcher, 2006). Emerging during a period of industrialization and growing numbers of working-class and immigrant children, this policy aimed to retain non-working school-aged youth in school through improved alignment between education and the workforce (Hayward & Benson, 1993). An added benefit was the production of a more informed citizenry, a skilled workforce, and higher economic productivity. Ultimately, this policy established vocational education as a distinct system with separate state boards, funding mechanisms, curriculum structures, teacher training programs, and professional organizations (Scott & Sarkees-Wircenski, 2004; Rojewski, 2002).

Following this formalization of vocational education in compulsory education, the United States has seen more than a century of continued support in this area. In 1963, the Vocational Education Act updated focal points of vocational training to align with the needs of the country during the Cold War and also sought to address social inequalities in the labor market more directly (Imperatore & Hyslop, 2017). In 1984, the first iteration of the Carl D. Perkins Vocational Education Act (Perkins I) maintained a focus on providing high-demand vocational programming and teacher training. The second iteration of the Perkins Act, authorized in 1990, expanded the objectives of vocational education to include preparing students more broadly for careers as opposed to instruction in specific work roles, thereby emphasizing the need to integrate curricula from both academic and vocational domains (Gordon & Schultz, 2020). This shift in vocational education's objectives expanded to include preparing students for postsecondary education, reflecting the acknowledgment that vocational education needed to evolve to meet the changing demands of the workforce (Rojewski, 2002; Castellano et al., 2003; Fletcher et al., 2013). This movement continued with Perkins reauthorizations in 1998, 2006, and 2018 (the most recent version). The 2006 version (Perkins IV) further reduced the historical divide between academic and vocational education by rebranding vocational education as career and technical education, emphasizing its role in preparing students for both college and careers (Imperatore & Hyslop, 2017; Plasman et al., 2020). Further, with each subsequent federal policy governing vocational (or career and technical) education, the federal government increased its emphasis on the importance of integrating technical education more fully with academic education to better promote college and career readiness for all students.

This integration allows students to move more freely between educational and workforce pathways. As such, though contemporary CTE is the most comparable form of education in the United States to vocational education and training in the international sense, it still falls short of a true vocational education and training (VET) system as delivered in many other countries.

Comparative VET: Vocational Education in the United States and Germany

While CTE represents the closest equivalent to VET in the international context, the U.S. system does not match the rigorous standards of many global VET frameworks. A comparison with Germany, for instance, highlights significant differences in the administration and structure of vocational education. Below, we outline key contrasts between the U.S. CTE system and Germany's comprehensive VET programs.

Although differences exist, it is important to first highlight some ways in which the U.S. system is comparable to VET systems internationally. In comparison with Germany, for instance, both countries face challenges in adapting vocational education to technological advancements and evolving labor market demands (Köhler & Drummer, 2018; Solga et al., 2014; Stringfield & Stone, 2017). Both also work to enhance the image of vocational education, attract more students, and address skills gaps within their employment systems (Hamilton, 1999; Lerman, 2008). However, the methods of vocational education delivery reflect broader differences in educational priorities and systemic organization.

Key differences relate to apprenticeship and governance (Boesel, 2011; Hamilton, 1999; Lerman & Rauner, 2011). The United States has historically viewed apprenticeship as a minor component of training, with formal apprenticeships largely unlinked to high schools and postsecondary institutions. Efforts to integrate apprenticeships into secondary education through youth apprenticeship programs have faced significant challenges, including concerns from labor unions and administrators about conflicts with existing adult apprenticeship systems, as well as coordination issues and federalism challenges (Glover & Bilginsoy, 2005; Commission on the Skills of the American Workforce, 1990; Hamilton, 1990; Lerman & Pouncy, 1990). In contrast, Germany's vocational education system is deeply rooted in the dual system, combining classroom instruction with on-the-job training to provide students with practical skills and a smooth transition into the workforce (Pilz, 2016).

Regarding governance, Germany's system involves strong central control, with federal government collaboration with business and labor partners to regulate apprenticeships and coordinate with state-controlled vocational schools (Boesel, 2011). In contrast, the United States has a much more decentralized governance structure, with decision-making power primarily in the hands of state and local authorities, resulting in a substantially more fragmented system (Boesel, 2011; Hamilton, 1999).

The evolution of vocational education away from a true dual system, combined with the history and development of CTE, cannot be divorced from the longstanding debates concerning the purposes of education in the United States. These debates have centered on the fundamental purposes of education, including whether its primary aim should be workforce preparation or broader societal and personal development.

The Purpose of Education in the United States

Labaree (1997) presents a valuable lens by which to examine the various objectives of education. Specifically, he identifies that the history of educational goals in the United States reflects a dynamic interplay between three distinguishable educational goals, namely democratic equality, social efficiency, and social mobility. Each tenet "has exerted considerable impact without succeeding in eliminating the others, and each of which has at times served to undermine the others" (Labaree, 1997, p. 41).

Democratic Equality

The principle of democratic equality has been a cornerstone of American education since the establishment of common schools in the mid-1800s. These schools were seen as essential for nation-building and democratic participation (Meyer et al., 1979). Horace Mann, a prominent figure in the public education movement, emphasized education as essential for fostering civic virtue and addressing capitalism's risks and social inequality (Mann, 1957), leading to the nineteenth-century expansion of public schooling to ensure elementary education for all children. Efforts to increase access to secondary and higher education followed, resulting in mass enrollment (Labaree, 1990). The pursuit of democratic equality has significantly shaped American education, influencing curriculum design, access to education, and opportunities for students. It has spurred efforts to combat discrimination, promote inclusivity, and guarantee that all students can fully engage in the educational system.

Social Efficiency

As Labaree (1997, 2009) noted, the concept of social efficiency in education has profoundly impacted American schools, primarily through the implementation of vocationalism and educational stratification. It reflects a pragmatic approach, aligning educational practices with the practical constraints embedded in the market-based structure of economic and social life. One significant outcome of this influence is the historical shift toward vocationalism, marked by the establishment of vocational programs at the secondary and community college levels to prepare students for specific occupations. These programs provide practical education to supply well-trained employees for specific jobs, but their focus on social efficiency has unintentionally undermined educational equality, creating a stratified system. Schools mirror the vertical job market structure, placing greater value on degrees from prestigious institutions. Within schools, ability grouping and curriculum tracking perpetuate inequality by providing varied educational experiences. Social efficiency limits opportunities in the name of economic necessity but also values education as a public good that benefits the whole society. This perspective underscores the importance of effective instruction and universal learning for enhancing productivity and economic growth.

Social Mobility

The social mobility goal in education focuses on providing students with the credentials necessary to advance within the existing socioeconomic structure (Labaree, 1997, 2009). In the pursuit of social mobility, education is seen as a private good, where individuals seek to accumulate educational "property" to gain advantages in social competition. This perspective fosters educational stratification, with hierarchical distinctions and qualitative differences between institutions fueling intense competition for credentials valued in the job market. Consequently, students prioritize credentials over knowledge, reinforcing a consumer-driven, meritocratic system that disproportionately benefits those in advantaged positions. Thus, the social mobility goal in education may perpetuate inequality by prioritizing individual status attainment over the production of human capital at the societal level. Human capital as a theory developed by Becker (1962) stipulates that investing in individuals' skills and capacities through education and training is essential for maximizing productivity. This theory is core to CTE and is discussed in further detail below.

As outlined by Labaree (1997, 2009), three distinct but intimately related goals—social mobility, social efficiency, and democratic equality—shape American education. Social mobility emphasizes individual advancement and upward mobility, treating education as a means to achieve personal aspirations, often at the expense of others. In contrast, social efficiency adopts a top-down perspective, prioritizing vocational training and practical skills to align students with societal structures and meet economic demands. This approach promotes a stratified education system, with pathways tailored to perceived future vocational roles, mirroring broader societal hierarchies.

Democratic equality, on the other hand, seeks inclusivity and social justice by expanding educational access and ensuring equal treatment. It emphasizes civic virtues and the common good, aiming to foster a more equitable society. While both social mobility and democratic equality advocate for expanding access to education, the former often undermines the latter by perpetuating inequalities tied to individual achievement rather than addressing systemic disparities based on class, race, and gender.

The tension among these goals underpins the challenges facing CTE, particularly in navigating stigmatization and perceptions of vocational training within the economy.

Vocational Education Stigmatization

The conflict between the aims of education has contributed significantly to the stigmatization of vocational education in the United States, rendering it often perceived as an inferior pathway compared to more strictly academic pursuits. Thus, vocational education has been marginalized due to its association with limited social mobility and its perceived role in preparing individuals for lower-status jobs. This perception of vocational education as a less academically rigorous alternative, especially when compared to traditional academic pathways, has been reinforced over time (Lewis & Cheng, 2006; Shavit & Müller, 2000). It has been portrayed as a "dumping ground" for students deemed less academically inclined, priming them for low-wage jobs rather than higher education or prestigious careers (Blank, 1999).

One key factor contributing to the stigmatization of vocational education is tracking—a practice by which students are separated into different educational paths based on perceived academic ability or future career prospects (Fletcher & Zirkle, 2009). This practice, which is deeply rooted in historical inequalities related to race, class, and gender, has significantly influenced the negative perception of vocational education (Hodge et al., 2020). Tracking perpetuates inequalities in education and occupational opportunities, particularly among low-income, minority, or underperforming students (Bowles & Gintis, 1976; Marks, 2006; Oakes & Saunders, 2008; Ozer & Perc, 2020; Reichelt et al., 2019). Specifically, students who underperform academically tend to come from low-income backgrounds, belong to racial minority groups, or have different learning needs and have historically been much more likely to be tracked into non-academic or lower-academic placements. This placement reduces their chances of attending university and entering high-prestige occupations (Eriksen & Dougherty, 2024; Marks, 2006; Reichelt et al., 2019; Woessmann, 2009).

Vocational tracks may inhibit educational attainment in several ways. First, lower-performing students in vocational tracks may be deprived of peer-learning benefits (Eriksen & Dougherty, 2024; Hanushek & Woessmann, 2006). Second, vocational tracks historically offered a more restricted curriculum, resulting in less opportunity to participate in advanced coursework (Oakes & Saunders, 2008). Third, historically, less selective tracks devoted less time to instruction, which was often delivered at lower complexity levels, resulting in students learning less and struggling on college admission tests (Metz, 1978; Brunello, 2004). Finally, placement in a perceived lower track often signaled to students that they were less worthy, dampening their future expectations and aspirations (Hanushek, 2019; Marks, 2006; Vanfossen et al., 1987; Woessmann, 2009; Zimmer, 2003). Unsurprisingly, secondary students in vocational tracks were then less likely to succeed in postsecondary education.

Broadly speaking, tracking in education has been criticized for perpetuating social inequalities by segregating students based on perceived ability and career prospects, stigmatizing vocational education (Bowles & Gintis, 1976; Woessmann, 2009; Hanushek, 2019). This practice limits access to a broad curriculum, reduces instructional time, and lowers expectations, hindering academic success and upward mobility, particularly for minority and lower-class students. Historically, vocational education exemplified these disparities.

Destigmatizing CTE: From Vocational Education to CTE

In the 1990s, vocational education underwent a significant transformation in response to the stigma attached to traditional vocational pathways and to address emerging skills demands (King, 2011). The Carl D. Perkins Act of 1990 shifted the purpose of vocational education to emphasize broader career preparation and integrate academic and vocational curricula (Gordon & Schultz, 2020). This shift aimed to better prepare students for postsecondary education, reflecting workforce demands (Castellano et al., 2003; Fletcher et al., 2013; Rojewski, 2002). It also prompted educators, administrators, and policymakers to reevaluate traditional curriculum structures (Rojewski & Kim, 2003). The 2006 reauthorization of the Perkins Act (Perkins IV) marked a pivotal moment by formally rebranding vocational education as "career and technical education" to reflect its evolving nature (Gordon & Schultz, 2020). This shift emphasized broader outcomes, combining technical skills with academic knowledge and general employability skills, and aimed to reduce stigmatization. Perkins IV sought to transform vocational education into a respected pathway, preparing students for both college and careers while enhancing their capacity to contribute to global economic competitiveness (Klein & Green, 2011; Plasman et al., 2020). The development of Perkins IV was driven by rapid technological advancements and concerns about American competitiveness, prompting a reassessment of education systems to enhance workforce skills and innovation (Friedman, 2005; Imperatore & Hyslop, 2017). Following the 2001 No Child Left Behind (NCLB) Act, Perkins IV shared a focus on academic achievement and success across all student subgroups (Gordon & Schultz, 2020). NCLB's emphasis on academics influenced Perkins IV, leading to a renewed integration of academic education with technical learning and hands-on experiences for a more holistic approach to student development in CTE.

Perkins IV significantly impacted secondary education by expanding CTE courses that emphasize competency-based applied learning, integrating academic knowledge with technical skills (Gottfried & Plasman, 2018). This approach aimed to strengthen connections between high school education, postsecondary education, and careers (Brand et al., 2013). Additionally, increased accountability measures sought to ensure CTE programs effectively prepared all students for further education and meaningful careers (U.S. Department of Education, 2012).

The most recent version of Perkins, authorized in 2018 as the Carl D. Perkins Strengthening Career and Technical Education for the twenty-first-Century Act (Perkins V), continues to promote the integration of academic and technical education through enhanced partnerships and alignment across secondary education, postsecondary education, and industry (Plasman et al., 2023). Further, Perkins V builds upon the framework established by Perkins IV to provide access to high-quality CTE programming for traditionally underrepresented students (Gordon & Schultz, 2020). Unlike Perkins I–III, which primarily focused on vocational education and delineated five clusters (i.e., agriculture, construction, manufacturing, transportation), Perkins IV and V encompass sixteen clusters, including information and communication technologies, engineering, health, education, and business, again reflecting the evolving nature of vocational and technical education and the labor market demands (Klein & Green, 2011; National Association of State Directors of Career and Technical Education, 2006).

Benefits of CTE for Students

Research highlights the substantial benefits of CTE participation, offering both monetary advantages in the labor market and nonmonetary benefits, such as improved self-worth and engagement (Dougherty, 2018; Kelly & Price, 2009). These benefits extend across secondary education, postsecondary education, and career outcomes. At the secondary level, CTE participants exhibit higher self-efficacy, better attendance, advanced STEM course enrollment, improved standardized test performance, and increased high school graduation rates (Bozick & Dalton, 2013; Gottfried & Plasman, 2018; Plank et al., 2008; Plasman et al., 2021). These effects are particularly pronounced among underrepresented populations, including low-income students, who report higher school engagement, and students with learning disabilities, who experience improved odds of high school completion and college enrollment (Plasman & Gottfried, 2018, 2021, 2023).

Beyond high school, CTE coursework is also linked to postsecondary education outcomes. For instance, Dougherty (2018) found that students completing CTE pathways are more likely to apply to and enroll in postsecondary education, a trend supported by Staklis and Klein's (2010) findings linking proficiency on CTE-related assessments to college enrollment, particularly at the 2-year level for students earning certifications (Glennie et al., 2017). Similar to outcomes at the secondary level, traditionally marginalized populations benefit significantly from CTE participation in postsecondary education. Plasman et al. (2024) found that students with learning disabilities who earned CTE credits were more likely to apply to college and complete a federal application for student aid (FAFSA). Additionally, these students were less likely to need remedial coursework and more likely to declare STEM majors once enrolled in postsecondary education (Plasman & Myles, 2024). CTE participation is also strongly linked to labor market benefits. Plasman (2019) found that non-college-going students who participated in CTE earned higher wages. Similarly, a rigorous evaluation of career academies revealed a significant positive impact on annual earnings, with much of the effect observed among male participants (Kemple & Willner, 2008). Additionally, Dougherty et al. (2019) identified increased employment rates and higher wages as key outcomes of CTE participation. Overall, students with access to structured skills and experiences that facilitate transitions to college or careers gain significant benefits.

New CTE with New Opportunities

While research evidence is optimistic with respect to the benefits of CTE participation, there remains the necessity to interpret how and why CTE may produce these benefits. Human capital theory has been and continues to be an important perspective by which to explore the benefits of CTE, but the key points are differently nuanced (Becker, 1962; Gottfried et al., 2014). Additionally, new pushes to blur the lines between academic and vocational tracks have led to a new perspective by which to consider CTE—college and career readiness (Brand et al., 2013; Stone & Lewis, 2012). Through this new perspective, CTE fosters accessible and meaningful educational and career pathways, underscoring its role in broadening opportunities for individuals from diverse backgrounds. Below, these perspectives are discussed in more detail.

Human Capital Theory

The goal to provide participants with skills and knowledge to succeed in an occupation has been a common thread since the formalization of vocational education in the United States more than a century ago (Gordon & Schultz, 2020). This purpose aligns particularly well with Becker's (1962) human capital theory that suggests individuals can improve their capacity to be productive through additional education and skill training. Participation in CTE coursework likely promotes general and specific human capital development through three key mechanisms: reinforcement of academic skills, relevance and engagement, and new skill development (Plasman & Gottfried, 2018).

First, students in CTE coursework may benefit from opportunities to apply theoretical learning in practical settings, with the additional focus on academic skills through experiential activities potentially enhancing success in traditional academic contexts (Stone et al., 2008). Second,

CTE coursework connects traditional academic content to specific career-based applications, fostering engagement and relevance through hands-on experiences and applied contextual learning (Brand et al., 2013). Finally, CTE programs are deliberately designed to integrate academic knowledge with work-based practices, equipping students with the skills needed for both college and career readiness after high school graduation. These include industry-specific expertise and critical competencies such as critical thinking, reasoning, collaboration, research, and problem-solving (Brand et al., 2013; Oakes & Saunders, 2008; Stern & Stearns, 2006). Ultimately, these mechanisms are important in helping smooth the transition between secondary education and career and postsecondary education.

College and Career Readiness

Ensuring students receive the education, skills, and training needed for the modern economy requires employing a range of strategies, including increasing the rigor and relevance of teaching, improving preparation for college and careers, raising high school graduation rates, providing career exploration opportunities, and facilitating smoother transitions to postsecondary education (Brand et al., 2013; Balfanz et al., 2013). CTE has emerged as a transformative educational approach, equipping students with academic, technical, and employability skills essential for postsecondary success and lifelong learning (Partnership for 21st-Century Skills, 2010). Moving beyond the traditional vocational model of low-level courses and job-specific training, CTE now offers rigorous, integrated, and sequenced programs aligned with postsecondary pathways. These programs foster competencies such as critical thinking, collaboration, problem-solving, innovation, teamwork, and communication.

Additionally, CTE pathways provide tailored support for diverse learners, making them an inclusive and increasingly viable option for students of all abilities. By engaging a broader range of students, these programs enhance high school graduation rates, promote postsecondary success, and improve labor market outcomes (Brand et al., 2013).

Discussion

The Future of CTE

As we look to the future, the evolving demands of the workforce and economy necessitate a strategic focus on key areas vital for the development and success CTE. Addressing these areas will meet societal and economic needs while fostering inclusivity and equity in vocational education. These enhancements enable CTE to move beyond job preparation, offering pathways for personal and professional growth and expanded career opportunities. By fostering diverse learning experiences and skill development, CTE can serve as a robust alternative to traditional postsecondary education while building a foundation for lifelong achievement. As such, we propose a handful of considerations for CTE/VET as we move deeper into the twenty-first century: (1) workforce readiness and global participation; (2) access to programming; (3) alignment with local workforce needs; (4) secondary to postsecondary transitions; (5) integrated curricula; (6) broader awareness; and (7) lifelong learning. We conclude this chapter with a discussion each in greater detail below.

CTE for Workforce Readiness and Participation in the Global Economy

CTE is recognized as a crucial strategy to meet workforce development needs and enhance economic competitiveness (Whiston & Blustein, 2013). Quality career development programs are critical for raising wages and building a skilled workforce to compete in high-demand fields (Tyszko et al., 2014). In today's global economy, states are increasingly turning to CTE to improve college and career readiness and empower individuals to pursue careers essential for economic growth and stability (van Bruinswaardt et al., 2015). Integrating career development strategies that align personal aspirations with regional labor market needs, CTE fosters empowerment and economic growth (Goan et al., 2006; Plasman & Gottfried, 2022). Looking ahead, it is imperative to maintain focus on developing and enhancing CTE programs to ensure they remain effective in meeting the constantly evolving requirements of the workforce and economy.

Improving Access to CTE Programs

Improving CTE in general will require prioritizing the accessibility of CTE courses for all students, regardless of race, gender, class, or ability.

Students should have access to a wide range of courses without restrictive barriers. Key stakeholders, including administrators, teachers, and counselors, play a pivotal role in disseminating information about CTE courses, career pathways, and opportunities (Bartholomew, 2014). They should actively facilitate the enrollment process, ensuring it is straightforward and accessible, and provide students with the autonomy to select from a diverse range of CTE offerings. It is essential to educate students and families about local workforce demands, postsecondary education options, and the potential earnings associated with various fields, as this knowledge can broaden accessibility and bolster enrollment rates. To mitigate the risk of CTE classroom closures due to teacher shortages or insufficient funding, proactive measures must be taken at the federal and state levels.

Aligning with Local Workforce Demands

In the realm of CTE, alignment between educational programming and local workforce needs is a fundamental principle (Northern & Petrilli, 2019). Aligned programs ensure that CTE remains relevant and responsive to the demands of the economy, thereby facilitating a seamless transition for students from education to the workforce. By tailoring CTE programs to local workforce needs, students are equipped with the skills and knowledge that are directly applicable to the available job opportunities in their community. Additionally, aligning CTE with local workforce needs is essential for employers. It enables them to find well-educated and skilled personnel who can readily contribute to their organizations, thereby strengthening the overall economic development of the region. Aligning CTE with local workforce needs fosters a symbiotic relationship between educational institutions and employers. Institutions gain insights into evolving job market needs, enabling them to adapt programs more effectively to meet these needs. Employers, meanwhile, can have more influence on the development of the local workforce, ensuring the demands of the economy can be met.

Smooth the Transition from Secondary to Postsecondary

CTE programs offer a potentially impactful pathway for students by providing a bridge to postsecondary education for those who may not have otherwise continued their schooling (Cullen et al., 2013). However, there is concern that CTE programs could inadvertently track students into paths that could limit their future employment and educational opportunities. Therefore, a comprehensive understanding of how CTE participation affects students' educational outcomes is essential for shaping effective education policies. Evaluating CTE outcomes ensures programs remain beneficial pathways without narrowing student prospects.

Integrating Holistic Curricula
In shaping the future direction of CTE, a pivotal advancement involves the adoption of fully integrated curricula (Fletcher & Zirkle, 2009). Integrating vocational training with academic learning enhances opportunities, fosters academic excellence, and counters historical vocational education stigmas. Embracing such an approach not only tackles critical challenges within the education system but also sets the stage for a more inclusive and equitable future for all students.

Increasing Awareness Among Educators, Counselors, and Parents
Career counseling and guidance are integral components of CTE, forming the cornerstone of the career pathways approach (Mobley et al., 2017). By focusing on practical workplace knowledge and skills, CTE guidance practices offer students valuable insights into a range of career options. Recognizing the importance of counseling in CTE, schools must enhance these services—recent estimates put the students-counselor ratio at an unmanageable 376:1 (American School Counselor Association, 2023). Additional steps toward this prioritization include bolstering teacher education regarding CTE and increasing parental awareness of the opportunities offered by CTE. One means to help increase overall awareness is through increased transparency and accuracy of information. Data dashboards can play an essential role in this regard as they provide available, accessible, and easy-to-use data reporting and dissemination tools, meeting the needs of CTE learners, families, practitioners, policymakers, and the public. By addressing these challenges, schools can better support students in navigating their career pathways and achieving their full potential.

Fostering a Culture of Lifelong Learning
CTE contributes significantly to fostering a culture of lifelong learning through several key features. CTE programs enhance immediate job readiness while fostering continuous skill development and adaptability.

Moreover, these programs often integrate industry-recognized certifications and credentials into their curriculum, encouraging students to pursue them and emphasizing the importance of ongoing learning and professional development for competitiveness in the workforce (Dougherty, 2018). Additionally, CTE frequently provides work-based learning opportunities, such as internships, apprenticeships, and cooperative education experiences, offering students valuable industry exposure and instilling the expectation of continuous learning and skill improvement in a professional setting. Encouraging collaboration and knowledge sharing, CTE programs foster a culture of continuous learning. This approach not only prepares students for the workforce but also nurtures personal and professional growth beyond formal education.

Conclusion

As CTE in the United States continues to evolve, it faces a unique opportunity to redefine its role in preparing students for both immediate job readiness and long-term career growth. The preceding sections have highlighted CTE's historical roots and compared its implementation with international systems, such as in Germany, demonstrating both its progress and potential as a transformative force within the U.S. educational landscape. This chapter also traced the shift from traditional vocational training models to an inclusive, multi-faceted approach that aligns with contemporary goals of human capital development, college and career readiness, and personal empowerment.

The current shift in CTE signifies more than a response to workforce demands; it embodies a commitment to democratic education by enhancing accessibility and supporting diverse student backgrounds. These aims go beyond the economic framework, aiming to reduce social inequities and dismantle the stigmas long associated with vocational training. By expanding CTE pathways and integrating them with general education, CTE not only addresses social mobility goals but also contributes to a more inclusive approach to workforce readiness, making this model an invaluable asset to American education.

Moving forward, it is crucial for CTE to continue bridging gaps between education and industry through strategies that prioritize holistic curricula, alignment with local workforce demands, and collaborative partnerships among educators, students, and employers. This collaborative effort must also address accessibility challenges and improve students' transitions from secondary to postsecondary education, providing more flexible, career-relevant pathways that align with individual aspirations and economic needs.

To ensure the enduring relevance of CTE, a commitment to lifelong learning is paramount. By integrating industry-recognized credentials and emphasizing continuous skill development, CTE can foster a culture of learning that equips students not only for initial employment but for sustained growth across their careers. This aligns with broader educational purposes, positioning CTE as a forward-thinking model that reflects America's democratic ideals while meeting the demands of a complex, global economy.

The future of CTE lies in its potential to provide equitable, accessible, and adaptable education through expanding opportunity structures. As it continues to evolve, CTE can contribute to personal empowerment, social mobility, and economic prosperity, laying a strong foundation for the next generation of skilled and adaptable professionals.

References

American School Counselor Association. (2023). *National student to counselor ratio 1986–2023*. Author.

Balfanz, R., Bridgeland, J. M., Bruce, M., & Fox, J. H. (2013). *Grad nation: Building a grad nation, progress and challenge in ending the high school dropout epidemic*. 2012–2013 Annual update.

Bartholomew, S. (2014). Why we cannot afford to lose CTE in schools. *International Journal of Vocational and Technical Education, 6*(2), 7–12.

Becker, G. S. (1962). Investment in human capital: A theoretical analysis. *Journal of Political Economy, 70*(5), 9–49.

Blank, W. (1999). Future perspectives in vocational education. In *Workforce education: Issues for the new century* (pp. 281–289).

Boesel, D. (2011). Governing VET in the United States: Localization versus centralization. In *Work and education in America: The art of integration* (pp. 77–99). Springer Netherlands.

Bowles, S., & Gintis, H. (1976). *Schooling in capitalist America*. Basic.

Bozick, R., & Dalton, B. (2013). *Career and technical education and academic progress at the end of high school: Evidence from the Education Longitudinal Study of 2002*. RTI International.

Brand, B., Valent, A., & Browning, A. (2013). *How career and technical education can help students be college and career ready: A primer*. College and Career Readiness and Success Center.

Brunello, G. (2004). Stratified or comprehensive? Some economic considerations on the design of secondary education. *CESifo DICE Report, 2*(4), 7–10.

Castellano, M., Stringfield, S., & Stone, J. (2003). Secondary career and technical education and comprehensive school reform: Implications for research and practice. *Journal of Vocational Education Research, 73*(2), 231–272.

Cedefop. (2017). *The changing nature and role of vocational education and training in Europe. Volume 2: Results of a survey among European VET experts* (Cedefop research paper; No 64). Luxembourg Publications Office. https://doi.org/10.2801/548024

Commission on the Skills of the American Workforce. (1990). *America's choice: High skills or low wages!* National Center on Education and the Economy.

Congressional Research Service. (2018). *Reauthorization of the Perkins Act in the 115th Congress: The strengthening career and technical education for the 21st century* (Report No. R45446). Author.

Cullen, J. B., Levitt, S. D., Robertson, E., & Sadoff, S. (2013). What can be done to improve struggling high schools? *Journal of Economic Perspectives, 27*(2), 133–152. https://doi.org/10.1257/jep.27.2.133

Dougherty, S. M. (2018). The effect of career and technical education on human capital accumulation: Causal evidence from Massachusetts. *Education Finance and Policy, 13*(2), 119–148.

Dougherty, S. M., Gottfried, M. A., & Sublett, C. (2019). Does increasing career and technical education coursework in high school boost educational attainment and labor market outcomes? *Journal of Education Finance, 44*(4), 423–447.

Eriksen, J., & Dougherty, S. M. (2024). *Peer effects in vocational education and training. Working paper* (pp. 24–943). Annenberg Institute at Brown University. https://doi.org/10.26300/xmvk-mb25

Fletcher, E. C. (2006). No curriculum left behind: The effects of the No Child Left Behind legislation on career and technical education. *Career and Technical Education Research, 31*(3), 157–174.

Fletcher, E., & Zirkle, C. (2009). The relationship of high school curriculum tracks to degree attainment and occupational earnings. *Career and Technical Education Research, 34*(2), 81–102.

Fletcher, E. C., Lasonen, J. L., & Hernandez-Gantes, V. M. (2013). What is CTE?: Practitioners struggle to define their field in the United States. *International Journal of Adult Vocational Education and Technology, 4*(1), 16–30.

Friedman, T. (2005). *The world is flat: A brief history of the twenty-first century.* Farrar, Straus, and Giroux.

Glennie, E. J., Ottem, R. A., & Lauff, E. (2017). Examining the influence of the Florida Career and Professional Education Act of 2007: Changes in industry certifications and educational and employment outcomes. National Center for Innovation in Career and Technical Education.

Glover, R. W., & Bilginsoy, C. (2005). Registered apprenticeship training in the U.S construction industry. *Education+Training, 47*(4/5), 337–349.

Goan, S. K., Cunningham, A. F., & Carroll, C. D. (2006). *Degree completions in areas of national need, 1996–97 and 2001–02.* Institute of Education Sciences.

Gordon, H. R., & Schultz, D. (2020). *The history and growth of career and technical education in America.* Waveland Press.

Gottfried, M. A. (2015). The influence of applied STEM coursetaking on advanced mathematics and science coursetaking. *The Journal of Educational Research, 108*(5), 382–399.

Gottfried, M. A., & Plasman, J. S. (2018). From secondary to postsecondary: Charting an engineering career and technical education pathway. *Journal of Engineering Education, 107*(4), 531–555.

Gottfried, M. A., Bozick, R., & Srinivasan, S. V. (2014). Beyond academic math: The role of applied STEM coursetaking in high school. *Teachers College Record, 116*(7), 1–35.

Grubb, W. N., & Lazerson, M. (2011). The education gospel and vocationalism in US higher education: Triumphs, tribulations, and cautions for other countries. In *Work and education in America: The art of integration* (pp. 101–121). Springer Netherlands.

Hamilton, S. F. (1990). Apprenticeship for adulthood: Preparing youth for the future. The Free Press.

Hamilton, S. F. (1999). Germany and the United States in comparative perspective. *International Journal of Sociology, 29*(1), 3–20.

Hanushek, E. A. (2019). *Addressing cross-national generalizability in educational impact evaluation.* NBER Working Paper 25460. National Bureau of Economic Research.

Hanushek, E. A., & Woessmann, L. (2006). Does educational tracking affect performance and inequality? Differences-in-differences evidence across countries. *The Economic Journal, 116*(510), 63–76.

Hayward, G. C., & Benson, C. S. (1993). *Vocational-technical education: Major reforms and debates, 1917–present.* U.S. Department of Education, Office of Vocational and Adult Education.

Hodge, E., Dougherty, S., & Burris, C. (2020). *Tracking and the future of career and education: How efforts to connect school and work can avoid the past mistakes of vocational education.* National Education Policy Center.

Imperatore, C., & Hyslop, A. (2017). CTE policy past, present, and future: Driving forces behind the evolution of federal priorities. *Peabody Journal of Education, 92*(2), 275–289.

Kelly, S., & Price, H. (2009). Vocational education: A clear slate for disengaged students? *Social Science Research, 38*(4), 810–825. https://doi.org/10.1016/j.ssresearch.2009.05.002

Kemple, J., & Willner, C. (2008). *Career Academies: Long-term impacts on labor market outcomes, educational attainment, and transitions to adulthood*. MDRC.

King, J. (2011). Dilemmas of design: Education versus qualification in the US vocational system. In *Work and education in America: The art of integration* (pp. 15–32). Springer.

Klein, S., & Green, K. A. (2011). College for all: The American model for career and technical education. In *Work and education in America: The art of integration* (pp. 123–140). Springer Netherlands.

Köhler, T., & Drummer, J. (2018). Recent technological challenges in vocational education. In *Vocational teacher education in Central Asia: Developing skills and facilitating success* (pp. 3–14). Springer.

Labaree, D. F. (1990). From comprehensive high school to community college: Politics, markets, and the evolution of educational opportunity. In R. G. Corwin (Ed.), *Research on the sociology of education and socialization* (Vol. 9, pp. 203–224).

Labaree, D. F. (1997). Public goods, private goods: The American struggle over educational goals. *American Educational Research Journal, 34*(1), 39–81.

Labaree, D. F. (2009). *Educational research: Proofs, arguments, and other reasonings*. Springer. https://doi.org/10.1007/978-90-481-3249-2_4

Lerman, R. I. (2008). Are skills the problem? Reforming the education and training systems in the United States. In T. J. Bartik & S. N. Houseman (Eds.), *A future of good jobs* (pp. 17–80). W. E. Upjohn Institute.

Lerman, R. I., & Pouncy, H. (1990). The compelling case for youth apprenticeships. *The Public Interest, 101*, 62.

Lerman, R. I., & Rauner, F. (2011). Apprenticeship in the United States. In *Work and education in America: The art of integration* (pp. 175–193). Springer Netherlands.

Lewis, T., & Cheng, S. Y. (2006). Tracking, expectations, and the transformation of vocational education. *American Journal of Education, 113*(1), 67–99.

Mann, H. (1957). Twelfth annual report. In L. Cremin (Ed.), *The republic and the school: Horace Mann on the education of free men* (pp. 79–112). Teachers College Press.

Marks, G. N. (2006). Are between-and within-school differences in student performance largely due to socioeconomic background? Evidence from 30 countries. *Educational Research, 48*(1), 21–40.

Metz, M. H. (1978). *Classrooms and corridors: The crisis of authority in desegregated secondary schools*. University of California Press.

Meyer, J. W., Tyack, D., Nagel, J., & Gordon, A. (1979). Public education as nation-building in America: Enrollments and bureaucratization in the American states, 1870–1930. *American Journal of Education, 85*, 591–613.

Mobley, C., Sharp, J. L., Hammond, C., Withington, C., & Stipanovic, N. (2017). The influence of career-focused education on student career planning and development: A comparison of CTE and non-CTE students. *Career and Technical Education Research, 42*(1), 57–75.

Moodie, G. (2008). *EBOOK: From vocational to higher education: An international perspective*. McGraw-Hill Education (UK).

National Association of State Directors of Career and Technical Education. (2006). *Career clusters: A plan for education for a global economy*. http://www.careertech.org/Career_Cluster_Research_Brief_final.pdf

Northern, A. M., & Petrilli, M. J. (2019). Aligning CTE courses to local labor markets. *State Education Standard, 19*(3), 25–29.

Oakes, J., & Saunders, M. (2008). *Beyond tracking: Multiple pathways to college, career and civic participation*. Harvard Education Press.

Ozer, M., & Perc, M. (2020). Dreams and realities of school tracking and vocational education. *Palgrave Communications, 6*(1), 1–7. https://doi.org/10.1057/s41599-020-0409-4

Partnership for 21st Century Skills. (2010). *Up to the challenge: The role of career and technical education and 21st century skills in college and career readiness*. Author.

Pilz, M. (2016). Typologies in comparative vocational education: Existing models and a new approach. *Vocations and Learning, 9*(3), 295–314. https://doi.org/10.1007/S12186-016-9154-7

Plank, S. B., DeLuca, S., & Estacion, A. (2008). High school dropout and the role of career and technical education: A survival analysis of surviving high school. *Sociology of Education, 81*, 345–370.

Plasman, J. (2019). Linking occupational concentration to hourly wages for non-college going individuals. *Journal of Career and Technical Education, 34*(1), 29–51.

Plasman, J. S., & Gottfried, M. A. (2018). Applied STEM coursework, high school dropout rates, and students with learning disabilities. *Educational Policy, 32*(5), 664–696.

Plasman, J. S., & Gottfried, M. A. (2022). School absence in the United States: Understanding the role of STEM-related vocational education and training in encouraging attendance. *Journal of Vocational Education & Training, 74*(4), 531–553.

Plasman, J., & Myles, D. (2024). STEM pathway and college progression: The link between engineering CTE and postsecondary outcomes for students with learning disabilities. *Exceptional Children*. https://doi.org/10.1177/00144029241247034

Plasman, J. S., Gottfried, M. A., & Hutt, E. L. (2020). Then and now: Depicting a changing national profile of STEM career and technical education course takers. *Teachers College Record, 122*(2), 1–28.

Plasman, J. S., Gottfried, M. A., & Klasik, D. (2021). Do career-engaging courses engage low-income students? *AERA Open, 7*(1). https://doi.org/10.1177/23328584211053324

Plasman, J., Gottfried, M., & Oskay, F. (2023). Exploring the role of high school engineering courses in promoting science attitudes for students with learning disabilities. *Journal of Research in Science Teaching*. https://doi.org/10.1002/tea.21905

Plasman, J. S., Oskay, F., & Gottfried, M. (2024). Transitioning to success: The link between E-CTE and college preparation for students with learning disabilities in the United States. *Education Sciences, 14*(2), 116.

Reichelt, M., Collischon, M., & Eberl, A. (2019). School tracking and its role in social reproduction: Reinforcing educational inheritance and the direct effects of social origin. *The British Journal of Sociology, 70*(4), 1323–1348.

Rojewski, J. (2002). Preparing the workforce of tomorrow: A conceptual framework for career and technical education. *Journal of Vocational Education Research, 27*(1), 7–35.

Rojewski, J., & Kim, H. (2003). Career choice patterns and behavior of work-bound youth during early adolescence. *Journal of Career Development, 30*(2), 89–108.

Scott, J. L., & Sarkees-Wircenski, M. (2004). *Overview of career and technical education* (4th ed.). American Technical Publishers.

Shavit, Y., & Müller, W. (2000). Vocational secondary education, tracking, and social stratification. In *Handbook of the sociology of education* (pp. 437–452). Springer US.

Solga, H., Protsch, P., Ebner, C., & Brzinsky-Fay, C. (2014). *The German vocational education and training system: Its institutional configuration, strengths, and challenges* (No. SP I 2014-502). WZB Discussion Paper.

Staklis, S., & Klein, S. (2010). *Technical skill attainment and post-program outcomes: An analysis of Pennsylvania secondary career and technical education graduates*. National Research Center for Career and Technical Education.

Stern, D., & Stearns, R. (2006). *Combining academic and career-technical courses to make college an option for more students*. Graduate School of Education, UCLA.

Stone, J. R., & Lewis, M. V. (2012). *College and career ready in the 21st century: Making high school matter*. Teachers College Press.

Stone, J. R., Alfeld, C., & Pearson, D. (2008). Rigor and relevance: Enhancing high school students' math skills through career and technical education. *American Educational Research Journal, 45*(3), 767–795.

Stringfield, S., & Stone, J. R. (2017). The labor market imperative for CTE: Changes and challenges for the 21st century. *Peabody Journal of Education, 92*(2), 166–179.

Tyszko, J. A., Sheets, R. G., & Fuller, J. B. (2014). *Managing the talent pipeline: A new approach to closing the skills gap*. U.S. Chamber of Commerce Foundation.

U.S. Department of Education. (2012). Investing in America's future: A blueprint for transforming career and technical education. Author.

Van Bruinswaardt, C., Solberg, V. S., & Jarukitisakul, C. (2015). *Designing statewide career development strategies and programs*. National Collaborative on Workforce and Disability for Youth.

Vanfossen, B., Jones, J., & Spade, J. (1987). Curriculum tracking and status maintenance. *Sociology of Education, 60*, 104–122.

Whiston, S. C., & Blustein, D. L. (2013). *The impact of career interventions: Preparing our citizens for the 21st century jobs*. National Career Development Association and the Society for Vocational Psychology.

Woessmann, L. (2009). International evidence on school tracking: A review. *CESifo DICE Report, 7*(1), 26–34.

Zimmer, R. (2003). A new twist in the educational tracking debate. *Economics of Education Review, 22*(3), 307–315.

CHAPTER 9

Second Chance Schools in Spain. A Bridge Toward Continuity of Training and Employment for Young People at Risk of Social Exclusion

Davinia Palomares-Montero

Introduction

Development and Transformation of the Spanish Vocational Education and Training System

Since the mid-twentieth century, formal Vocational Education and Training (VET) in Spain has undergone various changes and reforms, primarily aimed at enhancing its quality and improving its social standing. Olmos-Rueda and Mas-Torelló (2019) offer an overview of the Spanish formal VET system and its evolution, which is updated with the recent legislative reform (Cedefop & Fundae, 2022). It is necessary to know this

D. Palomares-Montero (✉)
Department of Didactics and Scholar Organization, Faculty of Philosophy and Sciences of Education, University of Valencia, Valencia, Spain
e-mail: Davinia.Palomares@uv.es; davinia.palomares@gmail.com

E. Smeplass and C. Zirkle (eds.), *Opportunity Structures in Vocational Education and Training*, Palgrave Studies in Technical and Vocational Education and Training,
https://doi.org/10.1007/978-3-032-04410-5_9

development to understand how Second Chance Schools are currently, for the first time, explicitly mentioned in a law regulating vocational training in Spain.

The beginning of Spanish VET goes back to 1955 when the industrial sector emerged, and the first VET Act (Boletín Oficial del Estado-BOE, 1955) was promulgated. This law was ahead of its time and extremely ambitious: it institutionalized the figure of the industrial apprentice and the industrial master. In 1970, the General Educational Act (LGE) (Boletín Oficial del Estado-BOE, 1970) was introduced as part of an educational reform aimed at addressing the evolving demands of Spain's social, political, and labor landscape. VET was seen as an educational opportunity for students, rather than a marginal pathway. LGE stipulated that VET must prepare students to pursue their chosen profession while continuing their holistic education, being closely connected to the workplace. The VET system, regulated by Decree 995/1974, was formally integrated into the educational system and structured into levels. The first level served as the vocational pathway for students, particularly those who had not completed their compulsory education. The second level focused on specialized professional training aimed at facilitating entry into the workforce. It incorporated workplace experience for students on a trial basis, despite the system's strong academic orientation. This model remained in force until 1990, when the General Organic Law of the Educational System (LOGSE) (Boletín Oficial del Estado-BOE, 1990) was introduced. LOGSE is considered one of the most important reforms of the Spanish VET system. Vocational training was defined by a modular structure according to the European Qualification Levels for Vocational Training (levels 2 and 3, corresponding to the intermediate and advanced training cycles), in which on-the-job training became obligatory and a very important component. In addition, an alternative pathway was introduced for early school leavers—those without a certificate of compulsory secondary education—through basic training programs called Social Guarantee Schemes (corresponding to level 1 VET). Later, the Organic law of Education (LOE) (Boletín Oficial del Estado-BOE, 2006) was promulgated focusing on working toward the qualification improvement of people at risk of social and labor exclusion. VET was defined as a training model based on competencies, more flexible, adaptable, and accessible. Social Guarantee Schemes were replaced by the Initial Vocational Qualification Programs addressed to the same target but with relevant differences: young vulnerable people would equip with a Level 1 VET. The latter can be considered the first second chance training program in Spain for young people at risk, which plays a key role in promoting their social participation (although not yet formally).

Currently, the Organic Law on Education (LOMLOE) (Boletín Oficial del Estado-BOE, 2020), which amends the LOE, regulates the Spanish education system. It is committed to reinforcing the competency approach throughout the curriculum and to personalizing learning, with the aim that no student is disadvantaged at the end of compulsory education. This includes the modernization of vocational education and training in cooperation with all stakeholders. Of the two subsystems that make up Vocational Training in Spain, LOMLOE regulates the initial vocational training subsystem, which it calls vocational training in the educational system. However, the Organic Law 3/2022, of March 31, on the organization and integration of Vocational Training (LOOIFP) (Boletín Oficial del Estado-BOE, 2022) aims at a global transformation of the vocational training system, combining the previous dual system (vocational training within the education system and vocational training for employment) into a single vocational training system. This law provides a system in which all training can be accredited, accumulated, and capitalized. The training provision is structured in a vertical, graduated dimension, which establishes an ascending continuum according to the extent of each training program. The provision is organized from "micro-training" up to vocational training qualifications, depending on whether they include a single learning outcome, one or several vocational modules, or a complete package of vocational modules, while always maintaining their cumulative nature. All vocational training must be dual in nature, as it will be carried out jointly by the training institution and the company.

Figure 9.1. represents the Spanish education system which consists of several stages (Cedefop & Fundae, 2022): (i) Early childhood education (ISCED 0), (ii) Primary education (ISCED 1), which lasts 6 years (ages 6–12), (iii) Mandatory lower secondary education (ISCED 2), lasting 4 years (ages 12–16), (iv) Post-compulsory upper secondary programs (ISCED 3), (v) Higher vocational education and training (VET) programs (ISCED 5), and (vi) Higher education academic programs (ISCED 6, 7, 8). Initial vocational education and training (IVET) is offered at three levels: Basic VET programs (ISCED 353) for learners aged 15 and older, Intermediate VET programs (ISCED 354) for learners aged 17 to 18, and Higher VET programs (ISCED 554) for learners aged 18 to 19. At the basic level, IVET programs are mainly delivered in schools, while intermediate and higher-level VET programs may also offer more flexible learning options, such as distance education.

To serve students with special learning difficulties, curricular diversification programs are included from the third year of mandatory lower secondary education stage. Furthermore, to avoid early school leaving, open expectations for training and subsequent qualification and facilitate access to working life, initial professional qualification programs are established for students over 16 years of age (or, exceptionally at age 15) who have not obtained a graduate degree in compulsory secondary education.

Basic Vocational Education and Training as Youth Guarantee in Spain

International organizations and national governments are concerned about the situation of the young population without qualifications. Early leavers from education and training (individuals aged 18 to 24 who have not completed upper secondary education and are not enrolled in any formal or non-formal learning), students at risk of dropping out (those who, while still enrolled in education, display warning signs such as poor academic performance, low engagement, frequent absenteeism, or difficult socio-economic circumstances), and young people aged 15 to 29 not in employment, education, or training (NEETs) often overlap and face multiple socio-economic barriers, which limit their future opportunities.

According to the latest data provided by European Commission (2023) and the Spanish National Institute of Statistics (2024), early school leaving has improved in recent decades, but it remains a relevant issue compared to other European Union countries. In 2022, Spain‘s early school leaving rate was 13.9%, a figure significantly higher than the EU average, which stood at around 9.6% in the same year. Although the trend over the past 15 years shows a significant decrease—from 30% in 2008 to less than 14% in 2022—Spain remains one of the European countries with the highest rates of early school leaving. Moreover, although specific statistics on students at risk of dropping can vary widely between regions and systems, in Europe, early identification is key to reducing dropout rates. For instance, in Spain, the exact proportion of students considered "at risk" is harder to quantify as it fluctuates based on regional and local policies. In many European countries, early warning systems track these students based on attendance, behavior, and academic performance indicators. Finally, the NEET population faces multiple social and

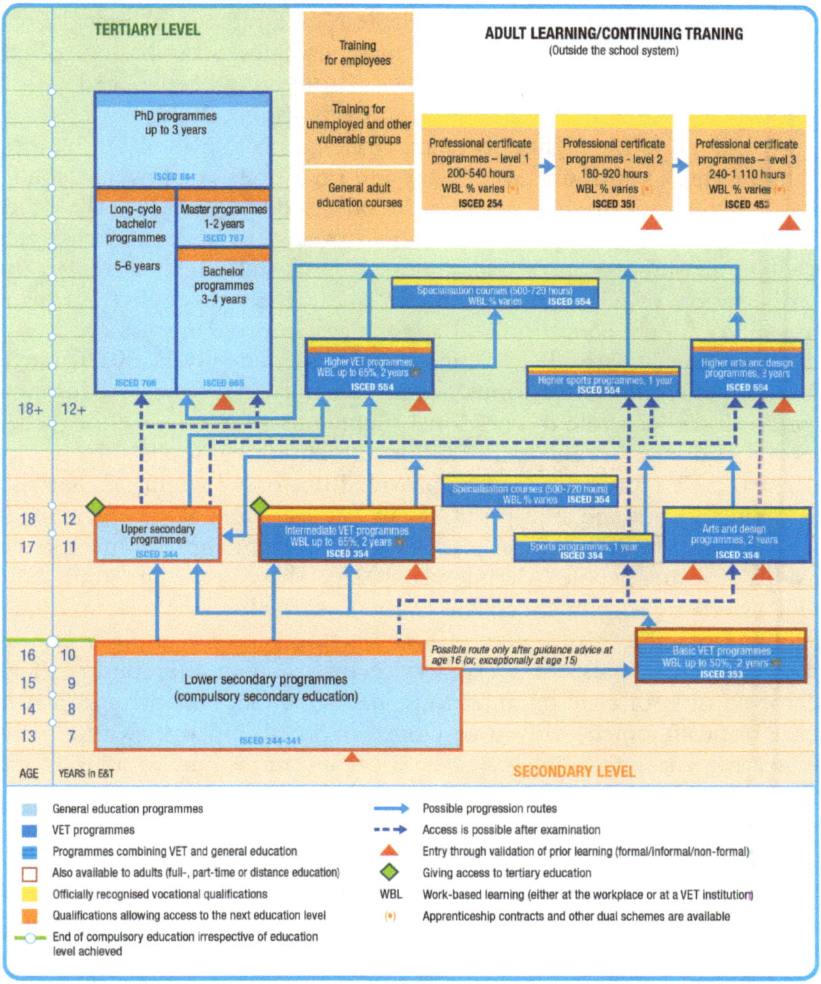

NB: ISCED-P 2011. The Spanish education system is not referenced to EQF levels.

Source: Cedefop, & State Foundation for Training in Employment (Fundae). (2023). Vocational education and training in Europe – Spain: system description. In Cedefop, & ReferNet. (2024). *Vocational education and training in Europe: VET in Europe database – detailed VET system descriptions* [Database]. www.cedefop.europa.eu/en/tools/vet-in-europe/systems/spain-u3

Fig. 9.1 Spanish education system. Source: Cedefop and Fundae (2022)

economic risks. In 2022, the EU average NEET rate was around 12.6%, with rates varying significantly across member states. Spain has one of the highest rates in the EU, with about 14.6% of young people aged 15–29 classified as NEETs in 2022 (European Commission, 2022).

This situation highlights structural challenges within the labor market, including high youth unemployment and precarious employment conditions, despite advancements in education (Cedefop, 2023). Like other EU member states, Spain is actively working to reduce these rates through a combination of policy measures. This chapter focuses on the flexibilization of educational pathways, particularly through the expansion of vocational education and apprenticeship programs. As previously discussed, the LOMLOE (Boletín Oficial del Estado-BOE, 2020) introduces more flexible trajectories within compulsory secondary education and reinforces the role of vocational training as an alternative for young people who may struggle with traditional educational models. In particular, basic VET programs can be provided not only by schools, but also by municipalities and non-profit organizations, often targeting individuals in vulnerable situations, including those from the aforementioned groups of young people (Marhuenda, 2023). Therefore, while some Basic VET programs are embedded within the formal education system, others emerge from the third sector of social action, non-for-profit organizations that are a pillar of social service provision in the country. This sector consists of civic organizations, neither public nor private (in the sense of profit aimed), driven by citizen or social initiatives with a focus on serving the public interest. These organizations aim to address the needs of young people who have left education and training prematurely, offering alternative support and opportunities (Butler & Payne, 2023).

Miller and Ohlin (2021) stated that modern community programs are crucial in tackling issues related to blocked opportunities, emphasizing the importance of "opportunity structures"in shaping youth outcomes. These programs strive to create environments that provide accessible resources and support, aiming to mitigate the socio-economic barriers that early school leavers, at-risk students, and NEETs face. By recognizing and addressing the structural factors contributing to youth disengagement, these initiatives align with the broader efforts to enhance educational pathways and improve labor market integration for vulnerable populations.

Spanish Second Chance Schools as Providers of Vocational and Educational Training for Young People at Risk

Second Chance Schools (SCS) were first conceived in the 1990s as part of a broader European Union initiative aimed at tackling youth unemployment and social exclusion (Commission on the European Communities, 1995). As a result, the European Association of Second Chance Schools[1] was officially created in Heerlen in the Netherlands in 1998. The aim was to create a network in which good practices and experiences are shared between members. Under this context, the Lisbon Strategy (European Council, 2000) was a crucial turning point, emphasizing the need to reduce early school leaving and to provide pathways for young people who had fallen out of the traditional education system. SCS offers an alternative to mainstream education, focusing on reintegrating youth through vocational training, basic education, and socio-emotional support, with an emphasis on improving employability. Today, SCS operates in various forms across Europe, adapting to the specific needs and policy environments of different countries (Blagoeva, 2022).

In Europe, SCS exists under different frameworks, reflecting varying national policies and educational structures. Countries like France, Germany, and the Netherlands have established robust models of SCS, with well-defined institutional support and integration with national education and employment services. France, one of the pioneers of the SCS concept, developed the "Écoles de la Deuxième Chance"[2] model, which has become a reference for other European countries. France's network is substantial, with over 146 schools across the country. Germany's SCS model is deeply embedded in its dual vocational training system. Young people who have dropped out of school are reintegrated into vocational education through partnerships between the education system and industries. While there is no single "second chance" school model, Germany's approach focuses on apprenticeships, combining work experience with formal education, offering a strong pathway to the labor market. The Dutch model integrates SCS within the broader framework of vocational education. Here, SCS functions as part of a system that supports early school leavers, combining personalized learning programs, vocational training, and work-based learning opportunities. This approach

[1] https://e2c-europe.org/

[2] https://reseau-e2c.fr/

is highly flexible and adapts to the specific needs of youth who face educational and social challenges. These countries share common goals—re-engagement of youth, development of professional skills, and social integration—yet they differ in execution, governance, and focus.

Other European countries, such as Austria, Belgium, and Portugal,[3] also have similar initiatives tailored to their national contexts, each focusing on re-engaging young people who are at risk of long-term unemployment or social marginalization. For instance, Glorieux et al. (2011) examined the effectiveness of a flexible system, in which students prepare themselves to do exams, to know the effect on diminishing the large inequalities in the Belgian educational system.

In Spain, SCS are relatively recent compared to their European counterparts. They emerged in response to the high rate of early school leaving, which has historically been higher than the European Union average. Spain's SCS system aims to offer tailored education and vocational training programs for young people aged 16 to 29 who have dropped out of the formal education system without completing secondary education. It offers a flexible and personalized learning approach, with a strong focus on basic skills development (literacy, numeracy, and digital skills), vocational training, and social and emotional support to foster self-confidence and motivation. The Spanish model shares similarities with the French and Dutch systems in its emphasis on integrating vocational training with education, but it also faces distinct challenges, such as limited institutional support and funding inconsistencies across different regions. The Spanish Association of SCS[4] was founded in 2016 as a response to the growing demand for educational alternatives for young people at risk of exclusion. The association was established with the goal of creating a national framework that would provide visibility, recognition, and support for SCS across Spain. Since its inception, it has played a critical role in advocating for the integration of SCS into the national education policy and in seeking official recognition for these schools as part of the formal education system. By 2024, the Association had grown to include 47 accredited schools across various regions of Spain, serving more than 8000 students annually.

[3] https://www.segundaoportunidade.com/

[4] https://www.e2oespana.org/

This chapter examines the vocational training offered by SCS in Spain and the characterization of the young people who require these programs. It results of a research project ref. AICO/2021/254,[5] named "Second Chance Schools: Reactions in response to needs changed by the effects of Covid-19", funded by the Regional R&D Program of the Government of Valencian Autonomous Region (Spain). The objective is to explore how the adjustment in this equation (training offered and profile demand) facilitates the enhancement of personal, social, and vocational skills, promotes educational engagement, and raises expectations for further training or reintegration into education.

Method

The methodological approach of this study is mixed, using both qualitative and quantitative techniques to collect data from different stakeholders involved in the SCS. This comprehensive approach aims to address the main objective in a holistic manner. The following sections outline the methodological design, including a description of the participants, the instruments used to collect data, the procedures for administering these instruments, and the analysis of the data collected.

Methodological Design

The methodological design involves a triangulation of three different approaches: i) an evaluative approach, which seeks to understand the nature of the training offered and the profile of the young people targeted from the perspective of the professionals and young participants in the SCS; ii) a descriptive approach, which does not aim to generate generalizable theory but to provide a detailed account of how the SCS delivers vocational training to groups in vulnerable situations; and iii) a survey-based approach, which focuses on collecting data from young people participating in the SCS training pathways. This methodological framework requires the use of tailored data collection and analysis strategies for each approach, as well as the definition of different participant groups.

[5] Two technical reports have been published as products of the project, presenting the main results of the supply and demand analysis (https://hdl.handle.net/10550/96789 and https://hdl.handle.net/10550/96790). The progress in understanding the fit between the two is presented in this chapter.

Table 9.1 Participants distribution in the SCS study

Instrument	*Agents*		*N*
Questionnaire	Enrolled students[i]	1st round	1119
		2nd round	1779
Discussion groups[ii]	Teachers		46
	Enrolled students		53
Document analysis	SCS (management teams and/or teachers)		17

Source: Own elaboration

[i]Enrolled students come from 28 SCS. The total number of enrolled students was 7855 in the academic year 2021–22 and 8493 in the academic year 2022–23 (https://www.e2oespana.org/)

[ii]Teachers and enrolled students come from 25 SCS (during the project period, 2021–2023, there were 43 accredited SCS in Spain)

Participants

The methodological triangulation in this study includes the triangulation of respondents. Specifically, two profiles of participants are distinguished: (i) professionals from the training teams of the SCS, and (ii) young people enrolled in the SCS programs (see Table 9.1.). In addition, a document analysis was carried out, examining accreditation reports, activity indicators, and other relevant documents either published on institutional websites or provided by SCS management teams.

Tools and Procedures for Data Collection

The fieldwork took place between 2022 and 2023, with questionnaires administered at the beginning of the course (October 2022) and another near the end of the course (May 2023). The focus groups were conducted between October and November 2023. And the document analysis was carried out throughout 2022 to serve as information to prepare the discussion groups. Survey data were analyzed using IBM SPSS Statistics, version 29.0. The content of the discussion groups was recorded and transcribed for subsequent analysis using MAXQDA Plus 2022. A document analysis was developed intuitively by four researchers using a template of relevant dimensions and categories.

Questionnaire

A customized questionnaire was designed and validated by three professionals with extensive practical experience in SCS. It was administered online at 28 SCS. The questionnaire collected data on students' social and academic backgrounds, their expectations at the beginning of the year (first phase), and their outcomes at the end of the year (second phase). A total of 1119 students participated in the first phase and 1779 in the second phase.

Discussion Groups
Two scripts were developed for the discussion groups, one for the young people and one for the teachers. The first recapitulates the experiences before and during the transition to the SCS and then comments on the teaching and learning process experienced in the SCS. The second addresses the processes of access and reception of students, the training processes, as well as mediation and/or referral at the end of the process, in addition to commenting on issues specific to the governance and management of the SCS. Here we will focus on the aspects that link the supply and demand of training in the SCS.

Document Analysis
Several sources of information were used in the document analysis. On the one hand, the information available on the websites of each of the institutions sponsoring the schools. On the other hand, prior consent for the transfer of data, information on indicators, and the accreditation criteria of the Spanish Association of Second Chance Schools for the courses analyzed. In addition, we had the opportunity to visit most of the schools to write a document about each of them, which in turn was validated by professional teams (between autumn 2022 and spring 2023). These reports served to configure the script for the discussion groups that were developed later (autumn 2023). The dimensions and categories were related to the contextualization of the SCS (location and profile of students), intervention strategies (selection, training, and derivation/intermediation), description of the teaching and learning process, and relationships with the context.

Results

This section presents the results obtained from the different instruments in a general and integrated way. It presents the relationships between the document analysis and the opinions of the different informants (teachers and students), through their evaluations in the discussion groups and the responses to the surveys. Thus, to respond to the proposed objective, three subsections are presented that respond to each of the elements raised therein: the analysis of supply, the description of demand, and the adaptation of both. This analysis is essential to assess whether SCS is meeting its integration goals, and whether the training programs it offers are effectively meeting the needs of its students.

Analysis of the Educational Offer of Second Chance Schools in Spain

Spanish Second Chance Schools as an Educational Offer

These schools are present in 10 autonomous communities in Spain and serve young people between the ages of 15 and 29 who, for the most part, have not managed to complete secondary education or lack other basic qualifications to access the labor market. SCS represents a well-defined socio-educational model that emphasizes diversity through the distribution of competencies. This model recognizes the different backgrounds and needs of students, making it adaptable to different educational contexts.

The training offered is structured around two main axes: academic and vocational. In the academic field, many SCS focus on preparation for the Compulsory Secondary School Certificate (ISCED 2) or the Basic VET programs (ISCED 353), essential qualifications to facilitate access to employment or to higher-level vocational training, such as the Intermediate Vocational Training Cycle (ISCED 354). This type of regulated training is available in some SCS, although its implementation largely depends on regional regulations and the accreditation that the schools have been able to obtain. On the other hand, in vocational training, SCS offer unregulated training that includes Level 1 and, to a lesser extent, Level 2 vocational certificates. Vocational training in these programs targets sectors such as commercial services, hospitality, maintenance, and other high-demand occupations in urban areas, where most SCS are located.

Internships in companies are a fundamental part of the training programs, as they allow young people to gain real work experience, develop technical skills, and facilitate their job placement. These practices involve collaborating companies, which see SCS as an opportunity to train future employees with a practical approach tailored to their needs.

SCS operate with a degree of autonomy that allows them to manage their staff, spaces, and relationships with local businesses. This autonomy is critical to creating a tailored educational environment that is responsive to the unique needs of their student population.

The core principles that guide SCS include individualization, diverse pathways, and flexible pathways. This focus ensures that each student's educational journey is personalized, considering their specific circumstances and aspirations. SCS maintains a stable and recognized identity, characterized by an educational vocation that values collaboration and community engagement. Their flexible management style, collaborative culture, and distributed leadership foster an environment where educators and students can thrive together, reinforcing their mission to provide meaningful second chances to those in need.

Spanish Second Chance Schools as an Agent in the Territory
SCS play a crucial role as agents in their respective territories, responding to the unique historical, territorial, and legal contexts in which they operate. However, the effectiveness of these schools is often affected by the absence of comprehensive regulations, such as those established by LOOIFP (Boletín Oficial del Estado-BOE, 2022) This lack of a coherent legal framework results in fragmented funding, with sources of funding scattered among employment services, the education sector, social services, and private entities.

The SCS landscape is populated by various organizations, including foundations, associations, and cooperatives, which collectively contribute to the creation and operation of these schools. Their presence is particularly strong in regions like the Basque Country, Catalonia, Andalusia, Madrid, and Aragón, where they have managed to integrate effectively into local communities. In contrast, the presence of SCS is limited in regions such as Castilla y León, Castilla-La Mancha, the Canary Islands, Navarra, and the Valencian Autonomous Region. In some regions, such as the Balearic Islands, Extremadura, Asturias, Cantabria, Galicia, La Rioja, Murcia, Ceuta, and Melilla, there are currently no Second Chance Schools. Geographically, SCS tend to thrive in urban environments where the economic landscape is often dominated by commerce and tourism. This urban character not only facilitates access to resources and employment opportunities for students, but also encourages community engagement through volunteer initiatives. The strong local ties that SCS cultivates in their areas enhance their ability to meet the unique needs of their students, making them important players in the educational and social fabric of their regions.

Spanish Second Chance Schools as Articulators of Functions

SCS primarily serves an educational function, characterized by a flexible and extended curriculum designed to meet the diverse needs of its students. This approach allows for a personalized learning experience, where students can progress at their own pace and focus on their individual abilities. Flexibility in both curriculum and pace of learning is essential to fostering an environment that supports a variety of learning styles and backgrounds.

In addition, SCS acts as a bridge between educational policy and active employment policy, facilitating pre-work exploration and aligning educational offerings with market demand. This alignment ensures that students not only obtain educational qualifications, but also acquire the skills and experience necessary to meet employer requirements, thereby enhancing their employability.

In addition to their educational role, SCS serves an important social function, particularly for young people who may lack family support structures. In cases where students reside in residential centers or supervised housing, SCS provide essential services to regularize their legal status and address other social needs. This comprehensive support includes organizing recreational activities and leisure opportunities, contributing to the holistic development of students and fostering a sense of community and belonging.

Finally, as institutions that operate at both the second and third levels of organization, SCS play a crucial representational role. By acting as intermediaries between students, local communities, and the broader education and employment systems, SCS advocates for the needs and rights of marginalized youth. Their representative role underscores their importance not only as educational institutions, but also as key actors in the social and economic fabric of their communities.

Profile and Characteristics of Young People in Second Chance School in Spain

The first round of the questionnaire (in which all participants answer the socio-demographic questions—n =1119) is used to describe the student profile of the SCS. The student body has a larger proportion of boys, a trend linked to the well-known phenomenon of higher dropout rates among males (Martínez García, 2011). Additionally, 75% of the students are aged 20 or younger, highlighting that the chances of joining these programs decrease with age. Almost half of the students are immigrants, underscoring the significant disparity in dropout rates based on student origin. Many students come from families facing economic and cultural hardship, with over half reporting severe financial challenges. Nevertheless, a smaller group of students have parents with university degrees and a stable financial background, suggesting that educational disengagement is not solely a problem for low-income families (Valls, 2020). For more details, see Table 9.2.

Many of these students have had negative previous educational experiences, which strongly influence their expectations. Nearly 70% have

Table 9.2 Descriptive variables of the youth profile

Variables	*N*	*%*
Gender		
Male	806	72.0
Female	283	25.3
Other/Don't know/no answer	30	2.7
Age		
15–17 years	395	35.3
18–20 years	450	40.2
21–25 years	198	17.7
26–30 years	48	4.3
Invalid/no answer	28	2.5
Country of birth		
Spain	591	52.8
Other	517	46.2
Don't know/no answer	11	1.0
Family educational level		
Basic education or none	244	21.8
Intermediate education	444	39.7
Higher education	177	15.8
Don't know /no answer	207	18.5
Other/not specified	47	4.2
Family economic situation		
No difficulties	257	23.0
Some difficulties	250	22.3
Moderate difficulties	260	23.2
Significant difficulties	230	20.6
Severe difficulties	99	8.8
Don't know / no answer	23	2.1

Source: Palomares-Montero et al. (2024)

repeated at least one grade, 40% have been expelled at least once, and 30% were recurrent or chronic absentees. Additionally, over half received support to address learning difficulties, and almost 60% felt they had little or no attention from teachers. Many of the students in SCS have experienced situations that could affect their disengagement from the system (Table 9.3.).

Most young people entering SCS have very clear expectations regarding their future training and employment. 52% of students confirm that their main goal is to train to get a job, while 39.5% hope to obtain a Compulsory Secondary School Certificate (Table 9.4.). These expectations reflect an urgent need for access to vocational training that will

Table 9.3 Descriptive variables of prior educational experience

Variables	*N*	%
Attendance		
Every day	489	43.7
Almost every day	188	16.8
Missed occasionally	219	19.6
Rarely attended	131	11.7
Don't know/no answer	92	8.2
Educational support		
Yes	580	51.8
No	449	40.1
Don't know/no answer	90	8.1
Expulsion		
Never	522	46.7
Warning without expulsion	115	10.3
Once	131	11.7
More than once	321	28.6
Don't know/no answer	30	2.7
Good teacher attention		
Yes	476	42.5
Partially	380	34.0
No	234	20.9
Don't know/no answer	29	2.6
Grade repetition		
Yes	750	67.1
No	344	30.7
Don't know/no answer	25	2.2

Source: Palomares-Montero et al. (2024)

Table 9.4 Descriptives of expectations before entry

Expectations before entry	*N*	%
Get compulsory secondary school certificate	447	39.5
Entrance exam/intermediate level (ISCED 354)	178	15.9
Continue studying	255	22.8
Train to get a job	582	52.0
Get to know yourself better/improve personal relationships	214	19.1
I don't know/I don't answer	40	3.6

Source: Palomares-Montero et al. (2024)

allow them to enter the labor market in the short or medium term. Furthermore, a significant percentage of these young people come from families with economic difficulties, which increases the pressure to find employment and contributes to the family's livelihood.

As far as the internships[6] are concerned, the evaluations of the students who have completed them are very positive (Table 9.5). For all items, the ratings are very high, with the maximum response (strongly agree) being over 55%, which generally shows that training in enterprises is well perceived by students and that it is training that reinforces the skills acquired at school and expands the work expectations of young people. It is also worth noting that between 10% and 15% of students give a negative assessment (disagreeing with the statement or not agreeing much with it), so there is room for improvement in practice. Almost 15% say that the work placement did not help them to find a job, 16% did not receive support from their colleagues, 10% did not think that the evaluation carried out was useful, and 9% said that they did not receive support from the company tutor.

SCS plays a crucial role in addressing the educational needs of young people, especially during key transition periods in adolescence. These transitions are influenced by a variety of personal, social, cultural, and academic factors, and SCS adopts an ecological approach to development, offering a holistic view of the individual. This comprehensive perspective allows them to address the diverse and interconnected challenges faced by students. The social background of the students significantly impacts

[6] The data are from the 2nd round of the questionnaire, considering the students who have done internships.

Table 9.5 Description of the assessment of the quality of practices

Variables	*Scale*	*N*	*%*	*Variables*	*Scale*	*N*	*%*
... have improved skills learned in E2O	1	13	4,2	... her/his tutor is or was always available	1	16	5,1
	2	25	8,0		2	12	3,8
	3	64	20,4		3	83	26,5
	4	194	62,0		4	188	60,1%
	NR/DK	17	5,4		NR/DK	14	4,5
	Total	313	100		Total	313	100
... have applied what they have learned in E2O in internships	1	18	5,8	... felt helpled by his/her co-workers	1	21	6,7
	2	23	7,3		2	30	9,6
	3	77	24,6		3	66	21,1
	4	179	57,2		4	182	58,1
	NR/DK	16	5,1		NR/DK	14	4,5
	Total	313	100		Total	313	100
... have improved their chances of finding a job	1	19	6,1	... consider the evaluation helpful	1	12	3,8
	2	27	8,6		2	20	6,4
	3	72	23,0		3	76	24,3
	4	182	58,1		4	183	58,5
	NR/DK	13	4,2		NR/DK	22	7,0
	Total	313	100		Total	313	100
... value the training in E2O more	1	12	3,8	... have helped him/her to be aware of their need of being educated	1	12	3,8
	2	23	7,3		2	21	6,7
	3	76	24,3		3	80	25,6
	4	183	58,5		4	173	55,3
	NR/DK	19	6,1		NR/DK	27	8,6
	Total	313	100		Total	313	100
... have learned professional skills	1	12	3,8				
	2	18	5,8				
	3	80	25,6				
	4	186	59,4				
	NR/DK	17	5,4				
	Total	313	100				

Scale: 1. Strongly disagree, 2. Somewhat disagree, 3. Somewhat agree, 4. Strongly agree, NR/DK No response/Do not know
Source: Own elaboration

their ability to enter and remain in the education system, as well as their overall achievement. SCS responds to these needs through individualized support and flexible educational pathways, ensuring that students receive personalized attention tailored to their unique circumstances, helping them to overcome barriers and succeed in their transitions. Young people of immigrant origin are particularly vulnerable, with a higher likelihood of following educational dropout paths or enrolling in Basic VET.

Matching Training Provision to Young People's Needs

The analysis of the SCS training offer in relation to the needs and expectations of the young people who take part in it shows that in many respects, there is a reasonably good match, but also points to areas where improvements are needed. By recognizing that young individuals face varying degrees of access to legitimate opportunities, the SCS can better tailor its programs to address the systemic barriers that influence these youths' training experiences and employment outcomes (Cloward & Ohlin, 1960).

Job-Oriented Training

One of the areas in which SCS's offerings appear to be most responsive to the needs of young people is its focus on employment. With 52% of young participants citing training for employment as their primary goal, SCS has focused much of its programming on job training. Level 1 and 2 Certificates of Professionalism are the most suitable programs at SCS, providing young people with opportunities to acquire technical and work skills that are valued in the job market. These certificates are particularly attractive to young people who do not have the Compulsory Secondary School Certificate, as they do not require prior educational qualifications.

In addition, internships in companies are a central component of SCS training programs and are considered an effective tool to facilitate young people's transition into the workforce. According to the study, a high percentage of SCS students participate in internships, and the companies that partner with SCS highly value the training students receive in social and professional skills. These internships not only give young people the opportunity to gain work experience, but also increase their motivation and commitment to the training process. A concrete example is the case of SCS in regions such as the Basque Country and Catalonia, where collaboration with companies in sectors such as hospitality and retail has facilitated the employment of a high percentage of students.

Flexibility in Training Pathways

Another positive aspect of SCS provision is its flexibility, which allows young people to personalize their training pathways according to their needs and circumstances. This flexible approach is particularly important for young people who face socio-economic barriers, such as the need to work to support their families or difficulties related to their immigration status. SCS offers training pathways of different lengths, ranging from short programs of less than 1 year to longer programs of 2 years or more. This allows students to adapt their participation in the programs according to their personal situation.

However, this flexibility also poses certain challenges. Our study shows that despite SCS's efforts to offer long and structured pathways, many young people leave the program before completing them. In our study, we asked students whether they had considered dropping out of their studies. Most students answered negatively to this question (72%), while 19% admitted that they had thought about it but ultimately did not drop out. A small percentage (3%) had dropped out but returned later (Palomares-Montero et al., 2024). Young people from more vulnerable socio-economic backgrounds are more likely to drop out of longer pathways due to the pressure to find immediate employment. This suggests that while SCS has achieved a good match between its provision and young people's employment expectations, there is still a need to strengthen retention strategies to prevent students from dropping out prematurely.

Limited Access to Regulated Training

One of the main challenges facing SCS training provision is the limited access to regulated programs. Although many SCS offer non-regulated training, the possibility of obtaining official qualifications, such as the Compulsory Secondary School Certificate, the Basic VET, or the Intermediate VET, varies from region to region. In autonomous communities such as the Basque Country, SCS has signed agreements with local administrations to provide intermediate-level vocational training, giving students wider access to regulated education. However, in other regions, such as Andalusia, SCS is not authorized to provide this type of training, which limits young people's opportunities to obtain recognized qualifications.

This unequal access to regulated training has a negative impact on young people who wish to continue their education beyond the basic level. Those who wish to pursue intermediate or higher vocational training are often forced to leave SCS and seek alternatives in other educational institutions. This can be a significant obstacle for vulnerable young people, who often lack the financial and social resources to make the transition to other educational centers. It is therefore crucial that SCS expand their capacity to provide regulated training, particularly in the Autonomous Communities where they are not currently authorized to do so.

Gender Inequality in Training Offerings
Another major challenge is the persistence of gender inequality in SCS training provision. Although women make up around 23% of the student body, their participation in vocational training programs remains skewed toward traditionally female-dominated occupations such as personal care and childcare. Despite efforts to include women in vocational training for traditionally male-dominated sectors such as carpentry or electrical work, progress has been limited. Conversely, men are under-represented in female-dominated occupations, reflecting persistent gender stereotypes in access to training and employment. VET programmes continually struggle to create inclusive environments that support women and challenge masculinized fields and workplace norms. However, a gap exists between theory and VET practices (Ray & Zarestky, 2022). Breaking these stereotypes is essential to improve equity in SCS training provision and to provide equal opportunities for men and women to access different career fields. SCS should continue to promote programs that challenge these barriers and provide real opportunities for women to train in male-dominated fields and vice versa.

Conclusions

The SCS, with its offer of basic vocational training, is configured as a particularly relevant path for young people who, having completed secondary education without obtaining a diploma, are looking for an alternative training path. As Boeren and Whittaker (2018) stated, second chance education would be a new category of adult education provision available for the lowest educated adults in society. In Spain, the LOMLOE emphasizes the flexibility of training pathways, and the LOOIFP establishes a more open and accessible structure for VET, with programs that allow access to training from the age of 15 through to Basic VET (ISCED 353). These programs are designed to provide a level 1 vocational qualification, recognized in the National Catalogue of Professional Qualifications (CNCP), and offer the possibility of progressing to more advanced levels of training. In addition, young people can obtain the Compulsory Secondary School Certificate at the same time as studying at the Basic VET, underlining the system's second chance approach. In this context, the Spanish Association of Second Chance Schools plays a crucial role in the implementation and consolidation of second chance education in Spain. These schools provide an environment adapted to the needs of young people who have left the education system without qualifications, offering personalized training programs that combine education, vocational training, and career guidance. Therefore, SCS could be considered as a supportive school environment for alternative education youth (Edgar-Smith & Palmer, 2015).

Furthermore, the SCS model is in line with the objectives of LOMLOE and LOOIFP by providing training that promotes both the development of basic skills and access to vocational qualifications. SCS focus not only on academic training but also on socio-emotional support, facilitating young people's reintegration into the education system and improving their employability prospects. Personalized guidance and tutoring programs help to reduce early school leaving and motivate students to continue their education. SCS also work closely with companies and vocational training centers to ensure that young people receive practical, labor market-oriented training. This direct link with the productive fabric is essential to improve the integration of young people into the labor market, as envisaged by the LOOIFP in its strategy for the integration of vocational training. These centers contribute to the Spanish government's commitment to reducing early school leaving and improving the employability of young people. This conclusion is in line with Bloom's (2010) conclusion: that some second chance programs generate significant increases in employment or earnings in the short term.

Over time, several attempts can be observed to improve Spanish vocational training, but it should be noted that if something has characterized Spanish vocational training since its inception and until recently, it has been its low social value and recognition. At present, this recognition is much greater, improving the social image of these basic VET programs, increasing the enrolment numbers and the recognition of third-sector social organizations for job placement and second chance centers providing specific training for vulnerable groups. Recognition of prior learning is strongly dependent on the status of VET qualifications in education systems and labour markets (Maurer, 2022). Although

SCS are not ruled nor recognized by law and they are mostly promoted by non-for-profit organizations, for the first time, the formal system of education recognizes its participation and collaboration as providers of VET, commonly Basic VET.

The concept of opportunity structures is particularly relevant here, as it helps to explain how economic and political factors shape the pathways available to young people who face limited options within the traditional education system (Cloward & Ohlin, 1960). The flexibility and accessibility of SCS programs create an alternative structure of opportunities for these youth, especially for those in vulnerable situations. The SCS are, for the most part, a good match between the training on offer and the needs of young people in vulnerable situations. The placement focuses of the training, the flexible approach to training schedules, and the cooperation with companies have enabled many young people to acquire useful skills and find employment. However, important challenges remain, particularly regarding access to formal training, retention in longer programs and overcoming gender inequalities in training provision. In addition, SCS faces ongoing challenges in securing stable funding and achieving full legal recognition within the Spanish education system. While the demand for SCS continues to grow, the Spanish Association of SCS is advocating for more robust policy support at both regional and national levels. Looking ahead, the association aims to further develop its network and increase collaboration with European partners to share best practice and improve outcomes for young people across Spain.

The SCS's role in offering alternative pathways exemplifies how opportunity structures can be shaped to help disadvantaged youth access training that aligns with their socio-economic context, thereby increasing their chances of future success.

References

Blagoeva, E. (Ed.). (2022). *Framework for second chance schools in Europe. Report*. New Bulgarian University Press. https://publishing-house.nbu.bg/bg/elektronni-izdaniq/knigi/framework-for-second-chance-schools-in-europe-report

Bloom, D. (2010). Programs and policies to assist high school dropouts in the transition to adulthood. *Future of Children, 20*(1), 89–108. https://doi.org/10.1353/foc.0.0039

Boletín Oficial del Estado-BOE. (1955). *Ley, de 20 de julio de 1955, sobre Formación Profesional Industrial*. https://www.boe.es/buscar/doc.php?id=BOE-A-1955-10412

Boletín Oficial del Estado-BOE. (1970). *Ley 14/1970, de 4 de agosto, General de Educación y Financiamiento de la Reforma Educativa*. https://www.boe.es/buscar/doc.php?id=BOE-A-1970-852

Boletín Oficial del Estado-BOE. (1990). *Ley Orgánica 1/1990, de 3 de octubre, de Ordenación General del Sistema Educativo*. https://www.boe.es/buscar/doc.php?id=BOE-A-1990-24172

Boletín Oficial del Estado-BOE. (2006). *Ley Orgánica 2/2006, de 3 de mayo, de Educación*. https://www.boe.es/buscar/doc.php?id=BOE-A-2006-7899

Boletín Oficial del Estado-BOE. (2020). *Ley Orgánica 3/2020, de 29 de diciembre, por la que se modifica la Ley Orgánica 2/2006, de 3 de mayo, de Educación*. https://www.boe.es/buscar/doc.php?id=BOE-A-2020-17264

Boletín Oficial del Estado-BOE. (2022). *Ley Orgánica 3/2022, de 31 de marzo, de ordenación e integración de la Formación Profesional*. https://www.boe.es/diario_boe/txt.php?id=BOE-A-2022-5139

Boeren, E., & Whittaker, S. (2018). A typology of education and training provisions for low educated adults: Categories and definitions. *Studies in the Education of Adults, 50*(1), 4–18. https://doi.org/10.1080/02660830.2018.1520017

Butler, P., & Payne, J. (2023). Employer engagement with third-sector activation programmes for vulnerable groups: Interrogating logics and roles. *Journal of Social Policy*, 1–19. https://doi.org/10.1017/S0047279423000211

Cedefop. (2023). *Vocational education and training in Spain. Short description*. Publications Office of the European Union. http://data.europa.eu/doi/10.2801/221364.

Cedefop & Fundae. (2022). *Vocational education and training in Europe - Spain: System description* [From Cedefop; ReferNet. Vocational education and training in Europe database]. https://www.cedefop.europa.eu/en/tools/vet-in-europe/systems/spain-u2

Cloward, R. A., & Ohlin, L. E. (1960). *Delinquency and opportunity: A study of delinquent gangs*. Routledge.

Commission on the European Communities. (1995). *White paper on education and training. Teaching and learning. Towards the learning society*. Publications Office of the European Union.

Dubet, F. (2010). The crisis of transmission and the decline of the institution. *Política y Sociedad, 47*(2), 15–25.

Edgar-Smith, S., & Palmer, R. B. (2015). Building supportive school environments for alternative education youth. *Preventing School Failure: Alternative Education for Children and Youth, 59*(3), 134–141. https://doi.org/10.1080/1045988X.2013.865587

European Commission. (2022). *Youth guarantee: National implementation plan*. Publications Office of the European Commission.

European Commission. (2023). *Education and training monitor 2023. Comparative report*. Publications Office of the European Commission.

European Council. (2000). *Presidency conclusions: Lisbon European council 23 and 24 March 2000*. https://www.europarl.europa.eu/summits/lis1_en.htm.

Glorieux, I., Heyman, R., Jegers, M., & Taelman, M. (2011). Who takes a second chance? Profile of participants in alternative systems for obtaining a secondary diploma. *International Journal of Lifelong Education, 30*(6), 781–794. https://doi.org/10.1080/02601370.2011.627472

Marhuenda, F. (2023). Informal learning in the vocational education and training of vulnerable populations. In P. Matthias (Ed.), *Informal learning in vocational education and training. Illuminating an elusive concept* (pp. 91–110). Springer. https://doi.org/10.1007/978-3-658-44341-2

Martínez García, J. S. (2011). Género y origen social. Diferencias grandes en fracaso escolar administrativo y bajas en rendimiento educativo. *RASE Revista de Sociología de la Educación, 4*(3), 270–282.

Maurer, M. (2022). Recognizing prior learning in vocational education and training: global ambitions and actual implementation in four countries. *Comparative Education, 559*(1), 1–17. https://doi.org/10.1080/03050068.2022.2132445

Miller, A. D., & Ohlin, L. E. (2021). Introduction to delinquency and community: Creating opportunity and controls. In J. McCord & J. H. Laub (Eds.), *Contemporary masters in criminology: The plenum series in crime and justice* (pp. 19–40). Springer. https://doi.org/10.1007/978-1-4757-9829-6_19

Olmos-Rueda, P., & Mas-Torelló, O. (2019). Building up a VET system: Formal VET. In F. Marhuenda-Fluixá (Ed.), *The school-based vocational education and training system in Spain. Achievements and controversies* (pp. 21–40). Springer.

Palomares-Montero, D., Merino, R., Olmeda, E., García-Gracia, M., & Córdoba, A. I. (2024). Second chance schools' students: New engagement, recovering learning, reconnecting pathways. *Revista Complutense De Educación, 35*(3), 553–564. https://doi.org/10.5209/rced.85965

Ray, S., & Zarestky, J. (2022). Gender in vocational education and training: an integrative review. *European Journal of Training and Development, 6*(9), 876–893. https://doi.org/10.1108/EJTD-12-2019-0196

Spanish National Institute of Statistics. (2024). *Encuesta de población activa*. Instituto Nacional de Estadística. https://www.ine.es

Valls, O. (2020). *Attitudes, choices and inequalities in the transition from compulsory to post-compulsory school* [Unpublished doctoral thesis]. Universidad Autónoma de Barcelona.

CHAPTER 10

The Benefits, Challenges, and Opportunities for Technical & Vocational Education & Training in Zimbabwe

Davison M. Mupinga and John Chakamba

Introduction

Many developing countries face unemployment challenges (Global People Strategist, 2024; Olowe, 2024). For the youth, the unemployment rate remains high. In 2022, the unemployment rate in the Middle East and North Africa was 24.8% while Latin America had 20.5% and Sub-Saharan Africa was at 12.7% (Akileswaran et al., 2023). In Zimbabwe, the youth unemployment rate in 2023 was approximately 14.29% (Federal Reserve Economic Data, 2024). To reduce unemployment, many of the countries

D. M. Mupinga (✉)
Kent State University, Kent, OH, USA
e-mail: dmupinga@kent.edu

J. Chakamba
Department of Art Design and Technology Education, University of Zimbabwe, Harare, Zimbabwe

E. Smeplass and C. Zirkle (eds.), *Opportunity Structures in Vocational Education and Training*, Palgrave Studies in Technical and Vocational Education and Training,
https://doi.org/10.1007/978-3-032-04410-5_10

have adopted education and skill development programs such as technical and vocational education (Kennedy, 2020; Malle, 2016). According to UNESCO (2015), technical and vocational education and training (TVET) programs, also known as career and technical education (USA), are a combination of formal and informal learning that helps people develop the skills and knowledge they need to succeed in the workplace. These programs can be used for workforce development, that is, for acquiring practical skills, upskilling and reskilling employees, studying technologies and related sciences, developing proper workplace attitudes, and gaining knowledge about various occupations (California Workforce Development Board, 2025). In many countries, including Zimbabwe, the programs are offered in informal and formal settings such as public/private schools, tertiary institutions, and workplaces.

Prior to gaining political independence in 1980, Zimbabwe had the "F2 secondary schools" that primarily focused on practical and vocational skills, often considered to be of lower quality, and designed for students deemed less academically capable (Mugomba, 1980; Mutumbwa, 2023). The schools were later abolished after independence due to their negative connotations of segregation and limited opportunities for Black students. Currently, all high school students in Zimbabwe are expected to study at least one technical subject, as per Ministry of Education Policy Circular # 77 of 2006 (Zimbabwe Government, 2006). Tertiary institutions such as agricultural colleges, polytechnics, and other private colleges have always offered TVET programs. The TVET subjects offered in high school vary according to the geographic location of the school (urban versus rural) and what the school selects to offer. Given the disparities in resources between urban and rural schools in Zimbabwe, one can expect programs like construction and agriculture to be offered in the rural areas (due to shortages of electricity) while subjects like Metal Technology and Design and Wood Technology and Design to be offered in urban areas. In the lower levels of school systems, TVET programs tend to focus on awareness of careers and not on technical skill development (Mupinga et al., 2005). Depending on the set-up at the high school, the programs may emphasize career awareness, even though the expected outcome is skill development.

Zimbabwe's education had a reputation of being one of the most advanced in Africa, with high literacy rates and significant expansion across all levels (Mutisi, 2024). Despite operating with limited resources and unclear program guidelines (Sisimayi & Chikasha, 2016),

Zimbabwe's TVET programs are making notable achievements in students' lives and the country's economy. In addition, TVET practitioners in other countries, especially developing countries, who are interested in adapting educational programs to adverse economic conditions can learn from Zimbabwe's experiences. Furthermore, the ongoing initiatives to improve the quality of TVET programs in Zimbabwe have positive implications for researchers and practitioners. This chapter will focus on the benefits, challenges, opportunities, and suggestions for TVET programs in Zimbabwean high school, as perceived by program implementors.

Literature Review

Technical and vocational education focuses on practical, hands-on skills that align with market demands, and it is increasingly becoming a transformative force in Africa's educational landscape (Zwane, 2024). While it has always been a challenge for TVET programs to keep up with the demands of the twenty-first-century workplace (Winter, 2014), the programs have added benefits of developing transferable skills. Several African countries, such as South Africa, Nigeria, and Kenya, have reported better job prospects for TVET graduates due to their job-ready skills (African Development Bank, 2023). Furthermore, the International Labor Organization [ILO] (2021) noted that TVET graduates enjoy better earnings compared to their peers who have not undergone vocational training. Globally, employers point to difficulties in hiring skilled workers due to mismatch of skills between what new employees have and what employers are looking for (Englehart & Mupinga, 2020). TVET programs seem to be central in addressing skill shortages in the workplace.

While TVET programs provide valuable skills for immediate employment and are therefore essential to economic development, not everyone views them positively. Criticism leveled against TVET, according to Morley (2024), includes channeling of students into specific trades early in high school, thereby limiting their career choices, a big disadvantage in fast-changing workplaces that are characterized by technological advancements. Besides, assigning students to available TVET programs ignores students' occupational interests, values, ability, and personality traits, factors that should be considered when choosing careers. In addition, the programs often lack funding and institutional support. This leads to the use of outdated training equipment and materials (Shereni, 2019), insufficient and poorly prepared teaching staff, and a lack of access to the latest technologies. This situation affects the quality of TVET graduates. Furthermore, there is stigma associated with TVET programs. Though the negative attitude is changing, the programs continue to be unpopular and viewed as geared for "academically less able" students (Mhlanga et al., 2021) and as a second-tier option for students not college bound (Trafficanda, 2016). In Zimbabwe, negative attitudes are a carry-over from the now defunct F2 secondary schools. The criticism tends to present recruitment challenges for TVET programs. The question to ask is, how can TVET programs be implemented so they can be reputable and attract all students?

The implementation of vocational education programs as part of the school system has generated a lot of debate. At the center of the debate are the following critical questions: Should TVET programs be part of general education and what should the ideal curriculum be? What is the best structure for TVET programs in developed or developing countries? Who should be enrolled in TVET programs? What level of skill proficiency should TVET programs develop in high school students? The debate between philosopher John Dewey and Education Commissioner of Massachusetts, David Snedden, highlights these challenging issues. Dewey argued for "education in service of democracy" while Snedden advocated for social efficiency (Gordon & Schultz, 2020). According to Bates (2018), vocational education was the backbone of Snedden's social efficiency, and he regarded vocational education as the ideal alternative to a traditional liberal arts curriculum.

Dewey, on the other hand, was opposed to technical efficiency and "advocated vocational exploration as a means to acquire practical knowledge, apply academic content and examine occupational and societal values" (Gordon & Schultz, 2020, p.42). He felt that vocational education would place the cost of technical training on the public and ultimately have training that might be rendered obsolete due to technological changes (Bates, 2018). It is important to understand that Dewey's view of vocational education, as is the case in Zimbabwe's high school TVET programs, would not produce graduates who are very skilled in their trades.

From the debate comes two models for offering TVET: (1) vocational education programs that are simply providing vocational awareness of the trades, and (2) training in conditions that mimic the industry such that trainees are proficient in their skills upon graduation. It's always an advantage for students to learn in an environment closely modeled after the workplace. The students will acquire skills related to their trade as well as employability skills and workplace safety awareness. Employability skills are not job specific, but they are skills that cut horizontally across all industries and vertically across all jobs, from entry level to chief executive officer (Cotton, 1993).

Throughout the world, these two models for vocational education have been part of education systems. Therefore, as educators decide on the nature of TVET programs, the purpose needs to be considered. For most African countries who regard TVET programs as a panacea to unemployment, it behooves them to identify the causes of unemployment and decide on what is the best model for TVET programs.

Unemployment is a term that refers to employable individuals looking for a job but are unable to find a job (Corporate Finance Institute [CFI], 2024). It also refers to people in the workforce who may be working but not in an appropriate job. There are different forms of unemployment, and one needs to understand the type of unemployment to identify an appropriate solution to the problem. According to the CFI (2024), there are four types of unemployment: (1) demand deficient, (2) frictional, (3) structural, and (4) voluntary unemployment. Demand-deficient unemployment happens when companies experience a reduction in demand for their products or services (CFI, 2024), and the organization is forced to scale back production or services. Even though the workforce has the required skills, employees are furloughed due to a shortage of work. Frictional unemployment refers to workers who are in-between jobs and are trying to find a job most suitable for their skills. Kenton (2023) describes structural unemployment as a situation where workers lack the requisite job skills or live in a geographic area where their specific jobs are not available. Simply put, this happens when there is a serious mismatch between the skills companies' need and skills the workers possess. Voluntary unemployment happens when a worker decides to leave a job because it is no longer financially compelling, for example, a worker's salary is less than his or her cost of living. Unfortunately, TVET has become the default solution to unemployment in many countries even when a country's economic situation suggests otherwise.

As can be observed from the various types of unemployment, TVET programs may not address every form of unemployment. For example, TVET cannot address demand deficient unemployment since this is caused by market conditions. In addition, TVET programs have no role in voluntary unemployment. However, TVET programs can address structural and frictional unemployment. For example, in Zimbabwe, unemployment is mostly due to non-availability of job opportunities (ILO, 2019). Likewise, some Zimbabwean workers do not possess the skills that the workplaces need (ILO, 2021) and TVET programs provide them with opportunities to learn new skills so that they can perform a different job or role in the workforce (reskilling).

Just like many developing countries, Zimbabwe has been experiencing a huge problem of unemployed high school leavers, the majority of whom have pursued a largely academic education through the formal school years (ILO, 2019). Due to factors like economic decline, company closures, and limited investment, the remaining industries in Zimbabwe are failing to absorb the youth (Gukurume, 2018), leading to high unemployment rates among recent graduates. According to data from Statista and Trading Economics, Zimbabwe‘s unemployment rate in 2023 was estimated to be around 8.76% (Statista, 2024) and that for youth was 14.29% (Federal Reserve Economic Data [FRED], 2024). Many experts have pointed to the need for technical education to provide the graduates with technical skills needed in industry. When many industries are closed or will close soon in Zimbabwe, the problem is not so much a shortage of workers who have the skills needed by industries but not having jobs (Akileswaran et al., 2023).

In summarizing the consequences of unskilled workforce on economic development, Thambyrajah (2024, para 6) said:

> A workforce that lacks the necessary skills is a drag on the economy. Productivity declines, innovation stalls, and businesses face higher costs as they pour resources into recruitment and training just to stay afloat. It's not just a business problem; it's a community problem. The consequences of this gap are far-reaching. We're seeing decreased productivity, lower economic output, and higher costs for businesses as they struggle to train and recruit talent.

Geographic Location, Population, Industries, and Education System in Zimbabwe

Zimbabwe is a landlocked country that shares borders with South Africa to the south, Mozambique to the east, Botswana to the west, and Zambia to the north. The country has a population of 16.6 million people, per

2024 census (Worldometer, 2024) and has a diversity of races, cultures, and religions. Approximately 98% of the population consists of Africans, while the remaining are Asians, people of mixed race, and Europeans. Zimbabwe's major industries include agriculture, mining, and production of steel, clothing and footwear, chemicals, fertilizer, foodstuffs, beverages, transportation equipment, and wooden artifacts (World Atlas, 2025). The major export earnings come from agricultural produce such as tobacco, cotton, and sugar (International Trade Administration [ITA], 2024a) and from minerals such as gold, platinum group metals, diamonds, nickel, and chrome (ITA, 2024b). The mining sector is the primary driver of Zimbabwe's export economy.

Education System in Zimbabwe

Zimbabwe's school system uses a 2-7-4-2-3+ structure, beginning with 2 years of early childhood development, 7 years of primary school, 4 years of lower secondary cycle, also known as Ordinary Level (O-Level: forms 1–4 or grades 8–11), 2 years of upper secondary, also known as Advanced Level (A-Level: forms 5–6 or grades 12–13), and 3 years or more of tertiary education at the undergraduate level, depending on the degree program. The secondary school is divided into three 2-year levels.

As per Ministry of Education, (Policy Circular Number P77 of 2006), all secondary schools in Zimbabwe are mandated to offer technical and vocational subjects (Zendera, 2013). The policy outlines a "Two Pathway Education" system in which students are divided into academic and vocational pathways based on their abilities and interests (Mhlanga et al., 2021). It requires a student to pursue one of the following options:

> Option 1: General/Academic core subjects including Computer Studies where a student may choose to pursue Business/Commercial as a major (at least two subjects) and one subject from Technical/Vocational as an elective.
>
> Option 2: General/academic core subjects including Computer Studies where a student pursues at least 2 Technical/Vocational subjects as major and one subject from Business/Commerce as an elective.

While this policy offered opportunities for offering TVET courses to more students, Pedzisai et al. (2014) found the implementation to be flawed due to lack of human and material resources, congested school timetables, lack of knowledge on ways of channeling pupils into the two options, and the removal of Form 2 national examination (Zimbabwe Junior Certificate examination) whose results would have been used to determine appropriate options moving forward into Form 3.

Examples of TVET courses offered in secondary schools include Metal Technology and Design, Wood Technology & Design, Home Economics & Design, Food Technology & Design, and Building Technology & Design. Schools do not offer the same TVET subjects.

Teachers for TVET programs are primarily trained in teachers' colleges such as Belvedere Technical Teachers College, and polytechnics such as Gweru Technical College. The colleges focus on both theoretical knowledge and practical skills.

Research Questions

With reference to high school TVET programs in Zimbabwe, this chapter addresses the following questions:

1. What are the benefits of high school TVET programs?
2. What challenges do the TVET programs currently face?
3. What opportunities, if there are any, exist for the high school TVET programs?
4. What recommendations can be provided to improve the implementation of high school TVET programs in Zimbabwe?

Method

Data were collected from reviews of literature, an examination of TVET policy documents, and an open-ended survey of high school TVET instructors. A convenience sample of 12 (8 male and 4 female) high school TVET instructors in a professional development workshop at a tertiary institution in Zimbabwe responded to the survey. The 12 teachers had teaching experiences between 1 and 20 years and represented three technical subject areas: Building Design and Technology (8), Food Technology and Design (3), and Wood Technology and Design (1).

Benefits and Positive Gains of TVET Programs

TVET programs provide numerous benefits to students and workplaces. The TVET teachers who responded to the survey noted the following benefits: (1) provision of in-demand industry-related technical and workplace skills (80%); (2) increase job opportunities for students (60%); (3) creation of jobs and reduction of unemployment (50%) due to development of entrepreneurial skills. Additional benefits for TVET programs include exposure of students to career opportunities, potential to earn higher income (ILO, 2021; Morley, 2024), improved graduation rates for TVET students (Ecton & Dougherty, 2024), alignment of training to industry (Morley, 2024), empowers individuals from disadvantaged backgrounds, and provides entrepreneurial opportunities (Top Africa News, 2023).

In-demand Technical & Workplace Skills

TVET programs provide students with technical skills and hands-on training which make the graduates more employable (Vijayakumar, 2023). Most industries in Zimbabwe desire employees who possess technical skills. For instance, employers desire digital literacy and information and communication technology skills, such as proficiency in software tools, programming languages, and data analysis, and these are now part of TVET programs. In engineering and manufacturing sectors, employers seek design, prototyping, and production skills, and these have been included in applicable TVET programs. Similarly, in Zimbabwe's construction industries, skills in bricklaying, plastering, roof tiling, plumbing, and electrical installations are in demand and TVET programs have adjusted accordingly.

In addition to equipping students with skills in demand, most tertiary formal TVET programs in Zimbabwe offer hands-on training via internships, co-op programs, and apprenticeships that provide practical, and real-world experience. Furthermore, there are on-site labs, maker spaces, and innovation hubs for immersive learning and fostering technical skill building.

Other efforts in building a skilled workforce in Zimbabwe include key education reforms and initiatives that support the country's industrialization and modernization goals. One of the initiatives is the implementation of The Nziramasanga Presidential Commission of Enquiry (Nziramasanga, 1999). This initiative informed curriculum reforms across education levels and resulted in the introduction of competency-based education and training for TVET and curriculum reviews for teacher education.

Depending on the purpose of the technical program and the way they are set up, most students graduating from high school TVET programs in Zimbabwe only possess entry-level skills in their respective trades (Mupinga et al., 2005). The students acquire basic skills to function in their respective trade areas and would only require little training compared to students who do not receive basic trade skills. In addition to technical skills, TVET programs provide graduates with employability (transferable) skills such as communication, decision-making, teamwork, multicultural sensitivity and awareness, planning, organizing and management, and leadership (SkillsUSA, 2018; Vijayakumar, 2023). These non-technical skills are unique and transferable from one workplace to another. Ironically, these are the skills the employers are looking for in new employees.

Exposure to Careers & Increased Job Opportunities

TVET programs provide students with career opportunities which include career awareness, exploration, and development. This is an expectation for all technical education programs, however, with limited career choices in each industry, this becomes difficult. High school students in Zimbabwe have limited career choices and study whatever technical subject is available at their school. They are still exposed to various career pathways within the "chosen" field and expected to select and pursue one of the pathways. In contrast, in the United States, high school students go to local technical high schools (career centers or joint vocational schools) that offer various TVET programs from which students choose subjects that align with their interests. Unfortunately, for students in developing countries, the career center option is not feasible due to logistics issues and transportation costs. This model would be a challenge for Zimbabwe under current economic conditions.

Reduce Unemployment

Many employers have lamented over the widening gap between the skills their workforce possesses, and the skills demanded by industries (Englehart & Mupinga, 2020; Rao, 2024). This trend has contributed to unemployment and underemployment rates (Morley, 2024), especially among youth in developing countries (Rao, 2024). According to Morley

(2024), TVET can reduce unemployment by providing tailored skillsets that industries are actively seeking. Furthermore, he noted that countries that heavily invest in vocational education, such as Germany and Switzerland, have low youth unemployment rates.

Due to rapid technological advancements and brain drain (ILO, 2019), many industries in Zimbabwe face critical shortages of skilled workers. In 2018, it was estimated that applied arts and humanities industries had the least skill deficit of 18% while the rest of manufacturing, engineering and technology, medical and health sciences, and natural and applied sciences fields had more than 90% skill deficits (Manyukwe, 2024). The TVET programs in Zimbabwe will play a key role in providing workers with job-specific skills that align with industrial demands. Ultimately, providing technical training in school settings will reduce training costs for companies and increase the quality of the workforce. However, to reduce unemployment in Zimbabwe, there is need to provide the technical skills to students as well as improving the country's economic conditions, otherwise, the brain drain is not likely to stop.

Pathways to Employment for Individuals from Disadvantaged Backgrounds

In addition, TVET programs provide pathways to employment for individuals from disadvantaged backgrounds who might not have access to higher education (St. Claire, 2023) and might be unemployed. Zimbabwe's Ministry of Primary and Secondary Education (MoPSE) has implemented a new curriculum for primary and secondary education designed to facilitate the development of twenty-first-century competencies while ensuring inclusivity in the pursuit of development and empowerment. With most households struggling financially in Zimbabwe (World Bank Group, 2024), students from lower socio-economic backgrounds would be unable to continue their education beyond high school, and TVET programs provide them with a chance to be employed. There are several examples of students who did not pass the required five "O" level subjects but went on to use their TVET skills in the informal sector industries. In addition, short technical courses are offered to empower socio-economically disadvantaged men and women who do not possess any formal education (Murima, 2023).

Entrepreneurial Opportunities

TVET programs offer entrepreneurial skills that graduates are likely to use to start their own companies and create jobs. The Siyaso-Magaba home industries in Mbare, Harare, is part of the informal sector of the economy and provides employment opportunities for people who may not have access to formal employment. This industrial area is filled with entrepreneurs, most of whom are TVET graduates, running informal businesses of various trades, ranging from welding, tinkering, motor mechanics to selling of building materials and catering (Zinyemba & Changamire, 2014). Many of these entrepreneurs have technical backgrounds or have acquired technical skills through on-the-job training. At a time when most formal industries in Zimbabwe have closed due to harsh economic conditions, the informal businesses are thriving. Industries of this nature have sprung up in other parts of Harare and other cities in Zimbabwe, for example, Glen View Area 8 Home Industries Complex, and Mabvuku Home Industries in Harare. Similar forms of informal sector industries can be found in other African countries, for example, the "Jua Kali" in Kenya (Wiens, n.d.), Kasi economy in South Africa (CNN, 2023), and "Jua-kali" in Uganda (ChimpReports, 2023).

Several success stories of TVET graduates who have started their own ventures exist in Zimbabwe. For example, private TVET colleges registered with the Ministry of Higher and Tertiary Education, small-scale farming enterprises, and companies in cosmetology, construction, and catering. Furthermore, three tertiary TVET graduates used their computer science skills to establish a web development company that customizes websites and mobile apps (The Zimbabwe Herald, 2023). The company has quickly gained recognition in the industry. Therefore, the impact of technical education courses is evident in Zimbabwe's entrepreneurship sector, where the number of entrepreneurs is vastly growing in the industry (Murima, 2023).

Potential for Earning a Salary

Studies have shown that graduates from TVET programs often earn higher salaries than those who were in general education programs (Ecton & Dougherty, 2024). For instance, in low-income countries, TVET graduates experience a 15% higher increase in earnings than those who only receive a general education (Morley, 2024). It may be difficult to attribute high salaries to TVET programs in countries with high unemployment rates such as Zimbabwe, but it is evident that there are economic benefits to such programs. Even though there was no available data to make direct salary comparison between TVET graduates and students who attended the academic stream in Zimbabwe, there is great demand for graduates from technical training programs (Oak Foundation, 2024). For instance, Harare Polytechnic offers several short-term technical course programs to youth and adults without high school graduation qualifications. The graduates from these programs have the potential of earning a living from utilizing the acquired skills.

Training Aligned with Industry
The alignment of TVET programs with industries guarantees, to some extent, relevant technical training. In Zimbabwe, education programs collaborate with companies to align training with specific job market demands, especially at the post-secondary school level. Furthermore, the TVET programs use Higher Education Examination Council (HEXCO) curricula, which are designed in close consultation with industry making all National Foundation Certificates (NFC) courses offered in secondary schools relevant to industries (Zendera, 2013). An additional component that could add relevance to TVET programs is the use of advisory boards whose purpose is to provide input on industry's needs, expert advice, guidance, and explore new opportunities. This component should be a requirement for all technical programs to stay relevant.

Overall, TVET programs have numerous benefits to students in Zimbabwe, especially those who find hands-on education appealing. They provide basic technical and transferable workplace skills, expose students to careers within a given trade area, increase graduation rates and job opportunities, with the potential for graduates to earn high salaries. Similarly, industries benefit by saving money and time on training. In addition, productivity is likely to go up because employees will have skills that align with the needs of the industry. Furthermore, entrepreneurial skills will enable graduates to start their own businesses, resulting in creation of jobs and reduction of unemployment.

Challenges for Technical Education in Zimbabwe

TVET programs are not immune to challenges and in Zimbabwe, these include negative attitudes toward TVET (Sisimayi & Chikasha, 2016), poor funding, lack of infrastructure (Shereni, 2019), outdated training materials, and irrelevant curriculum (Murinda, 2014). In a study by

Mhlanga et al. (2021), TVET teachers in Zimbabwe cited additional challenges: (1) TVET teachers lack expertise to teach their programs; (2) inadequate training materials and equipment (textbooks, specialist rooms/labs/workshops); (3) insufficient training time, and (4) lack of opportunities for TVET graduates to gain industrial experiences. These challenges mirror those noted by the TVET instructors who responded to this study survey.

Teachers Lack Technical Expertise

The lack of technical expertise among TVET teachers is a concern that affects the quality of graduates the programs produce in Zimbabwe. The majority of the current TVET teachers were trained in teacher education programs that focused on pedagogical skills and very little on technical skills, resulting in producing teachers who have good instructional methods but were weak in technical skills. Zimbabwe now requires technical teachers to have a technical qualification, such as a National Diploma in a respective trade area and then a diploma in teaching. This is an effort to correct the technical skill deficiency among the TVET teachers. As the country focuses on providing infrastructure for implementing TVET programs, the quality of the teachers in charge of implementing the programs should not be ignored. Setting up and running TVET programs is expensive considering the high capital costs of equipment, expensive consumables, and at times, accreditation requirements (Reed, 2015) therefore, entrusting implementation of programs to unqualified teachers would be counterproductive.

Inadequate Funding

When it comes to funding, TVET programs require a large capital outlay to purchase equipment, tools, and training materials. Many of the schools in Zimbabwe are owned by the government and based on the current harsh economic conditions in the country (UNESCO, 2024), it's unlikely there would be much funding for TVET programs. The same economic situation is also likely to affect private schools. Furthermore, TVET programs require specialized training equipment and workshops, and all these rely on adequate funding. The funding challenges are even more pronounced in Zimbabwe's rural schools because they are government owned but funded partly by local communities that often face financial difficulties due to limited resources and the high cost of maintaining infrastructure (Ndongwe & Kangai, 2023).

Lack of Training Materials and Equipment

The lack of training equipment and materials is related to poor funding for the TVET programs. Some Zimbabwe TVET programs such as Wood Technology and Design and Metal Technology and Design require expensive equipment for training, and without adequate funding to purchase and or maintain equipment, the laboratories/workshops are poorly equipped. One teacher indicated that many TVET programs have "workshops and equipment in need of repair" due to neglect. Unfortunately, with ongoing power shortages in the country, schools struggle to operate TVET machinery. Another teacher reported shortages of "training materials such as supplies for [their] workshops/laboratories, tools, textbooks, and have broken equipment that is in dire need of repair." Yet another teacher reported outdated infrastructure resulting from inadequate funding.

Insufficient Training Time

Typically, a high school day in Zimbabwe is about 8 hours (starting at 7:15 am and ending at 3:00 pm), with a break between morning and afternoon sessions. The day consists of around 6–7 class sessions covering core subjects like English, Mathematics, local language, Science (Biology, Chemistry, Physics), History, and Geography. Depending on the type of school and student focus, there are additional class sessions for practical subjects like Agriculture, Home Economics, or other technical skill courses (USAPGlobal, 2015). Students are often scheduled to study a technical subject for 4.5 hours (1.5 hours theory and 3 hours laboratory). Roughly, this equates to 45 hours per term (10 weeks) for the technical subject and for the labs, the time includes entry and exit procedures, approximately 30 min per day.

Lack of Opportunities for TVET Graduates to Gain Industrial Experience

On making the industrial experience mandatory for all high school students in Zimbabwe, The Secretary for MoPSE announced that a five-month industrial attachment, after completion of "O" level examinations, started January 2016 (The Sunday Mail, 2017). While the intent is significant, there are some logistical challenges when it comes to implementing the ministerial directive (Mutisi, 2024). Not only do teachers have to find sites that are willing to take their students, but the teachers also need to schedule and supervise the industrial visits. Currently, "university students are struggling to secure places for unpaid internship or attachment," noted one comment to the newspaper article on the Secretary's announcement. Lack of transportation to visit the students at the industrial sites becomes another challenge for the teachers. Opportunities for industrial visits may also be limited or non-existent for students in rural areas because most industries are in urban areas.

Negative Attitudes Toward TVET

Negative attitude toward TVET is not unique to Zimbabwe, as this has been experienced in other countries as well (Krupnick, 2017). Over the years, vocational education has been regarded as a field for the "academically less able" (Dougherty, 2016; Trafficanda, 2016), as such, it has been treated as "second-class" education. In Zimbabwe, most parents do not perceive technical education as a route to success in the same way as the academic subject route (Chiweshe et al., 2013), partly because these courses are not counted among the required courses a student must pass at "O" level. Throughout the world, vocational education programs have been able to reinvent themselves and remove some of the stigma. In the USA, for instance, the programs changed in name from vocational education to career and technical education to improve image issues. Lately, the public is warming up to career and technical education as they see economic value in the programs. In Africa, a youth employment and skills study conducted across six African countries showed that TVET enrolments are on the rise, but parents and prospective students still see this path as a "fallback option" (African Center for Economic Transformation, 2022). In contrast, a Social Market Foundation (2021) survey in UK, on the public's view of vocational education, found that most respondents

(85%) believe a vocational qualification is valuable for those seeking a "good job."

Recommendations/Opportunities for Improvement

High school TVET programs in Zimbabwe are operating under difficult conditions, namely, severe economic conditions, unqualified teachers, shortage of funding, lack of training facilities, and outdated infrastructure. However, on the positive side, the programs continue to champion their mission despite the meager resources they must operate with. The program implementors should be commended for not sitting back and waiting for the situation to improve; they are putting into practice the innovative skills they are developing in TVET students.

In the Mhlanga et al. (2021) study, TVET teachers suggested the following improvements for implementing programs: (1) training of more technical teachers (to meet the current shortages and providing incentives for TVET teachers in rural areas); (2) improving infrastructure (such as access to water and electricity) in schools to enable them to offer TVET programs; (3) making TVET compulsory for all students; and (4) encouraging greater collaboration between schools and with industry. TVET teachers who responded to this study's survey reported similar suggestions and added two more ideas: (1) improving the relevance of technical subjects and (2) increasing funding for TVET programs.

Train More Technical Teachers

The fact that Zimbabwe has expanded the training of technical teachers to technical colleges, instead of just teachers' college, is an effort to increase the number of trained teachers. Regarding the rural schools, one teacher noted, "The Ministry of Primary and Secondary Education needs to recognize the critical shortage of TVET teachers in the rural areas" and suggested "providing incentives [extra remuneration] for teachers who work there." However, the idea of extra remuneration for rural TVET teachers will not be feasible under the prevailing harsh economic situation in the country.

An opportunity for Zimbabwe to improve the supply of TVET teachers is to utilize retired people with expertise in relevant industries to teach trade skills. Such individuals would be assigned to work with regular TVET teachers and focus on teaching trade skills. These individuals could be licensed through 'alternative licensure' programs (Teachers of Tomorrow, 2024). These programs are not new, as is the case in the United Sates. They offer a non-traditional pathway to becoming a certified teacher for individuals who have a minimum basic academic education (e.g., high school diploma) and technical expertise (e.g., five years of full-time work experience or equivalent) in a given trade area. In general, these programs allow individuals to start teaching under a provisional teaching license while completing a structured licensure program.

Improving Funding Levels

Adequate funding is critical to the success of TVET programs. However, given the current economic challenges in the country, improvement of financial support is not likely to happen soon. The lack of funding should not be a hinderance to training students. TVET programs can collaborate with the community and share resources. Schools can offer their services to communities in exchange for expertise, tools, or materials that they do not have. For instance, agriculture teachers and students can utilize a farmer's piece of land for training purposes. This is a win-win situation for the students and farmers, in that the farmer benefits from the students' labor and students gain skills. In construction programs, the students can acquire skills by building houses for community members using members' building materials. The same would apply to other technical programs.

Another suggestion is to operate the TVET program as a profitable business, for example, an agricultural program selling produce or livestock to the community and use the profits to fund the program. The business model has added benefits in developing entrepreneurial skills as the students are responsible for product ideas, development, and marketing. Funds generated from the program activities can also be used to improve TVET infrastructure.

Improving Relevance and Attitudes Toward TVET

The issue of negative attitudes toward TVET programs affects recruitment of both students and teachers and needs to be addressed urgently. Suggestions to improve the perception toward TVET programs include (1) giving the technical subjects an equal weighting to the academic subjects, for example, as one teacher said: "counting [TVET subject] as one of the five subject passes at 'O' level," and (2) showing relevance of the programs through advisory committees and business-education partnerships.

Collaboration Between Schools and Industries

Throughout the world, employers are facing skill gaps, that is, skills the employees possess are not the skills that the employers are looking for. This presents an opportunity for strengthening collaboration between education and industries. While these collaborations take various forms, common strategies include advisory committees, internships, and business-education partnerships. The use of advisory committees (made up of representatives from industry) is another strategy to ensure TVET programs develop the right workplace skills. Generally, advisory committees assist TVET programs by suggesting authentic learning experiences for students through mentoring, job placement, guest speaking engagements, and donations. Advisory committees are not there to give directives to schools but to suggest what skills the TVET programs should focus on.

Business-education partnerships add relevance to TVET programs by having technical programs partner with a local industry for the purpose of sharing research and development ideas, industrial attachment for the instructors, in-service training of employees at the partner TVET institution (Froese-Germain & Moll, 1997). This strategy is a win-win for industry, educators, and students. As noted earlier, these partnerships are an excellent strategy to provide workplace skills to prospective employees and opportunities to share resources such as equipment and other non-cash resources. Employers will have direct input in the student training and will have confidence in knowing that graduates of TVET programs have the skills employers require.

It should be noted that TVET programs serve the industry, and therefore, they are not solely the solution to unemployment issues. However, as Akileswaran et al. (2023) noted, it is important to provide the youth with skills in-demand but, doing so without creating the jobs will not reduce unemployment issues.

Despite the many challenges faced by Zimbabwe's TVET programs, they are beneficial to students and the economy. The benefits include, but not limited to, provision of an alternative to purely academic education, development of the technical and workplace skills needed in the workplace, exposure to various career pathways, and reduction of unemployment. In essence, TVET programs play a role in inclusive and sustainable economic development (ILO, 2021).

Overall, a lot needs to happen to improve the situation of TVET programs in Zimbabwe, starting from changing the public's mindset, crafting realistic education policies, and providing adequate support to the programs. In the case of Zimbabwe, Morley (2024) succinctly describes what needs to happen:

> There must be a fundamental shift in societal attitudes towards vocational education to overcome these challenges. Governments, employers, and educational institutions must work together to raise the profile of TVET as a high-quality, future-proof option. Better funding models are essential to ensure that TVET institutions can offer up-to-date, industry-relevant training. At the same time, stronger partnerships with industries will help align curricula with the latest market needs... Only with these changes can TVET reach its full economic potential and solidify its role as a key driver of workforce development in a rapidly changing global economy (para #13).

References

African Center for Economic Transformation. (2022, October 30). *The paradox of technical and vocational education in Africa*. https://acetforafrica.org/research-and-analysis/insights-ideas/articles/the-paradox-of-technical-and-vocational-education-in-africa/.

African Development Bank. (2023). *Technical education transforming rural towns and livelihoods in Kenya*. African Development Bank Group.

Akileswaran, K., Mazumdar, J., & Perez Albertos, A. (2023). Youth unemployment in the developing world is a jobs problem. *Stanford Social Innovation Review*. https://doi.org/10.48558/C6V0-YB62

Bates, C. (2018, Fall). Our divided education system: Vocational training—to the near-exclusion of all else—now sometimes begins in ninth grade. It makes a kind of sense, but ... really? *Democracy: A Journal of Ideas, Fall 2018*(50). https://democracyjournal.org/magazine/50/our-divided-education-system/#:~:text=Snedden%20argued%20for%20education%20in,education%20in%20service%20of%20democracy

California Workforce Development Board. (2025). *What is workforce development?* https://cwdb.ca.gov/faqs/#:~:text=Workforce%20development%20is%20a%20term,industry%2Drelevant%20certification%20or%20licensure

ChimpReports. (2023, March 4). *Ministry of Gender Issues Business Toolkits to Jua-kali Groups*. https://chimpreports.com/ministry-of-gender-issues-business-toolkits-to-jua-kali-groups/

Chiweshe, M., Edziwa, E., Cryton, J., & Chakamba, J. (2013). The gloomy outlook of practical subjects in Zimbabwe: A case of history perpetuated? *International Journal of Asian Social Science, 3*(4), 890–898.

CNN. (2023, Jan 9). *How informal business thrives in South Africa*. https://www.cnn.com/videos/business/2023/01/09/marketplace-africa-kasi-informal-economy-spc.cnn

Corporate Finance Institute [CFI]. (2024). *Unemployment*. https://corporatefinanceinstitute.com/resources/economics/unemployment/

Cotton, K. (1993). *Developing employability skills*. School Improvement Research Series. https://educationnorthwest.org/sites/default/files/DevelopingEmployabilitySkills.pdf

Dougherty, S. M. (2016). *Career and technical education in high school: Does it improve student outcomes?* Thomas B. Fordham Institute.

Ecton, W. G., & Dougherty, S. M. (2024). Heterogeneity in high school career and technical education outcomes. *Educational Evaluation and Policy Analysis, 45*(1). https://doi.org/10.3102/01623737221103842

Englehart, M., & Mupinga, D. M. (2020). Essential employability skills as perceived by employers: Implications for technical and vocational education and training. *International Journal of Vocational Education and Training, 25*(1), 29–42.

Federal Reserve Economic Data [FRED]. (2024, July 2). *Youth unemployment rate for Zimbabwe*. https://fred.stlouisfed.org/series/SLUEM1524ZSZWE

Froese-Germain, B., & Moll, M. (1997). *Canada: Business-education partnerships a troubling trend*. Education Monitor. https://www.corpwatch.org/article/canada-business-education-partnerships-troubling-trend

Global People Strategist. (2024, July 11). *Challenges for employment in developing countries*. https://globalpeoplestrategist.com/challenges-for-employment-in-developing-countries/

Gordon, H. R. D., & Schultz, D. (2020). *The history and growth of career and technical education in America* (5th ed.). Waveland Press.

Gukurume, S. (2018). *Navigating precarious livelihoods: Youth in the SME sector in Zimbabwe*. Institute of Development Studies. https://doi.org/10.19088/1968-2018.176

International Labor Organization [ILO]. (2019). *State of skills-Zimbabwe*. https://researchrepository.ilo.org/esploro/outputs/995384591102676

International Labor Organization [ILO]. (2021). *Technical and vocational education and training in Africa: A review*. wcms_881406.pdf (ilo.org).

International Trade Administration [ITA]. (2024a). *Zimbabwe - Country commercial guide: Agricultural sectors*. https://www.trade.gov/country-commercial-guides/zimbabwe-agricultural-sectors

International Trade Administration [ITA]. (2024b). *Zimbabwe - Country commercial guide: Mining and minerals*. https://www.trade.gov/country-commercial-guides/zimbabwe-mining-and-minerals

Kennedy, U. E. (2020). Curbing youth unemployment for achieving sustainable development through global partnership in TVET. *International Journal of Vocational Education and Training, 25*(1), 19–27.

Kenton, W. (2023, January 06). *Structural unemployment: Definition, causes, and examples*. https://www.investopedia.com/terms/s/structuralunemployment.asp#:~:text=Structural%20unemployment%20occurs%20because%20workers%20lack%20the,companies%20need%20and%20what%20workers%20can%20offer

Krupnick, M. (2017). *After decades of pushing bachelor's degrees, U.S. needs more tradespeople. The Hechinger Report*. PBS News hour. https://www.pbs.org/newshour/education/decades-pushing-bachelors-degrees-u-s-needs-tradespeople

Malle, A. Y. (2016). Inclusiveness in the vocational education policy and legal frameworks of Kenya and Tanzania. *Journal of Education and Learning., 5*(4). https://doi.org/10.5539/jel.v5n4p53

Manyukwe, C. (2024, December 5). *Government to align scholarships, critical skills shortages*. https://www.universityworldnews.com/post.php?story=20241204041725265#:~:text=The%20engineering%20and%20technology%20skills%20cluster%20showed,in%20the%20sector%20are%20not%20formally%20skilled

Mhlanga, M., Tenha, J., & Ndlovu, F. (2021). Technical and vocational education and training policy in Zimbabwean secondary schools: Teachers' views. *International Journal of Humanities Social Sciences and Education, 8*(7), 111–118. https://doi.org/10.20431/2349-0381.0807013

Morley, S. (2024, Sept 24). *Skills, jobs, and prosperity: Unpacking the economic impact of vocational education*. World of TVET (Blog). https://www.worldoftvet.com/blog/economic-impact-of-vocational-education#:~:text=Challenges%20and%20Criticisms%20Despite%20its%20many%20benefits%2C,their%20flexibility%20in%20an%20ever%2Dchanging%20job%20market

Mugomba, A. T. (1980). *Independence without freedom: The political economy of colonial education in Southern Africa*. ABC: CLIO.

Mupinga, D. M., Burnett, M. F., & Redmann, D. H. (2005). Examining the purpose of technical education in Zimbabwe's high schools. *International Education Journal, 6*(1), 75–83.

Murima, S. (2023, November 5). *The power of short courses in empowering knowledge and skill proficiency*. The Sunday News, November

5, 2023. https://www.sundaynews.co.zw/the-power-of-short-courses-in-empowering-knowledge-and-skills-proficiency/

Murinda, E. (2014). *Skills training and the employment situation of young people in Zimbabwe*. Presentation at Providing Africa's Youth with Skills and Training for Jobs conference. https://tvsd.adeanet.org/sites/default/files/documents/zimbabwe_country_report.pdf

Mutisi, J. K. (2024, October 1). *The state of Zimbabwe's education system and the need for industrial placements for unexperienced lecturers*. TechZim. https://www.techzim.co.zw/2024/10/zimbabwes-education-system-industrial-placements-lecturers/#:~:text=Zimbabwe%27s%20education%20system%20has%20long%20been%20regarded,producing%20some%20of%20the%20continent%27s%20best%2Deducated%20citizens

Mutumbwa, A. (2023). Vocationalizing the school curriculum for Blacks in pre-independent Zimbabwe: A weak or firm foundation for current voc-tech education. *International Journal of Humanities Social Sciences and Education, 10*(12), 60–62. https://doi.org/10.20431/2349-0381.1012006

Ndongwe, E., & Kangai, C. V. (2023). Challenges faced by Zimbabwe rural district council primary school heads in applying the legal framework that guides financial management in schools. *International Journal of Humanities Social Sciences and Education, 10*(11), 29–41.

Nziramasanga, C. T. (1999, August). *Zimbabwe: Report of the presidential commission of inquiry into education and training*. https://africasocialwork.net/wp-content/uploads/2022/10/Nziramasanga_Final.pdf

Oak Foundation. (2024). *Empowering students through technology education: An uncommon story*. https://oakfnd.org/empowering-students-through-technology-education-an-uncommon-story/

Olowe, M. O. (2024). Vocational and technical education: A functional education for sustainable development in a global turbulent era. *International Journal of Vocational and Technical Education Research, 10*(1), 1–11.

Pedzisai, C., Tsvere, M., & Nkhonde, M. (2014). The Zimbabwe two-pathway education curriculum: Insights into policy implementation challenges and opportunities. *International Journal of Advanced Research in Management and Social Sciences, 3*(5), 162–173.

Rao, S.A. (2024, Jul 8). *Empowering the future: Why developing countries need vocational guidance and technopreneurship for a self-sustainable TVET system*. LinkedIn: https://www.linkedin.com/pulse/empowering-future-why-developing-countries-need-tvet-huoye/

Reed, M. (2015, September 08). *Vocational programs cost more: A key point on the economics of higher education*. Inside Higher Education. https://www.insidehighered.com/blogs/confessions-community-college-dean/vocational-programs-cost-more

Shereni, N. (2019). The role of technical and vocational education and training (TVET) in restoring hospitality sector specific skills in Zimbabwe: A students' perspective. *Journal of Hospitality & Tourism Education, 32*(4), 1–9.

Sisimayi, G. W., & Chikasha, D. A. S. (2016). The politics of vocational and technical education curriculum: An analysis of the implementation of vocational and technical education in Zimbabwe. *Journal of Educational Research, 1*(5), 1–11.

SkillsUSA. (2018). *Overview of SkillsUSA*. https://www.skillsusa.org/

Social Market Foundation. (2021, April). *Not just other people's children: What the public thinks about vocational education*. https://www.smf.co.uk/wp-content/uploads/2021/04/Public-attitudes-to-vocational-education-April-2021.pdf

St. Claire, R. (2023). *Bridging the gap: The transformative power of TVET colleges and applied skills training*. https://publicsectorleaders.co.za/tvet-colleges-transforming-the-employment-landscape/

Statista. (2024). *Zimbabwe: Unemployment rate from 2004 to 2023*. https://www.statista.com/statistics/809092/unemployment-rate-in-zimbabwe/

Teachers of Tomorrow. (2024, September 4). *Alternative Teacher Program Guide in 2025*. https://www.teachersoftomorrow.org/blog/insights/alternative-teacher-program-guide/

Thambyrajah T. (2024, September 30). *Bridging the skills gap: The role of TVET in economic development*. https://www.linkedin.com/pulse/bridging-skills-gap-role-tvet-economic-development-thambyrajah-t-8cdwc/

The Sunday Mail. (2017, June 11). *O-Level industrial attachment begins*. https://www.sundaymail.co.zw/o-level-industrial-attachment-begins

The Zimbabwe Herald. (2023, Sept 8). *Ex NUST students establish a web firm*. https://www.herald.co.zw/ex-nust-students-establish-web-firm/

Top Africa News. (2023). *15 reasons why TVET education is important in today's economy*. https://www.topafricanews.com/2023/04/10/15-reasons-why-tvet-education-is-important-in-todays-economy/

Trafficanda, J. (2016, August 9). *Common challenges and new opportunities for career and technical education programs*. Work-Based Learning. https://coloradosucceeds.org/work-based-learning/common-challenges-opportunities-cte-programs/

UNESCO. (2015). *Recommendation concerning technical and vocational education and training (TVET)*. https://unevoc.unesco.org/home/TVETipedia+Glossary/show=term/term=Technical+and+vocational+education+and+training#start

UNESCO. (2024, January 24). *Zimbabwe: Education country brief*. https://www.iicba.unesco.org/en/zimbabwe

USAPGlobal. (2015). United States Student Achievers Program (USAP). https://usapglobal.org/zimbabwe/2015-usap-zimbabwe-application-form.pdf

Vijayakumar, B. (2023, August 30). *Advantages of technical education system*. https://www.linkedin.com/pulse/advantage-technical-education-system-vijayakumar-b/

Wiens, M. (n.d.). *Jua Kali—The informal Kenyan sector for "Git Er Done"*. https://migrationology.com/jua-kali-kenyan-informal-labor-sector/#comments

Winter, D. (2014, December 10). *How will education keep up with the demands of 21st century work?* https://www.linkedin.com/pulse/how-education-keep-up-demands-derek-winter/

World Atlas. (2025). *The biggest industries in Zimbabwe*. https://www.worldatlas.com/articles/the-biggest-industries-in-zimbabwe.html

World Bank Group. (2024). *The World Bank in Zimbabwe*. https://www.worldbank.org/en/country/zimbabwe/overview

Worldometer, (2024). *Zimbabwe Population*. https://www.worldometers.info/world-population/zimbabwe-population/

Zendera, E. (2013). *Technical and vocational education and training policy implementation in secondary schools*. Zimbabwe Open University.

Zimbabwe Government. (2006). *Ministry of Education, sports art and culture* (Directors policy circular number P 77 of 2006). Harare.

Zinyemba, A. Z., & Changamire, T. (2014). Challenges and survival strategies for women in small scale catering business: A case study of Siyaso Magaba, Mbare. *International Journal of Business, Economics and Management, 1*(11), 372–381.

Zwane, N. (2024). *Technical and vocational education and training (TVET) as a driver of employment, entrepreneurship and economic progress in Africa*. TVET Journal. https://tvetjournal.com/tvet-systems/technical-and-vocational-education-and-training-tvet-as-a-driver-of-employment-entrepreneurship-and-economic-progress-in-africa/

PART II

Conceptualizing VET Success Between Individual Needs and Societal Development

CHAPTER 11

Contemporary Purposes for Vocational Education

Stephen Billett

Contemporary Purposes of Vocational Education

Educational programs and processes should be intentional in their design and enactment as directed to achieving the purposes for which they have been established. This includes finding ways to opening pathways, fostering accessibility and progression for all kinds and classifications of students. Often, those purposes are established by those who sponsor or administer those systems. Whilst the statements of intent that direct the development and enactment of specific programs are often viewed in terms of hierarchies of aims, goals, and objectives (Marsh, 2004), there are also broader purposes that shape the overall provision of vocational education from which the need for and overall focus of those more specific statements of intent are derived. This chapter advances five such purposes that are seen as being salient for contemporary vocational education and training (VET). Whilst far longer lists of those purposes exist

S. Billett (✉)
Griffith University, Brisbane, QLD, Australia
e-mail: s.billett@griffith.edu.au

E. Smeplass and C. Zirkle (eds.), *Opportunity Structures in Vocational Education and Training*, Palgrave Studies in Technical and Vocational Education and Training,
https://doi.org/10.1007/978-3-032-04410-5_11

(Billett, 2011b; Fuller, 2015), based on a program of research over the past three decades, these five sets of purposes are identified as being those that address current imperatives for VET. They comprise, firstly, engaging with young people and their parents/caregivers and informing them impartially about VET and the occupations it serves, thereby informing potential students of opportunities that can be provided by this educational provision. This is an important educational purpose in an era when VET is seen as having lower status than higher education and the occupations it serves. Secondly, inclusion efforts and provision of opportunities can assist individuals identify their preferred occupations (i.e., their 'calling') (Dawson, 2002; Rehm, 1990). Advising them about and aligning their interests and capacities with those occupations is an important VET purpose given, for instance, the high dropout rates and low retention in key occupations. Thirdly, effective initial occupational preparation is an important purpose for employers, governments, communities, and students alike, although approaches to it differ across nation states for historical, cultural, institutional, and resource-related reasons. Now, many governments are requiring VET graduates of all kinds are to be 'job ready', including those that do not have extensive work experience or placement components within them. So, opportunities are required to secure these kinds of outcomes. Fourthly, accessible and effective provisions of continuing education and training (CET) are needed to support learning across working life, particularly when they are negotiating changing work roles or new occupational practices. Fifthly, being prepared to be innovative in and through work needs to be promoted through VET in an era of constant changes in requirements for occupational competence and workplace performance. Working age adults need to continually learn to secure their employability and advancement and often need to be provided with access to opportunities for realising those outcomes (Hoyrup et al., 2012; Lemmetty & Billet, 2023). Added here are opportunities to contribute to their workplaces transforming work procedures and practices premised on innovations, many of which can be driven by employees. It is these that are described and discussed in this chapter and in ways explicitly aligned with the focus of this volume that is on opening pathways, fostering accessibility and progression for individuals through VET.

Purposes of VET

From a program of research in both Australia and other countries over recent decades, it is possible to consider and delineate what should be the purposes for VET for the contemporary era (Billett, 2024). In the following sections, as foreshadowed, each of these five purposes is discussed through a commentary on why they are salient and their implications for organisation and implementation of VET provisions. Some key considerations are that without adequate interest and engagement by sufficient numbers and kinds of students and their parents, the ability of VET to achieve its intended goals and potential is rendered perilous (Wolf, 2016). Moreover, personal and institutional investments will be squandered if VET graduates fail to identify their future occupations (i.e., vocations) through their VET programs. Moreover, given the changing occupational and workplace requirements, initial occupational preparation through VET needs to secure adaptability along with the canonical occupational knowledge they are acquiring (Stevenson, 1991). This adaptability is necessary to respond to new workplace challenges and changes in work practices (i.e., new technologies, occupational and work requirements, ways of working with whom work is conducted) (Billett, 2006). Furthermore, CET provisions aligned with working age adults' needs are becoming essential for them to navigate changes to occupational requirements and workplace performance across lengthening working lives. Importantly, supporting innovations at work is also essential for both future employability and workplace viability (Hoyrup et al., 2012; Lemmetty & Billet, 2023), as workers' skilfulness is central to both those imperatives.

As can be seen, these purposes are not wholly focused on individuals' needs, but also address broader workplace, community, and societal imperatives. These include the viability of public and private sector enterprises, in ways that are more educationally appropriate than merely being responsive to the needs of key national stakeholders (e.g., government, industry bodies, professional associations, unions). Each of these five purposes is now described and discussed.

Purpose 1. Enhancing Engagement in VET as a Worthy Educational Sector

Across countries with developed and developing economies, the standing of VET is low, leading to quests for initiatives to enhance its status (Clement, 2014; UNESCO-UNEVOC, 2018a). These efforts are important because the relatively low standing of VET leads to structural problems in developing the occupational capacities required by communities, enterprises and realising national economic and social goals. In an era of high aspiration, VET is often seen as an option for those with no other choice (Parliament of Victoria, 2018; UNESCO-UNEVOC, 2018b). One of the consequences of this societal sentiment is that governments, employers, and educational systems view VET as low priority and scant and, sometimes, ad hoc funding. Added here are concerns about VET not attracting sufficient kinds, calibres, and quantum of young people. This can fall short of the required numbers being prepared for the kinds of occupations that nation states, their communities, and individuals need (Deissinger, 2002; Wolf, 2016). Consequently, engaging young people and their parents/caregivers, and informing them impartially about VET and the occupations it serves, offers a means to redress this situation. This includes finding ways of engaging them in structured activities in workplaces, and considering how VET institutions can act to make attractive and purposive their courses and provisions (Billett et al., 2022a). Hence, engagement has become an important priority for VET.

From projects undertaken in Australia (Billett et al., 2020) and through engagement with colleagues in European, Asian, African, and American nation states (UNESCO-UNEVOC, 2018b), some strategies to achieve these outcomes have been identified (Billett et al., 2022a). In all, efforts to enhance engagement by young people, championing, and directly supporting VET and the occupations it serves need to arise through actions at national, state/territory, and local levels (Billett et al., 2020, 2022a). Nationally, this can comprise governments' championing VET and actions to elevate its attractiveness to young people, their parents/caregivers, and in their communities. Similarly, industry and professional bodies' taking responsibility for and championing the occupations they represent in an informed and active way is found to be helpful (Billett et al., 2024), particularly at the local level where advice is provided to young people and their parents/caregivers about specific occupational pathways and their potential qualities and benefits. Here, there is a role for local social partnerships and public and private enterprises to engage productively in those processes (Billett et al., 2024; Billett & Seddon, 2004). At this local level, there is a key role for schools in providing experiences in VET institutions and/or work settings and providing impartial advice to assist young people and familiars make informed choices (e.g., visits to VET institutes, advisory processes). Partnerships amongst education, government, and local enterprises that provide engaging experiences and guidance about career choices are potentially helpful for young people and in ways that bring insights and contributions to their decision-making that cannot be found within the 'school fence' (Billett et al., 2007). Added here is the importance of engagement with experiences in work settings and support for young people and their parents/caregivers. These include having 'walk throughs' in local workplaces, opportunities for work experiences in significant local workplaces and open days so that young people can come to experience and understand what occurs in those settings, and the kinds of work being undertaken and occupational practices that otherwise they would be unaware (Billett et al., 2022b). Much of the ways of achieving those purposes are likely required to be generated at the local level, as it is through engagement with social partners, parents, familiars, and enterprises locally that these processes need to be enacted.

Engaging Young People

A case study of such an initiative can be found in the Health Education to Employment Pathway (HEEP) project conducted to inform young people, particularly young Indigenous people, about allied health roles as employment options and identify engaging provisions of preparing and retaining them within the healthcare workforce (Billett et al., 2022b). This study comprised interviewing young people in metropolitan and regional communities and identified that forms of engagement were crucial for achieving these outcomes, and these need to occur at the local level through appropriate educational and community actions. First, it was the importance of having Engaging work conditions that had qualities attractive to the young people (e.g., stable work and employment, opportunities, good pay, making friends). They also reported wanting to participate in Engaging work: i.e., inspiring, caring for others, and comprising compassionate work. Often, knowledge of this kind was provided through family members and other familiars about the importance of helping the community, but also with learning new skills and having the capacities to work in workforces that were diverse in their cultural and ethnic composition. They also privileged Engaging support including opportunities for experiencing different kinds of work, having 'tasters' through work experience programs, and being supported through mentoring and engaged in cadetships. The latter was particularly important for young people from financial impoverished families who could not afford the cost of their children's study (e.g., transport, food, accommodation). Importantly, these young informants wanted an initial occupational preparation characterised as being an Engaging education. That is, the educational experiences provided applicable outcomes and understanding and certification (Billett et al., 2022b).

All this requires engaging and informing young people about pathways to those occupations and opening up possibilities that they and their parents/caregivers might not have envisaged. These engagements might include considering what comprises worthwhile occupational pathways, preparing them for those occupations, and then retaining them within the workforce. Much of the emphasis here is on localised and close or interpersonal engagements with familiars, local experts, and educators, which indicates the importance of localised administrative and educational infrastructure. Whilst this project largely focused on disadvantaged young people, often in regional communities, the findings are probably applicable to the wider population. What these young people were requesting emphasises localised manifestations of support and guidance. So, more than being reactive, these requirements demand that VET takes an active pre-emptive approach in seeking to achieve these goals rather than being responsive to what others suggest. Consequently, VET systems and institutions need to actively engage with other agencies and tertiary education institutions in informing young people and their parents/caregivers about VET at the local level in unison with government, industry, and professional bodies and through local partnerships.

Purpose 2. Assistance in Identifying 'Suitable' Occupations

In perhaps the most powerful treatise on VET, Dewey proposed in Democracy and Education that the purposes are twofold. The first is to assist young people identify the occupations in which they are interested in and suited to (i.e., make them their vocations) (Dewey, 1916). He argued that for young persons to find their 'calling' (i.e., suitable occupation) was important for a satisfactory life. He proposed that assisting them identify what would become their vocation was central to who they were and how they would identify themselves in adult life. He stated that the opposite of vocations was not leisure, but idleness, indolence, and lack of purpose. This concern resonates today when many young people are taking long and circuitous pathways to reach their preferred occupations and, perhaps, never achieving that goal, thereby prompting the prospect of unfulfilled working lives. Of course, not everybody will find and identify occupations that become their vocation. Nevertheless, this is a worthwhile educational goal for personal and societal reasons. For individuals, having interest in and commitment to their occupations will potentially enrich their work life and avoid them falling into uncongenial callings, about which Dewey (1916) cautioned. Moreover, for individuals needing to remain employable, an interest in and commitment to an occupation is central for them to invest effort across working life to sustain their employability (Bennett, 2016; Billett et al., 2023). Given the needs of enterprises and nation states for workforces to remain current, adaptable, and responsive, this is far more likely to be achieved when both young people and working age adults have interest in and commitment to their occupations and when their subjectivities (i.e., sense of self) are aligned with occupations that become their vocations.

Consequently, assisting individuals identify their preferred occupations through advising them about and aligning their interests and capacities with those occupations is an important purpose of vocational education. The risk is that young people may make ill-founded and uninformed decisions about occupations. This can lead to them engaging in and persisting with uncongenial work that leads to unfulfilling working lives (Dewey, 1916). Additionally, such uninformed choices can have consequences, within communities and nationally, for the development of the kinds of required occupational skills.

Hence, young people (and their parents/caregivers) may need guidance and assistance to identify the occupations to which they are suited to and aligned with their interests. This kind of guidance and support needs to occur at the local level and again comprises instances of where educational and administrative infrastructure is required.

Assisting Young People Identify Suitable Occupations

It follows that assisting young people identify occupations to which they are suited to and interested in is an important goal for VET and, to manage personal and institutional resources. Whilst information about occupations can be derived from centralised sources (e.g., government, professional bodies, industry bodies), guidance and advice are required at the local level to assist young people align that information with their interests and capacities (Billett et al., 2020). Young people report being overwhelmed with the extent and kinds of information they receive about occupations and need guidance and support locally to align it with their interests and abilities. This can best occur at the local level and through processes in which openness is possible and informed advice provided impartially. Increasingly, schooling systems have processes that assist considerations of post-school pathways, and these often involve teachers engaging directly with young people and their parents (Billett et al., 2020, 2022b). But, again, these processes need to be informed and impartial. It is also helpful to make occupations visible to young people so that they are aware of their existence and what they constitute. For instance, in the HEEP project (Billett et al., 2022a, b) mentioned above, it was found that most young people have had experiences with nurses and doctors yet were unaware of the range of occupations enacted within the healthcare system and at their local hospital. It was often only when family members were involved in those occupations that young people were aware of them. Therefore, more than being responsive to their needs, there is a role for VET in making them aware of the kinds of work that occurred in their communities, perhaps because these may not be visible in their urban communities.

Some specific strategies advising occupational choices include: (i) social partnerships amongst local schools, tertiary education, and enterprises offering experiences informing post-school pathways (Billett et al., 2024); (ii) prevocational and taster programs being wider in range and availability; (iii) visits to local VET institutions and enterprises by school students and their teachers (Billett et al., 2020); (iv) local communities of engagement (i.e., teachers, enterprises, etc); and (v) engagement with workplace experiences and support to make them more organised, engaging, and effective (addressing high apprentice attrition). Added to these are series of micro strategies that were derived from discussions with teachers and administrators from schools and tertiary education institutions based on the feedback provided by young students and their parents (Billett et al., 2020) that comprise aspects of administrative, social, and educational infrastructure.

So, there are both personal and institutional imperatives for assisting young people identify the occupations to which they are suited through processes of informed and impartial advice. These imperatives have perhaps been most explicitly stated in Germany in what is referred to as the 'War for apprentices' (Deissinger, 2002), the high attrition rate of apprenticeships in many countries, and Wolf's (2016) concerns about the lack of young people seeking to participate in UK VET and being prepared for the very occupations that are most needed in that country. As mentioned, those personal imperatives are for young people to identify occupations and working life in which they can find satisfaction, personal sense of worth, and with which they can identify. The ideal is for these occupations to become their vocations and become a part of how they view and describe themselves. The national imperative is to ensure that there is a sufficient quantum and quality of workers across required occupations that serve individual and community needs. Without those workers and their ability to be effective and remain employable, there will be inevitable shortfalls in fulfilling societal needs for those occupations. This concern is far wider than being concerned about enterprises' profitability, because as with the HEEP project (Billett et al., 2022b), it extends to public sector institutions that provide care for the sick, the aged, and environment. Part of developing the sense of vocations in individuals' occupation is the ability to practice that occupation with confidence and competence (i.e., competent begets commitment) (Mclaughlin & Marsh, 1978). It follows that the effectiveness of initial occupational preparation is an important purpose of VET. What that study suggests is, however, that there will be community-specific factors that need to be accommodated in strategies associated with informing young people objectively and impartially and engaging them in careful decision-making about the kinds of occupations to which they are drawn yet are aligned with their capacities.

To achieve these outcomes can require proactivity by educators to initiate interactions amongst teachers, parents, school age children, local enterprises, and industry representatives. This kind of social and educational infrastructure is required to provide informed and impartial advice and guidance, and in ways that seek to make visible the range of potential opportunities for young people, whilst not seeking to restrict their decisions. This all needs to be driven by localised initiatives that work with and build partnerships amongst local public and private sector enterprises, representatives of industry sectors, and government to bring these partners to assist in this important process. Educational leadership here is driving these agendas and taking a key role in the development of processes that assist young people make informed choices about their preferred occupations and how they might best be prepared for them.

Purpose 3. Initial Occupation Preparation

The second purpose of education for occupations to which Dewey (1916) referred was for young people being adequately prepared for the occupations to which they are suited. This has, perhaps, become the principle purpose of VET in many countries: i.e., to prepare students for their preferred occupations (Greinert, 2002). Whilst this appears to be a common purpose for VET, approaches to achieving it differ widely across countries based upon historical, cultural, institutional, and resource-related factors. For instance, some countries have the dual experiences in workplace and educational settings, whilst others just in the latter. This preparation is often seen as the sole main purpose of VET by employers, governments, community, and students. Indeed, the common mandate is for graduates from VET programs to be prepared to be competent in the occupations that are the focus of their courses. For some, such as those participating in apprenticeships, traineeships, internships, and cadetships, their preparation involves periods of work experience, which makes them aware of the requirements of working life as well as specific needs of occupations, and also how that relates to particular workplace requirements (Chan, 2013; Deissinger, 2002; Fuller & Unwin, 2003). Yet, many governments are now requiring that VET graduates of all kinds are also to be 'job ready', not just occupationally prepared. However, securing that 'job readiness' is a tough educational goal. This is because: (i) educators may not know the work requirements where VET graduates will be employed; (ii) there are quite different educational objectives and processes required than for 'occupational' preparation; and (iii) these extend to so-called Twenty-First-Century skills (World Economic Forum, 2015) of (a) complex problem-solving, (b) critical thinking, (c) creativity, (d) people management, and e) coordination (Nokelainen et al., 2018). Generating these outcomes without extensive engagement in work settings, developing critical insights to understand situated and canonical occupational requirements and how these are practiced is potentially problematic. Consequently, vocational educators may have to be proactive and adapt what they do and have licence to shape educational experiences to achieve these outcomes for students (Brennan-Kemmis & Green, 2013).

To be adaptable and move smoothly into the work setting, it is helpful for VET graduates to know some of the variations of occupational practice and have access to educational processes to accommodating those variations. Thus, more than focusing on readiness in terms of occupational specific outcomes, it is necessary that the initial occupational preparation also seeks to generate adaptability in its graduates. That is, going beyond responsiveness to occupational requirements and immediate or current work requirements of workplaces, the need to develop that adaptability is important for at least three reasons. Firstly, there is a need to transfer or adapt what has been learnt in the educational setting to the circumstances and work beyond and outside of it and which are unlikely to be known by the student. Secondly, there are numerous, and possibly endless, manifestations of how particular occupations are practiced as they are dependent on the specific work situations, their circumstances, and what passes as effective/ineffective performance and whether problem solutions are judged relevant or irrelevant. So, what constitutes responding to non-routine problem-solving as a hallmark of expertise is highly situated (Billett, 2001). Consequently, in moving across circumstances of practice, adaptability is often necessary and can be best developed situationally. Thirdly, as occupational and work requirements change, these need to be addressed and supported again at the local level. Hence, developing occupational adaptability becomes a key educational goal. Yet, governmental concerns often are focused on national statements of competence, rather than processes securing these kinds of outcomes (Hamalainen et al., 2018), which need to be addressed locally.

Workplace Learning Experiences

For both initial occupational preparation and CET, workplace experiences and their integration are important elements of VET provisions. Workplaces provide access to the knowledge required for both occupational capacities and situational performance that can be experienced through occupationally authentic activities and interactions in those settings (Chan, 2013). In essence, workplaces afford richly contextualised experiences (i.e., engage multi-sensory processes, provide clues, cues, etc.), purposive activities (i.e., directed to goals, engage in decision-making), practice (i.e., engage, refine, hone), episodic experiences (i.e., establishing causal and propositional links), and monitoring progress and outcomes (i.e., appraising and evaluating performance) (Billett, 2014).

Table 11.1 Enriching workplace experiences

Models of workplace curriculum	Learning curriculum, learning activities at work, learning pathways, parallel practice
Workplace pedagogic practices	Verbalisation, pedagogically rich activities, guided learning (proximal guidance), partially worked example/direct instruction and hands on, heuristics, mnemonics, artefacts
Personal epistemological practices	Ontogenetic ritualisation, embodying knowledge, deliberate practice, guided re-discover, active engagement and construction, observation, averting gaze, readiness, assent

However, they also have weaknesses and limitations, thus requiring deliberate guidance and pedagogic interventions to avoid or minimise such limitations, as overviewed in Table 11.1. This table refers to models of workplace curriculum, practice pedagogies, and personal epistemological practices, whose details are elaborated elsewhere (Billett, 2014).

As is the case for what occurs in educational institutions, indicated, there are both strengths and limitations to learning from the activities and interactions that individuals engage with in work settings. Indeed, many models of initial occupational preparation experiences in both education and work settings are provided, and at best integrated. One potential outcome of that integration is the development of knowledge that is adaptable to work settings and different than those in which the initial experiences have occurred (Cooper et al., 2010). Hence, the integration of these two sets of experiences can make important contributions to both initial occupational preparation and its further development across working life (i.e., CET). However, it requires educational and social infrastructure at the local level to provide relevant experiences and achieve desirable outcomes.

Providing and Integrating Workplace Experiences

So, beyond just providing experiences in work settings, educational worth of workplace experiences can be enhanced through intentionally integrating them with experiences within educational institutions (Grollman & Tutschner, 2006). This can be used to achieve the kinds of learning and adaptability required to make VET graduates ready for occupational practice. Consequently, wherever possible is the importance of providing and integrating those experiences within VET programs to optimise their contributions and redress limitations (Billett, 2007, 2011a). These intentional efforts to integrate the two sets of experiences exercise curriculum, pedagogic, and learners' personal practices. That is, with curriculum practices comprising the intended, enacted, and experienced curriculum, there have to be: (i) intentional organisation for workplace experiences to be provided and then integrated into the program; (ii) pedagogically practices enacted prior to, during, and, perhaps most importantly, after the students have had work experiences; and (iii) personal practices on the part of the students and how they come to engage with experiences in each setting, how they come to reconcile the two sets of experiences and construct meaning through and from them (*see* Table 11.2).

For each of these practices, there are models and approaches that can be enacted depending upon the circumstances and intended outcomes although there are some key canonical elements to them. Variations for the sequencing and organising of the experiences include what should come first (i.e., educational or workplace setting experiences), their duration, focus, number, and sequencing (Billett, 2015a). For pedagogic practices, there are considerations about how students might be prepared to be ready for (i) working life, (ii) particular occupational requirements, and (iii) ability to contribute productively in work settings. Then, how their engagement and learning can be supported during the work experience, either directly, or through withdrawing back to the educational institution for educational interludes (Tan et al., 2023) through virtual engagement or student-led group processes, etc. As noted, importantly, when the students either have had periods of work experiences or completed them, post-practicum events, such as those that allow students to compare, contrast, and critique their work experiences and relate those experiences to achieving the goals of their courses, are likely to enrich the

Table 11.2 Elements of integrating educational and workplace experiences

Practices	*Organisation of experiences*
Curriculum practices	intended, enacted, and experienced
Pedagogic practices	before, during, and after workplace experiences
Personal practices	how students come to engage, mediate, reconcile, and accommodate those experiences

outcomes and make them far more prepared to practice the occupation and, potentially, be adaptable in that application.

However, more than being advised by the local community or national bodies, governmental agencies, employer groups, and representatives of workers about this provision of initial occupational preparation, the emphasis is on vocational educators taking active role in oversighting the entire educational experience of their students. This includes engaging with, negotiating access to, and generating support within those workplaces for VET students to have effective learning experiences. All of which requires these educators to be proactive and take the initiative in engaging with enterprises and others to organise and enact both the provision of workplace experiences and the integration of them into the overall VET program, to enhance the overall learning experiences. There may be prescriptions at the system level about how these experiences might be sought, and their integration for students. But ultimately, it will be most efficacious for teachers to engage with their students in the local settings. This is because it is only these educators that are best positioned to organise and provide those experiences and then integrate them into their programs of teaching and education enquiry. All this emphasises the importance of having educational infrastructure which is of the kinds and be exercised.

In these ways, the provision of workplace experiences and their integration into the overall VET program are important elements of initial occupational preparation in most of the acknowledged effective models of preparation. These can best be realised through social infrastructure comprising education institutions and workplaces that collaborate, respect each other's contributions, and seek mutual goals and outcomes. However, these considerations can also apply to models of and approaches to CET.

Purpose 4. Provisions of Continuing Education and Training

Even the most effective initial occupational preparation is unlikely to be sufficient for a lifetime of work and workplace performance, as indicated earlier. This is because changes in what constitutes occupational performance constantly evolves, as does the requirements for workplace performance, how that work is conducted, with whom, and through what means necessitate working age adults to constantly develop further their occupational capacities and competence to perform their work effectively.

That is, workers need to learn to sustain their employability. Much of the ongoing learning across working life arises through everyday work activities and interactions (Billett et al., 2014; Choy et al., 2013). Both the routine and non-routine problem-solving that all workers need to engage with, regardless of what kinds and classifications of work they undertake, are generative of new knowledge and its reinforcements and refinement (Billett, 2015b). That learning is mediated by these workers' efforts and agency, as well as support from other and more experienced and informed workers. However, there is still a need for accessible and effective provisions of CET that can support workers' learning, particularly when they are negotiating changing work roles or new occupational practices (Leow et al., 2023). It is those transitions that workers report needing to be supported through provisions of CET that provide certification. That accessibility needs to account for adults often needing to reconcile three other elements of their lives (i.e., work, family, and community).

Consequently, the timing, locations, and means by which the CET is provided need to be aligned with the other demands made of working age adults' lives (e.g., family, community commitments), so that they can manage their time efficiently. Accordingly, the kind of arrangements that might work for young people participating full-time in VET programs, during the day and having lengthy sessions, are unlikely to be appropriate for working age adults. Models of provisions that include evening and weekend timings, the use of both online and face-to-face engagements, and educational processes that draw upon and focus on the adult students' interests, capacities, and needs become particularly salient (Leow & Billett, 2022). It follows that measures of effectiveness include the provision of experiences to develop the kinds of knowledge they want to acquire. Yet, they may need a particular emphasis on the applicability to their current or intended work roles as this focus is likely to be helpful for their engagement and applicability to their current work practices. Important here also is that given the very heterogeneous nature of the adult population, they will inevitably have very diverse kinds and levels of readiness to participate in these CET provisions.

Hence, there is a need for both the model of education and the actions of educators to be open, responsive to these needs, but also to challenge and extend these adult learners. This is because whilst it is important to cater to these adults' readiness, it is also important to press them into extending further what they know, can do, and value (Billett, 2001).

Some of the educational infrastructure required to provide effective CET provisions need to step outside of pedagogical practices that are hostage to what occurs in educational institutions. Curriculum processes and approaches that can be useful here include work-based projects, group-based activities, professional clusters, individual, and group-based activities to identify, consider, and generate innovations for work and their implementation. Common amongst all of these kinds of CET provisions is that they shift the focus away from the educator onto the activities and interactions required by the adult students or learners. The adults need to engage in the thinking and acting associated with these approaches, albeit guided and supported by adult educators, group processes they have established, and peer support processes established and enacted by students themselves. All of this suggests the importance of adult educators not being teacher-centric, but being tolerant of and using adult learners' contributions, and these educators have both occupational and pedagogical expertise that makes them credible for the students.

From the above, regardless of whether referring to the learning occurring in and through working life or CET provisions, efforts to enhance and augment work-based activities need to be understood and enacted locally. This is the case with efforts to integrate what is being provided and experienced through educational provisions with those in and through work. It is the guidance, facilitation, and interventions to support this learning that are necessarily based upon situational knowledge and familiarity with the students in the context in which they work and study. Hence, decision-making about the organisation and enactment of CET provisions needs to occur locally and be enacted in ways that are accessible to working age adults and provide educational experiences which are pertinent to their intended learning outcomes.

Whilst there is applicability arising from these learning experiences and outcomes that are directly associated with improving work performance, and the development of individuals' capacities, there is also a need to ensure that these kinds of activities directly support and utilise innovations that arise in and through workplaces and work practices. This requires the kind of educational infrastructure and social engagement referred to above, including an openness to what constitutes educational practices and the valuing of work and learning more broadly. As indicated earlier, all kinds and classifications of workers engage in non-routine problem-solving that is both generative of workplace innovations, but also higher-order learning outcomes for those workers. Yet, rather than viewing workers' engagement in innovations as peripheral or fleeting, they stand to provide opportunities for advancing both the goals of workplaces as well as individuals' learning and development. This then is the fifth contemporary purpose of VET.

Purpose 5. Initiating and Enacting Workplace Innovations

As noted, the changing requirements for the goods and services provided by public and private enterprises mean that enterprises continually need to transform their procedures and practices to maintain their viability. Hence, being innovative and enacting workplace innovations has now become a necessary requirement for both workers and workplaces (Hoyrup et al., 2012). At one level, it provides continuity of employment, opening of opportunities, and bases for advancing work practices and performance. At another level, it provides the basis by which workers can be engaged in ways that bring their contributions to these changes and learn through doing so (Billett, 2012). Consequently, the co-occurrence of learning and workplace innovations can become a basis for both workers' continuing employability and their workplaces' viability. Hence, addressing changing occupational requirements and those needed to provide the goods and services to secure workplace viability is an important goal for VET. Importantly, those innovations are central to meeting communities' economic and social goals and, collectively, those of nation states. Often, workers' learning, and work innovations are considered separately rather than co-occurring or reciprocal. Workplace innovations are unlikely to be initiated, developed, and adopted without engaging workers' capacities and their learning (Billett et al., 2021). Regardless of whether those innovations come from outside of the workplace, or initiated and enacted within it, inevitably workers will need to adapt and adjust, thereby potentially extending what they know, can do, and value (Lemmetty & Billet, 2023). Consequently, a focus for VET is not just realising personal goals, but those associated with the viability and advancement of their workplaces. That co-occurrence is best initiated, enacted, and evaluated at the local (i.e., situated) level because of the situatedness of these dualities.

Initiating and Enacting Workplace Innovations

Investigations into small to medium size enterprises have found that workplace innovations are of different kinds, scales, and scope, as is the potential learning that can come from participating in them (Billett et al., 2022b; Billett et al., 2021). Those investigations indicated that there are: (i) strategic innovations (i.e., substantial changes in the enterprise's focus and activities), (ii) work practice innovations (i.e., significant changes in the processes that the goods and services are generated), and (iii) procedural innovations (i.e., the kinds of processes used to provide those goods and services). The latter involved workers' initiating, enacting, and evaluating innovations, whereas the other two are more likely initiated by workplace owners or managers. Regardless, in all three kinds of innovations, workers are central to the processes of change. They are needed to engage and learn, and the more fulsome they engage, more likely the richer the learning outcomes occur (Billett et al., 2022b). The role for VET and its educators in such processes can range from encouraging, guiding, and providing support for workplaces' innovations and their enactment, through to providing experiences that can develop knowledge that is unavailable in the workplace and therefore cannot be accessed by workers. In those investigations, for instance, it was found that when new workplace processes were required or workplace lacked expertise, the provision of CET was essential. Here, the provisions of educational experiences and actions of educators with expertise in those fields were vital, and need to comprise an infrastructure which is available locally and accessible within work settings and accepted as such. Thus, there are ways in which vocational educators and VET systems can support innovations in and through work and emphasise the importance of using the identification, justification, and enactment of innovations as bases for rich learning amongst workers involved in the generation and advancement of those innovations.

Enacting VET Purposes: Localised Engagements and Learner Agency

What has been proposed is that there are five purposes for contemporary VET, and these go beyond the initial development of occupational capacities for which this educational sector is perhaps most known. Added here is the growing importance of being able to attract and engage young people to participate in VET, assist them in identifying which occupation they are most interested, and to which they are suited. Then, because of constant changes in occupational requirements and workplace practices, their continuing development across working life and the needs to support workplaces to respond to challenges necessitate both CET provisions and focuses on the combination of workplace innovations and workers' learning. All of this is aligned with the concept of opportunity structures, viewed within this volume as the organisational frameworks within education, training, and the workforce that open pathways, fostering accessibility and progression for individuals through VET (Chap. 1). The central premise is that these structures comprise policies, practices, and collaborative networks that together become a social/societal environment for learners to navigate their educational and career trajectories. These structures can facilitate the development of professionalised educational and training routes that not only lead to formal certifications, but also open new career and employment opportunities for newcomers and experienced working age adults.

All this represents significant challenges and tasks for VET. Yet, for these to be achieved, beyond national mandates, ordering of occupational standards, regulation, and legislation, there is a need for administrative, social, and educational infrastructure to be exercised at the local level. This infrastructure is important because there is always the inevitable risk that when faced with key national challenges and crises, governments and supra-government agencies will tend to emphasise top-down and highly controlled and regulated responses to these challenges. That is, seeking to tightly manage what is perceived to be key priorities. Yet, in the case of VET, this is counter to what is needed and required because such measures have repeatedly been shown to be ineffective and counter to achieving these goals. Much of what has been proposed can only be realised through localised activities and interactions at the local level. This includes having infrastructure and expertise at that level to develop effective partnerships, identify educational needs and appropriate processes, and organise and enact experiences to achieve required outcomes. With partnerships, for instance, their longevity is often premised upon partners believing they can achieve their goals, gain something, but not everything, from their commitment to those partnerships and this needs to be understood and managed at the local level.

Much of what is proposed here is about opening pathways, fostering accessibility and progression for individuals through VET Chap. 1. This can be realised through engagements, interactions, and partnerships outside of VET institutions, comprising the social infrastructure that brings together partners with shared interests and abilities to make distinct contributions in securing mutually understood goals. However, often these need to be initiated, enacted, and sustained through the auspices of VET institutions, some form of administrative infrastructure. For instance, vocational educators might actively seek to engage with their industry partners and build capacity and infrastructure at the local level. That is, engagements with workplaces, communities, and industry bodies beyond the 'school fence' through local partnerships.

References

Bennett, D. (2016). Developing employability and professional identity through visual narratives. *Australian Art Education, 37*(2), 100.

Billett, S. (2001). Knowing in practice: Re-conceptualising vocational expertise. *Learning and Instruction, 11*(6), 431–452. https://doi.org/10.1016/S0959-4752(00)00040-2

Billett, S. (2006). *Work, change and workers*. Springer.

Billett, S. (2007). Integrating contributions of workplace and college experiences in developing vocational knowledge. In R. McClean, D. N. Wilson, & C. Chinien (Eds.), *International handbook for technical and vocational education and training*. Springer.

Billett, S. (2011a). Integrating expereinces in workplace and university settings: A conceptual perspective. In S. Billett & A. Henderson (Eds.), *Developing learning professionals: Integrating experiences in university and practice settings* (pp. 21–40). Springer.

Billett, S. (2011b). *Vocational education: Purposes, traditions and prospects*. Springer.

Billett, S. (2012). Explaining innovation at work: A socio-personal account. In S. H. Pederson (Ed.), *Employee-driven innovation: A new approach* (pp. 92–107). Palgrave Macmillan.

Billett, S. (2014). *Mimetic learning at work: Learning in the circumstances of practice*. Springer.

Billett, S. (2015a). *Integrating practice-based experiences into higher education*. Springer.

Billett, S. (2015b). Work, discretion and learning: Processes of learning and development at work. *International Journal of Training Research, 13*(3). https://doi.org/10.1080/14480220.2015.1093308

Billett, S. (2024). Building VET systems to advance communities: Beyond responsiveness. *Ammattikasvatuksen aikakauskirja, 26*(1), 98–104. https://doi.org/10.54329/akakk.143479

Billett, S., Ovens, C., Clemans, A., & Seddon, T. (2007). Collaborative working and contested practices: Forming, developing and sustaining social partnerships in education. *Journal of Education Policy, 22*(6), 637–656. https://doi.org/10.1080/02680930701625288

Billett, S., Choy, S., Dymock, D., Smith, R., Henderson, A., Tyler, M., & Kelly, A. (2014). *Towards more effective continuing education and training for Australian workers*. NCVER, Adelaide.

Billett, S., Choy, S., & Hodge, S. (2020). Enhancing the standing of vocational education and the occupations it serves: Australia. *Journal of Vocational Education and Training, 72*(2), 270–296. https://doi.org/10.1080/13636820.2020.1751247

Billett, S., Choy, S., Gibbs, K., Le, A. H., McKay, L., & Hay, S. (2024). Initiating, enacting and sustaining partnerships to inform post-school pathways in *Journal of Education and Work 37*(1–4) pp. 149–164. https://doi.org/10.1080/13639080.2024.2383564

Billett, S., & Seddon, T. (2004). Building community through social partnerships around vocational education and training. *Journal of Vocational Education and Training, 56*(1), 51–67. https://doi.org/10.1080/13636820400200245

Billett, S., Yang, S., Chia, A., Tai, J. F., Lee, M., & Alhadad, S. (2021). Remaking and transforming cultural practices: Exploring the co-occurrence of work, learning and innovation. In K. Collin, V. Glaveanu, S. Lemmetty, & P. Forsman (Eds.), *Creativity and learning: Contexts, processes and impact* (pp. 219–244). Palgrave Macmillan.

Billett, S., Stalder, B. E., Aarkrog, V., Choy, S., Hodge, S., & Le, A. H. (2022a). *The standing of vocational education and the occupations it serves: Current concerns and strategies for enhancing that standing* (Vol. 32). Springer Nature.

Billett, S., Tan, J., Chan, C., Chong, W. H., & Keat, J. S. C. (2022b). Employee-driven innovations: Zones of initiation, enactment and learning. In W. O. Lee, P. Brown, A. L. Goodwin, & A. Green (Eds.), *International handbook on education development in Asia-Pacific* (pp. 1–19). Springer Nature. https://doi.org/10.1007/978-981-16-2327-1_67-1

Billett, S., Salling Olesen, H., & Filliettaz, L. (2023). *Sustaining employability through work-life learning: Practices and policies*. Springer Nature. https://doi.org/10.1007/978-981-99-3959-6

Brennan-Kemmis, R., & Green, A. (2013). Vocational education and training teachers' conceptions of their pedagogy. *International Journal of Training and Research, 11*(2), 101–121. https://doi.org/10.5172/ijtr.2013.11.2.101

Chan, S. (2013). Learning through apprenticeship: Belonging to a workplace, becoming and being. *Vocations and Learning, 6*(3), 367–383. https://doi.org/10.1007/s12186-013-9100-x

Choy, S., Billett, S., & Kelly, A. (2013). Engaging in continuing education and training: Learning preferences of worker-learners in the health and community services industry. *Australian Journal of Adult Learning, 53*(1), 68–89.

Clement, U. (2014). *Improving the image of technical and vocational education and training. Deutsche Gesellschaft fur Internationale Zusammenarbeit (GIZ).*

Cooper, L., Orrel, J., & Bowden, M. (2010). *Work integrated learning: A guide to effective practice.* Routledge.

Dawson, J. (2002). Vocation in adult education: widening the gaze. In *Proceedings of the Canadian Association for the Study of Adult Education.*

Deissinger, T. (2002). Apprenticeship systems in England and Germany: Decline and survival. In *Towards a history of vocational education and training (VET) in Europe in a comparative perspective: Proceedings of the first international conference, Florence.* Office for Official Publications of the European Communities.

Dewey, J. (1916). *Democracy and education.* The Free Press.

Fuller, A. (2015). Vocational education. In *International encyclopedia of the social & behavioral sciences* (Vol. 25, pp. 232–238).

Fuller, A., & Unwin, A. (2003). Fostering workplace learning: Looking through the lens of apprenticeships.*European Educational Research Journal, 2*(1), 41–55. https://doi.org/10.2304/eerj.2003.2.1.9

Greinert, W.-D. (2002). European and vocational training systems: the theoretical context of historical development. In *Towards a history of vocational education and training (VET) in Europe in a comparative perspective: Proceedings of the first international conference, Florence.* Office for Official Publications of the European Communities.

Grollman, P., & Tutschner, R. (2006). *Possible intended and unintended effects of European VET policies - The case of integrating work and learning.* In European research network in vocational education and training symposium, Geneva.

Hamalainen, R., Lanz, M., & Koskinen, K. T. (2018). Collaborative systems and environments for future working life: Towards the integration of workers, systems and manufacturing environments. In C. Harteis (Ed.), *The impact of digitalisation in the workplace: An educational view* (Vol. 21, pp. 25–38). Springer.

Hoyrup, S., Bonnafous-Boucher, M., Hasse, C., Lotz, M., & Molller, K. (2012). *Employee-driven innovation: A new approach.* Palgrave Macmillan.

Lemmetty, S., & Billet, S. (2023). Employee-driven learning and innovation (EDLI) as a phenomenon of continuous learning at work. *Journal of Workplace Learning, 35*(9), 162–176. https://doi.org/10.1108/JWL-12-2022-0175

Leow, A., & Billett, S. (2022). Sustaining the employability of working-age adults: A Singapore case study. In B. Ng (Ed.), *Graduate employability and workplace-based learning development: Insights from sociocultural perspectives* (pp. 45–66). Springer Nature. https://doi.org/10.1007/978-981-19-5622-5_4

Leow, A., Billett, S., & Le, A. H. (2023). Towards a continuing education and training eco system: A case study of Singapore. *International Journal of Training Research*, 1–17. https://doi.org/10.1080/14480220.2023.2203944

Marsh, C. J. (2004). *Key concepts for understanding curriculum*. Routledge-Falmer.

Mclaughlin, M. W., & Marsh, D. D. (1978). Staff development and school change. *Teachers College Record, 80*(1), 69–94.

Nokelainen, P., Nevalinen, T., & Niemi, K. (2018). Mind all machine? Opportunities and limits of automation. In C. Harteis (Ed.), *The impact of digitalisation in the workplace: An educational view* (Vol. 21, pp. 13–24). Springer.

Parliament of Victoria. (2018). *Inquiry into career advice activities in Victorian schools*. Victorian Parliment.

Rehm, M. (1990). Vocation as personal calling: A question for education. *The Journal of Educational Thought, 24*(2), 114–125.

Stevenson, J. C. (1991). Cognitive structures for the teaching of adaptability in vocational education. In G. Evans (Ed.), *Learning and teaching cognitive skills* (pp. 144–163). ACER.

Tan, N., Shien, C., Ong, C., & Billett, S. (2023). Promoting student readiness for work-life through internships: Challenges and support. *Australian Journal of Adult Learning, 63*(3), 343–367.

UNESCO-UNEVOC. (2018a). *Improving the image of TVET: Making TVET attractive to youth.*

UNESCO-UNEVOC. (2018b). *Short summary of the Virtual Conference on improving the image of TVET: Making TVET attractive to youth*

Wolf, A. (2016). *Remaking tertiary education: Can we create a system that is fair and fit for purpose.*

World Economic Forum (2015). *New vision for education - Unlocking the potential of technology*. Retrieved from https://widgets.weforum.org/nve-2015/index.html

CHAPTER 12

Investment in Youth Through Vocational Education and Training

Eli Smeplass

Introduction

This chapter explores how vocational education and training (VET) in Norway is shaped by the country's social and political context. Investing in youth through education is a long-established strategy in Norway, seen as key to nation-building, fostering democracy, and promoting broad societal participation. Given the growing global emphasis on education as a tool for social inclusion and economic development, understanding how a nation like Norway integrates VET within its broader policy framework can offer valuable insights for other nations seeking to balance education with labor market integration.

Norway, a Nordic welfare state, is known for its strong commitment to social equality, comprehensive public services, and an inclusive labor market that provides opportunities for its citizens, especially its youth

E. Smeplass (✉)
Department of Teacher Education, Norwegian University of Science and Technology (NTNU), Trondheim, Norway
e-mail: eli.smeplass@ntnu.no

E. Smeplass and C. Zirkle (eds.), *Opportunity Structures in Vocational Education and Training*, Palgrave Studies in Technical and Vocational Education and Training,
https://doi.org/10.1007/978-3-032-04410-5_12

(Esping-Andersen, 1990; Telhaug, 2006; Antikainen, 2006; Telhaug et al., 2006). In this context, education in general, and particularly vocational education and training, has been strategically positioned as foundation of what is described as *youth investment*. This term signifies the deep recognition of the importance of early investment in the upcoming generation, ensuring they have access to quality training, broad opportunities, and reduced risks related to educational choices and associated costs at the individual level. In the Nordic welfare states, like Norway, principles of equality and inclusivity have strongly influenced the development of educational systems, with VET providing a clear pathway for youth to contribute to both the economy and society at large (Nylund & Rosvall, 2019; Michelsen, 2018; Dobbins & Busemeyer, 2015). Norway is a country that has established supportive measures to promote successful training for their population.

The chapter will discuss how the integration of VET into upper secondary education, formalized in Norway in 1994, plays a role in attracting and equipping young people with the relevant skills and qualifications necessary for entering the labor market. At the heart of Norway's social and economic policies is the 'work-first principle' or '*arbeidslinja*', which emphasizes the importance of promoting active labor market participation while ensuring comprehensive social protection (Øverbye & Stjernø, 2012). These principles are embedded in Norway's education and welfare systems, particularly through policies that ensure access to education, training, and employment for all youth, regardless of social and economic background. By integrating VET within this framework, Norway supports career mobility and ensures that its youth have the opportunity to contribute meaningfully to society.

Norway's high public investment in education is reflected in its per-pupil spending, which significantly outpaces the OECD average. In 2019, Norway allocated 55% more per student at the primary level and 34% more at the secondary level than the OECD average (Udir, 2022). A key focus of Norway's education system has been reducing the share of young adults without an upper secondary qualification, both historically and also in recent years. From 2016 to 2023, the share of 25–34-year-olds without this qualification decreased by 4% points, now standing at 15%, just above the OECD average (OECD, 2024). This recent reduction aligns with a broader goal to improve youth employability and access to quality education, particularly through vocational education and training pathways (Norwegian Government, 2021c). The rationale is that those with an upper secondary qualification experience significantly higher employment rates, with 86% of individuals aged 25–34 employed, compared to only 68% of those without such qualifications (OECD, 2024). Norway's success with VET is also influenced by its integrated model of cooperation between the state, employers, and trade unions, known as 'the tripartite arrangement' (Dølvik et al., 2014). This collaboration ensures that the VET curriculum remains relevant to labor market needs while simultaneously providing strong protections and incentives for students and employers alike (Thinqvist et al., 2019).

This chapter will describe some of the key features of Norway's VET system as part of a broader welfare strategy to discuss how it can be seen as an investment in the coming generations. By examining how VET has become an attractive option for youth, it highlights how the system integrates with other policies to promote labor market participation and social inclusion, promoting VET as an interesting reliable choice. Through the double focus on education and active labor market policies, Norway's approach offers valuable insights into how vocational training can enhance opportunity structures for youth and their engagement and support social and economic mobility.

Participation and Education in the Welfare State

Norway has transformed from a relatively modest, agricultural-based economy into one of the wealthiest and most egalitarian nations in the world. The country's economic development is closely linked to its natural resources, particularly oil and gas, which have propelled its economy since the late twentieth century. Norway is currently ranked second in the world on the United Nations' Human Development Index, surpassed only by Switzerland (UN, 2024). The Human Development Index is a composite measure that assesses average achievement in three essential dimensions of human development: a long and healthy life, access to knowledge, and a sufficient standard of living. However, the foundation of Norway's prosperity lies not only in its natural resources but also in its investment in education, social infrastructure, and a well-developed welfare system. Over the years, Norway's economic policies have focused on ensuring high levels of labor market participation and equitable distribution of wealth, driven by the country's commitment to social democracy (Norwegian Government, 2021a). The Norwegian model is commonly understood as the interplay between a well-organized labor market, comprehensive public welfare systems, and economic policies (Norwegian Government, 2023). Successful coordination across these areas is considered key to fostering a society with high employment and low inequality. Additionally, the structure of welfare programs helps promote high employment rates, effective adaptation to changes, and a fair distribution of income.

Norway has a high labor market participation rate, with 73% of the population aged 15–74 employed in 2023 (Norwegian Government, 2023). This is partly due to a significant increase in female workforce participation since the 1970s. Norway‘s focus on education, health, and policies like the 2011 pension reform, which encourages older workers to stay employed longer, has also contributed to a high employment rate. Additionally, the country's relatively low unemployment rate, at around 4% in 2024 (Statistics Norway, 2024a, c), illustrates a stable engaged workforce.

If young people do not complete their education or become detached from the labor market, it can have long-term consequences for the individual. Unemployment early in life is linked to later labor market integration and wages and increases the risk of future unemployment (Hyggen, 2010). Welfare systems that promote work participation are essential. Furthermore, the National Insurance Scheme (*folketrygden*) is a mandatory social insurance system that provides financial support to individuals in case of sickness, disability, unemployment, retirement, and death. It also covers healthcare expenses and offers benefits for various life situations, such as pregnancy and family care. The system is financed through member contributions, employer contributions, and state funding, and is managed by The Norwegian Labour and Welfare Administration (*NAV*). It ensures that all residents of Norway, regardless of nationality, are covered by the scheme. Due to the expenses of the welfare system, policies must ensure a high workforce participation to guarantee that the welfare system is sustainable. NAV plays a key role in this system by providing services related to employment, social security, and welfare, safeguarding that individuals facing barriers to employment receive the necessary support. NAV's mission is to reduce welfare dependency while fostering social inclusion and enabling transitions into the workforce (Djupvik & Eikås, 2024; Lægreid & Rykkja, 2014). NAV is responsible for several targeted investments in youth by offering specific support for those facing barriers to employment. Young people registering with NAV are assessed and placed into one of three support categories: standard, situational, or specially adapted measures. These measures include job search assistance, vocational training, and, for those with more significant challenges, medical treatment or tailored employment programs.

The development of labor policy in Norway, known as the 'work-first principle' or *arbeidslinja*, is closely tied to the country's social policies that have developed over time. Initially the work-first principle was rooted in Christian values and the Protestant work ethic, work was viewed as a moral duty to serve God (See for example Øverbye & Stjernø, 2012). However, after the nineteenth century, it evolved into a broader social policy, emphasizing that social benefits, such as assistance for the poor, should not exceed the income from the lowest-paying jobs, thus reinforcing a notion that 'work always pays off'. Following the political dominance of the Norwegian Labour Party after the 1930s, the focus shifted toward increasing production to ensure that the working population supported those who were unable to work. The core idea was that a healthy labor market required that all capable individuals not only had the right to work but also the duty to contribute. This principle has since become ingrained in Norwegian culture, where an individual's vocation or occupation plays a crucial role in social recognition. A strong normative expectation to work and contribute to society through production, service, and paying taxes is evident in most social situations. For instance, work norms are reflected in the most common question asked when meeting someone: 'What do you do for work?'. In the 1950s and 1960s, there was a strong belief that education and schooling were essential tools for increasing productivity and promoting economic growth. The emphasis on equality, democracy, and tolerance in education was not only a social aspiration but also a pragmatic approach to nation-building. By fostering a well-educated and cohesive society, these policies laid the foundation for economic development and long-term productivity. Hence, the development of educational institutions has been part of a broader strategy to advance the country's economy.

Prior to today's integrated VET system, there was a complex system of various schools and responsibility levels. The development of vocational training in Norway prior to the Second World War was mainly initiated bottom-up as a result of initiatives from individuals and organizations based on their experienced needs in specific industries and companies (Nyen & Tønder, 2014). A manyfold of training offers and types of VET schools were established without any coherent plan or strategy from the governments side. Modernization and the growth of industry, as in many countries, entailed that more workers worked in industry than in traditional crafts. From 1945, vocational schools, including technical schools, workshop schools, agricultural schools, maritime schools, and business schools, provided qualifications in a number of different trades (Karlsen, 2006). Vocational schools were the predecessors to today's vocational training and played an important role in training youth for relevant professions. They could be municipal, county, or national institutions. Vocational schools were established to organize vocational training in a more systematic manner and offered programs that provided training before, during, and after the apprenticeship period. After 1976, vocational schools were merged with gymnasiums under the same legislation (Smeplass, 2024). Since the 1900s, the idea of the 'unified school' (*enhetsskolen*) emerged, aiming to standardize education and provide equal qualifications for all students. This reform was a response to the increasing diversity in schools and the need to harmonize the system. During this period, there was a growing political debate over the role of education, with the Labor Party seeing schools as a bridge between social classes. By 1936, the folkeskole laws were introduced, establishing a seven-year compulsory school for all, marking the beginning of a unified national education system (Helsvig, 2014). The primary goal was to ensure that all children, regardless of their social background, received a similar quality of education, further reinforcing the notion that education should be a collective good.

The expansion of compulsory schooling continued into the 1950s and 1960s. The strong belief in education as a public good was solidified by major reforms during this period, where educational policies sought to provide equal opportunities for all citizens. Due to the country's geography, there was huge differences in educational opportunities in urban compared to rural areas. In 1969, the introduction of the primary school law marked a pivotal point, extending the compulsory schooling to 9 years with an important goal to make it available to all children, not just those from privileged backgrounds. This expansion reflected the increasing political emphasis on education as a means of social mobility and a tool for reducing inequality. Furthermore, Norway's education system was strongly influenced by the idea that education should not only be about academic knowledge but also about fostering democratic values. The school reform *Mønsterplanen* of 1974, introduced a new curriculum, which emphasized this broad educational mission, encouraging student-centered learning and the development of a morally equipped and responsible citizen (Tønnessen, 2011). This reform shifted the focus from strict subject specialization to an approach that encouraged critical thinking, social responsibility, and furthermore creativity. By the 1980s and 1990s, education policies continued to evolve, with increasing attention paid to ensuring that students were equipped with both academic knowledge and practical skills for the labor market (Volckmar, 2008). Since the late 1980s, the Norwegian school curricula such as M87 (Curriculum for the 10-Year Compulsory School 1987), *R94* (Reform 1994), *R97* (Reform 1997), *LK06* (Knowledge Promotion Reform 2006), and *LK20* (Renewed Knowledge Promotion 2020) have all explicitly communicated a value orientation that aligns with social and political ideas of what education should offer young people and society (Smeplass & Haugseth, 2025). These curricula, which have been widely accepted in the population, emphasize democratic values, equality, and social responsibility, ensuring that education not only imparts knowledge but also encourages active, informed, and responsible citizens. A central component of this educational philosophy has been the concept of *bildung* (or *danning* in Norwegian), which refers to the holistic social formation of individuals. This concept has been a core element of educational goals for decades and is widely accepted as an equally important objective. Understanding this context is fundamental for comprehending the current status and structure of Norwegian VET.

The Integration of VET into Upper Secondary Education

Since 1994, VET has been an integrated part of the Norwegian upper secondary system. Various vocational schools were now incorporated into the general educational framework. Students, starting at the age of 15, can enroll in a school-based program where they receive both academic and vocational training, with opportunities to specialize further in subsequent years. With the 1994 reform, all students also gained a legal right to an upper secondary education. The development of Norway‘s new upper secondary school system was driven by both utilitarian and democratic considerations, as there was a strong belief in the value of general and higher education as tools for economic growth (Nyen & Tønder, 2014; Persson Thunqvist et al., 2019). A key aspect of the debate was the recognition that modernization and the rapid pace of change in the labor market required students to acquire more theoretical knowledge and be better prepared for lifelong learning (Olsen, 2008). The reform also aimed to create more pathways into higher education for all students, regardless of background (Nyen & Tønder, 2014). During the 1980s, changing competence demands in industry and services led to many young people facing a shortage of employment opportunities. For many, the school system became an essential alternative, especially for those without formal qualifications, as jobs for untrained youth became more scarce. The VET reform of 1994 was quickly seen as a success, as it allowed more students to access their preferred programs. The government made significant investments to expand training options, resulting in an immediate increase in VET enrollment, with the share of youth entering vocational education rising from 45% to 55% of the youth cohort (Telhaug, 2006: 101). However, shortly after implementation, it became clear that the 2+2 model (2 years of general education followed by 2 years of vocational training) required more flexibility. To address this, the reform introduced expanded opportunities for students with low school motivation to receive more training through company-based apprenticeships (Legislative Amendment 80, 1992–1993). Additionally, the reform established the Follow-up Service (*Oppfølgingstjenesten*, or OT), which was tasked with providing additional support and guidance for students at risk of dropping out. This service aimed to ensure that students struggling with academic challenges or disengagement received personalized assistance to successfully complete their education. The Follow-up Service played a key role in improving coordination between primary education counseling, various upper secondary schools within a county, and municipal pedagogical and psychological services (Ibid.).

Thirty years later, the service continues to exist and has expanded its scope, helping students up to the age of 24 who have not applied for upper secondary education, interrupted their training, or lost their right to education. Functioning at the county municipal level—the same as VET provision, the services have various forms of collaboration with NAV. Through various efforts and project, these public services work jointly to promote youth employment when possible. The Follow-up Service facilitates access to alternative learning pathways, job placements, and other support services, striving for inclusivity and accessibility for youth at risk of non-completion. This ongoing service plays a vital role in ensuring that all youth, regardless of their challenges, have the opportunity to succeed and transition into either the labor market or further education.

Features of the Tripartite Collaboration

An essential feature of the Norwegian system is the tripartite collaboration, a longstanding model where the government, labor unions, and employers' organizations work together to shape labor policies, wage agreements, and social welfare initiatives. This collaboration has been crucial in developing the Norwegian welfare state and ensuring its sustainability. In Norway, more than half of all working are organized in a union. There are four main federations for workers which collectively represent a large number of unions and their members. These federations play a key role in negotiating workers' rights, wages, and working conditions, with most unions being affiliated, although some remain independent. Labor unions, particularly vocational unions, have played a pivotal role in developing the VET model, not only in advocating for workers' rights but also in shaping significant elements of Norway's education system (Nyen & Tønder, 2014; Teige & Stuart, 2013). They have been instrumental in developing VET curricula and ensuring fair working conditions. Over the years, unions have collaborated closely with government agencies and employers to create a balanced system that supports both economic productivity and social welfare. It is through this collaboration Norway has maintained low inequality and high labor market participation.

Landsorganisasjonen i Norge (LO), Norway's largest and most influential trade union confederation, was established in 1899. LO represents a broad range of workers from various sectors, including industry, public services, and the private sector. The organization was founded to unify trade unions and advocate for workers' rights, with the goal of improving working conditions, wages, and social benefits for its members. LO's close ties with the Labour Party (Arbeiderpartiet) have historically contributed to promoting social democratic policies and the development of the Norwegian welfare state. Although LO's primary mission is to represent the interests of its members, it also plays a significant role in shaping broader political and social policy in Norway.

On the employers' side, the Confederation of Norwegian Enterprise (NHO), founded in 1989 (from three organizations with the oldest dating back to 1886), represents a wide range of businesses and industries across Norway. NHO's membership includes over 33,000 companies, employing more than 700,000 workers (NHO, 2025). It plays a central role in shaping labor market policies, negotiating collective agreements, and advocating for employers' interests in various sectors. NHO works closely with LO, particularly through the 'Main Agreement', a key framework governing labor relations in Norway. This agreement is often referred to as the constitution of Norwegian labor relations and is a vital tool for ensuring stability in the labor market. NHO also represents Norwegian business interests within the European Union through its office in Brussels and promotes sustainable business practices. NHO's influence extends beyond labor relations, encompassing workplace safety, training, and skill development, all of which contribute to the competitiveness and sustainability of Norwegian industries (Norwegian Government, 2021b).

The tripartite collaboration ensures that VET in Norway remains relevant to the needs of the labor market while promoting inclusivity and equal opportunity. During the development of the 1994 VET reform, both LO and NHO were heavily involved in ensuring that the educational pathways aligned with industry needs and provided sufficient opportunities for all youth. They were also concerned with creating pathways for adults to obtain qualifications based on prior experience, through measures such as the assessment of competence pathway. The collaboration also highlighted the importance of creating clear transitions from VET to higher education, ensuring that students could continue their learning if they wished to do so (Nyen & Tønder, 2014).

A significant challenge in expanding VET in the 1990s was the need to secure enough apprenticeships and training placements in businesses. To address this, LO and NHO worked together to promote more robust apprenticeship schemes and make them less dependent on economic fluctuations. By committing to a joint information campaign, both parties aimed to raise awareness of the importance of VET to businesses and its long-term benefits, not only for the economy but also for job security. LO and NHO also set concrete goals for increasing the number of apprenticeships and practical placements in various sectors (Nyen & Tønder, 2014).

Moreover, the 2+2 model, which forms the basis of the modern Norwegian VET system, was developed with input from both LO and NHO. The model for the most part comprises 2 years of school-based education followed by 2 years of workplace-based learning, where 1 year is considered as training and the other as value creation. The government provides financial subsidies to businesses that offer apprenticeship placements, thus reducing the cost of providing vocational training. This ensures that businesses share responsibility for the apprenticeship process, while the government supports the educational component through subsidies (Norwegian Government, 2021b).

The Structure and Regulation of Norwegian VET

Today, there are ten vocational education clusters[1] in upper secondary school that lead to 196 different trade and journeyman certificates. These clusters cover a range of sectors, from agriculture, fishing, and forestry to building and construction, and from healthcare, childhood, and youth development to technological and industrial production. In each sector, students are trained for specific roles in the labor market, combining classroom learning with practical experience.

The system is structured as a sequential process, with each stage building on the knowledge and skills developed in the previous one (Smeplass & Schmees, 2023). In the first year (called Vg1), students receive a foundational education that combines general subjects (such as Norwegian, mathematics, and English) with program-specific subjects. During this year, students are introduced to the basic tools, techniques, and work processes relevant to their chosen vocational field. One key element of Vg1 is Vocational specialization (*yrkesfaglig fordypning* or YFF), which gives students a chance to explore specific trades in a practical setting. In YFF, students are placed in companies to gain experience with the content, tasks, and work methods within various professions. The VET teacher helps students try out professions in which they may potentially secure an apprenticeship after the second year. Through the subject, students can make connections with potential apprenticeship companies, and many find apprenticeships through vocational specialization. The subject also helps them gain insight into their future profession

[1] Agriculture, fishing and forestry, Building and construction, Crafts, design and product development, Electrical engineering and computer technology, Hairdressing, floral, interior and retail design, Healthcare, childhood and youth development, Information technology and media production, Restaurant and food processing, Sales, service and tourism and Technological and industrial production

and make more informed decisions about their next steps. In the second year (Vg2), students narrow their focus by selecting a specific program area within their vocational track. The education continues to be a blend of theoretical and practical learning, with a deeper focus on the competencies needed for the chosen occupation. YFF continues in this year, providing more opportunities for hands-on experience and engagement with potential future employers. The third year (Vg3) typically focuses on completing the certification or further specialized education. Students either continue in school or enter an apprenticeship in a company. This stage builds on the skills and knowledge gained in Vg1 and Vg2, leading to the completion of a journeyman's certificate or trade qualification. Students apply their learning in a real work environment, gaining direct experience in their chosen field. This sequential structure allows students to gradually progress from general knowledge to specialized skills, with opportunities to engage with the workforce at various stages, ensuring that education aligns with both industry needs and individual career development. The YFF helps students make informed decisions by offering practical insights into the work they will eventually pursue, providing them with experience from various vocations prior to making decisions about apprenticeship.

The VET system is regulated primarily through the Education Act (*Opplæringsloven*), which ensures that the system maintains high quality standards. This act outlines the responsibilities of public authorities, including county authorities, in ensuring that VET meets the required educational standards. The integration of VET into upper secondary has created a strong and popular system: since 2015, the proportion of students in VET has remained stable at around 50%, with a gradual increase from 49.9% to 51.7% in 2023 (Statistics Norway, 2024a, b, c). The Norwegian Directorate for Education and Training (Utdanningsdirektoratet, also known as Udir) is a government agency under the Ministry of Education and Research. Established in 2004, Udir is responsible for overseeing and developing Norway‘s educational system, including vocational training. The public authorities, including county authorities, ensure the quality of VET by appointing *vocational training committees* that advise on programs and approve training companies. These committees include representatives from employers, employees (through unions), and the government, with each party working collaboratively to ensure that VET programs meet industry standards and benefit apprentices. The Collaboration Council for VET (*Samarbeidsrådet for yrkesopplæring*, SRY) provides national advice and initiatives for VET, while county-level training boards and vocational training committees help oversee and coordinate apprenticeships and training opportunities at the local level.

The curriculum for VET (*Læreplanverket*) describes the fundamental approach to teaching, outlining key values, skills, and competence goals for the subjects. The curriculum consists of the general section (*Overordnet del*), subject and time distribution, and subject-specific curricula. The subject curricula describe the competencies that students are expected to achieve in each subject, including the relevance and values of the subject, core elements, interdisciplinary themes, and basic skills. They also outline the competence goals for different stages of education, descriptions of formative and final assessments, and regulations regarding final evaluations. The competence goals express what students and apprentices are expected to master upon completing their education at various stages of the training process. These documents are formal regulations under the education act and serve to govern the content of VET, ensuring it aligns with both national standards and industry needs. The curriculum guides the work of VET teachers as well as companies and their instructors, though they have the autonomy to design their teaching methods to fit the individual needs of students. Adapted education is required, ensuring that each student's training is tailored to their progress and challenges (LOV-2023-06-09-30). This is supported by initiatives like the mentioned Follow-up Service, which provides support to ensure that all students complete their education successfully.

In terms of apprenticeships, the Completion Reform (Norwegian Government, 2021c) aims to improve opportunities for all students to complete upper secondary education, ensuring that students who do not secure apprenticeships have access to alternative options. The reform includes expanded rights to apprenticeships, increased support for immigrant students, and stronger measures to address students' mental health and wellbeing. County authorities oversee training companies, ensure compliance with the Education Act, and provide financial support for apprenticeships. Training agencies, which are collaborative bodies involving multiple companies and industry organizations, assist in recruiting and supporting apprentices, ensuring that the apprenticeship system runs smoothly and meets educational standards.

The social partners, including labor unions and employer organizations, have significant representation in advisory bodies at both the national and county levels. These include the Collaboration Council for VET (SRY) at the national level, vocational training councils for specific trades, county vocational training boards, and trade-specific examination boards, all of which play a critical role in shaping and ensuring the quality of VET in Norway.

Attractivity of the Norwegian System

The Norwegian VET system offers a range of attractive features for students, making it a compelling choice. A key benefit is the system's flexibility and security. Since VET is integrated into upper secondary education, students are assured that they can complete their education, even if they later decide to pursue a different path (Norwegian Government, 2021c). This way, the system provides legal rights for students to continue their education, allowing them to transition into academic tracks after the first 2 years without extending their time in school. Supplementary studies qualifying for higher education can be pursued either before or after an apprenticeship. This reduces the risk of making an irreversible decision, providing a safety net for those unsure about their future career direction. In 2024, 53% of applicants to upper secondary education chose VET as their first preference, marking the highest proportion since 2012 (Udir, 2024a). This trend is particularly notable among girls, whose preference for vocational tracks has reached its highest level in over a decade. While traditionally male-dominated fields like technology and industrial subjects continue to attract significant interest, there is also a steady rise in the number of young women entering these areas, contributing to the diversification of vocational pathways.

Apprenticeships also offer significant advantages, not only as an engaging way to gain professional skills but also by providing financial support. Apprentices earn a salary while receiving on-the-job training, which is a strong motivator for many students. In Norway, the salary for apprentices is structured to reflect their progress in the training process. For example, in the early stages of their apprenticeship, apprentices earn about 30% of the full wage for skilled workers in their field, with this percentage increasing over the course of the training. By the end of their apprenticeship, apprentices can earn as much as 80% of a fully qualified worker's salary.

Moreover, the Norwegian system includes strong protections for employees, which benefits apprentices as well. Due to the historical influence of labor unions, employees in Norway enjoy a regulated workweek, paid holidays, and tariff-based wages (LOV-2005-06-17-62). Over time, these protections have expanded to include agreements on pay, pensions, and job security. The social partnership between employers and unions plays a crucial role in ensuring fair conditions for workers across various sectors, including apprenticeships.

For many students, the prospect to earn money while learning a trade is an important incentive. Apprenticeships in high-demand sectors, such as the aquaculture industry, offer substantial earnings. For instance, apprentices in aquaculture can earn over 34,000 NOK per month (approximately €3020), which is nearly double the salary of apprentices in fields like hairdressing, where the average monthly salary is around 17,820 NOK (approximately €1580) (Statistics Norway, 2024b). The variation in salary is partly due to the difference in wage levels across industries, but also reflects the responsibilities and expertise required in certain sectors. Students also have an option to supplement the apprenticeship salary with a student loan, with slightly different rules for those under and over 21 years.

In addition to the wages apprentices receive, they are supported by subsidies to businesses that take them on. These subsidies help cover training costs and ensure that apprentices are given the appropriate guidance and opportunities for development. The Norwegian government, through the Norwegian Directorate for Education and Training (Udir), provides financial support to approved training companies, helping to maintain high-quality education and training standards. The government provides a monthly subsidy for apprentices following the main training model, with varying amounts depending on the type of apprenticeship (Udir, 2024b). For the most common path, which includes 2 years of school-based learning and 2 years in the workplace, the subsidy is 7328 NOK per month (approximately €650). For apprentices following an extended training model, with 2 years of practical learning, the subsidy is higher, at 9771 NOK per month (approximately €865).

There are additional subsidies for apprentices in fields with high cultural value or those over 21 years of age, ensuring that no apprentice is left without support during their training. Apprentices working in traditional craftsmanship fields, considered crucial for cultural preservation, are eligible for an additional subsidy of 68,196 NOK per year (approximately €6000). For apprentices over 21, who may have already exhausted their right to further secondary education, the government provides a lower subsidy of 5728 NOK per month (approximately €510) (Statistics Norway, 2024b).

The financial support and salary structure for apprenticeships are designed to ensure that apprentices can support themselves during training but also to make apprenticeship a viable and appealing option for young people entering the workforce. This support system plays an essential role in Norway's high employment rate among those with vocational qualifications. In fact, Norway has one of the highest employment rates for individuals aged 25–34 with vocational qualifications at the upper secondary level, with around 93% of these individuals employed, compared to the OECD average of 83% (Stengrundet & Nygård, 2024).

Norway's labor market, highly regulated and focused on worker welfare, further contributes to the attractiveness of the VET system. The Working Environment Act (*Arbeidsmiljøloven*) guarantees workers a safe and healthy work environment. This law covers not only physical safety but also mental health, with provisions for workplace accommodations, working hours, dismissal protections, and more. Workers are entitled to paid leave, including parental, sickness, and vacation leave, ensuring they are supported during health issues or life events.

Another key feature of the Norwegian labor market is the high rate of union membership, which plays a pivotal role in negotiating favorable working conditions. Unions are actively involved in collective bargaining, securing fair wages and benefits for workers across sectors. The network of unions in Norway, alongside strong legal protections, ensures that the labor market remains secure and equitable for all workers, including apprentices. As this is highly institutionalized in society, students and their families have high trust in the apprenticeship arrangements and the legal support that is provided throughout the training. The quality assurance systems provided by the county municipalities and the Norwegian Labour Inspection Authority also strengthens this view that VET is a safe and attractive choice.

Through this combination of flexible education options, financial support, strong labor protections, and active union involvement, the Norwegian VET system offers an attractive model for students, ensuring both personal development and financial security as they choose a pathway in upper secondary training.

A Strategic Investment in Youth Through VET

Norway stands out for its significant investment in education, allocating substantial resources to ensure that every citizen, particularly the youth, has access to equitable and relevant education. The emphasis on VET as a central component of the upper secondary education system has become a key factor in Norway's strategy to reduce youth unemployment, increase social mobility, and align educational outcomes with labor market demands.

On a critical note, the Norwegian VET system faces several challenges, similar to those observed in other contexts. Completion rates for VET are slightly lower than for general academic tracks (Halvorsrud, 2017), highlighting the need for further research into strategies for supporting students, particularly those from vulnerable groups. Dropout rates are linked to factors such as social class, gender, and disability, indicating that despite existing measures, the system still struggles to fully address these disparities. Additionally, the high cost of maintaining the system continues to draw political scrutiny, reflecting the significant expectations placed on its outcomes and efficiency.

One of the most attractive features of the Norwegian VET system is its flexibility. Unlike in many other countries where choosing a vocational track could limit future options, Norway offers a system that has standardized certain components through the curriculum, providing a main model with potential for adjustments and furthermore allows students to transition between academic and vocational pathways. This contributes to reducing the individual's risks associated with making wrong choices at an early age. For students and families, this is likely an important feature promoting VET as a choice. The integrated structure of the sequential VET system, where the first 2 years are a mix of theoretical and practical learning, means that students are not locked into one career path from the outset. They can also explore different fields through the vocational specialization subject (YFF) and furthermore change direction if they decide that another path suits them better, without necessarily adding substantial time to their educational journey. The training is in any case covered by the welfare state.

However, this flexibility also extends beyond academic mobility and contributes to the development of a professional identity and pride. The system can nurture a sense of belonging and identity for young people as they begin to see themselves as part of a specific profession or trade at a young age. Professional pride is cultivated not only through the mastery of skills but also through the deeper understanding of one's role within a larger professional community. Whether through work-based learning, apprenticeships, or specialized training, young people can gain an integrated sense of purpose and commitment to their chosen professions. This pride is reinforced by the recognition that they are contributing to industries that benefit society and the economy, thus strengthening their commitment to quality and service in their chosen trade.

Companies are also highly motivated to participate in the apprenticeship system, as it offers them the opportunity to recruit future employees who are trained in the specific skills they need. By providing apprenticeships, businesses can ensure a continuous supply of qualified workers, while also having a direct role in shaping the competencies and qualifications that students acquire during their training. This active participation in VET helps companies secure the relevance and quality of training, ensuring that the skills taught align with the real-world demands of the industry. As such, the apprenticeship system becomes a mutually beneficial arrangement: students gain valuable work experience and financial support, while companies can foster a workforce that meets their operational needs and contribute to the ongoing development of the VET system.

Furthermore, the apprenticeship component of the system ensures that students are gaining the theoretical knowledge needed for their chosen profession while also acquiring hands-on experience in real work environments. This practical training is a crucial aspect of the system's success, as it allows students to apply their learning in context, making them more prepared for employment once they complete their education. In this respect, VET in Norway serves educational goals and plays a vital role in ensuring that youth are adequately equipped to enter the labor market with the necessary skills and qualifications.

The Norwegian VET system also offers attractive financial incentives for students. Since apprentices are paid during their training, the apprenticeship pathway is both attractive and feasible. Norway‘s government further supports this system by offering subsidies to businesses that take on apprentices, ensuring that the cost of training is shared between employers and the state. This structure enhances the accessibility of apprenticeships, especially in sectors where the supply of training placements may be limited. Variation in apprentice salaries across different industries highlights how the system adapts to the unique demands of each sector. For example, apprentices in high-demand industries like aquaculture often receive higher wages due to the specialized skills and market demand, while sectors like hairdressing, which are also essential and skill-based, offer salaries that reflect industry standards. This flexibility in wage structures demonstrates how the system supports a wide range of professions, each contributing to the economy in important ways. For many young people, the opportunity to earn while learning, with salaries that reflect the value of the profession and the skills being developed, is likely a key motivator for choosing the apprenticeship route.

At the same time, the strong legal protections provided to apprentices and workers in Norway add an additional layer of security. Labor unions, a significant force in Norway's political and social landscape, have played an instrumental role in ensuring that workers' rights are upheld, including those of apprentices. This protection is embedded in Norway's comprehensive labor laws, which guarantees safe working conditions. For apprentices, knowing that they are covered by such protections provides peace of mind as they enter the workforce and begin their careers.

I argue that the concept of *youth investment* is central to Norway's approach to vocational education and training, which aims to equip young people with the skills and opportunities necessary for active participation in the workforce and society. While financial support and security certainly make the apprenticeship route attractive, its appeal also lies in its integration with broader social policies that emphasize participation, productivity, inclusivity, equal opportunities, and career mobility. A key pillar of Norway's vocational education and training system is the state's commitment to ensuring that all young people, regardless of background, have access to high-quality education and training. This commitment reflects the broader social contract, where education, labor market participation, and social protections are closely interconnected. As shown, the system's attractiveness stems from its alignment with Norway's wider welfare and economic policies. By combining education with labor market participation and strong social protections, the system provides young people with a secure entry point into the workforce and opportunities for further education and career advancement. This comprehensive approach ensures that VET as a pathway to employment is a long-term investment in youth as active contributors to the economy and society.

References

Antikainen, A. (2006). In search of the Nordic model in education. *Scandinavian Journal of Educational Research, 50*(3), 229–243.

Djupvik, A. R., & Eikås, M. (2024). Activation policy and discretion in the Norwegian Labour and Welfare Administration (NAV)—introducing a multi-layered explanatory model. *European Journal of Social Work*, 1–13. https://doi.org/10.1080/13691457.2024.2405849

Dobbins, M., & Busemeyer, M. R. (2015). Socio-economic institutions, organized interests and partisan politics: the development of vocational education in Denmark and Sweden. *Socio-Economic Review, 13*(2), 259–284.

Dølvik, J. E., Fløtten, T, Hippe, J. M & Jordfald, B. (2014). *Den nordiske modellen mot 2030. Et nytt kapittel? Rapport 2014:46*. Fafo. https://www.fafo.no/images/pub/2014/20393.pdf

Esping-Andersen, G. (1990). *The three worlds of welfare capitalism*. Polity.

Halvorsrud, K. (2017). Student dropout in upper secondary education in Norway: A challenge to the principles of the welfare state? *London Review of Education, 15*(2), 302–316. https://doi.org/10.18546/LRE.15.2.12

Helsvig, K. G. (2014). *1814-2014 Kunnskapsdepartementets historie*. Ministry of Education and Research.

Hyggen, C. (2010). *Livsløp i velferdsstaten, NOVA Rapport, 6/2010*. Norsk institutt for forskning om oppvekst, velferd og aldring. https://docplayer.me/3296462-Livslop-i-velferdsstaten.html

Karlsen, G. E. (2006). *Utdanning, styring og marked*. Universitetsforlaget.

Lægreid, P., & Rykkja, L. H. (2014). One-stop-shops for social welfare: The NAV office in Norway. In *Organizing for coordination in the public sector* (Public sector organizations) (Vol. 12, pp. 78–90). Palgrave Macmillan. https://doi.org/10.1057/9781137359636_7

Legislative amendement 80. (1992–1993). Innst. O. nr. 80. (1992–93) Innstilling fra kirke- og undervisningskomiteen om lov om endringer i lov av 21. juni 1974 nr. 55 om videregående opplæring og i lov av 23. mai 1980 nr. 13 om fagopplæring i arbeidslivet. (Ot.prp. nr. 31.). https://www.stortinget.no/no/Saker-og-publikasjoner/Stortingsforhandlinger/Lesevisning/?p=1992-93&paid=6&wid=ba&psid=DIVL641&pgid=ba_0702

LOV-2005-06-17-62. Lov om arbeidsmiljø, arbeidstid og stillingsvern mv. (arbeidsmiljøloven). Ministry of Labour and Social Inclusion. https://lovdata.no/dokument/NL/lov/2005-06-17-62

LOV-2023-06-09-30. Lov om grunnskoleopplæringa og den vidaregåande opplæringa (opplæringslova). Ministry of Education and Research. https://lovdata.no/dokument/LTI/lov/2023-06-09-30

Michelsen, S. (2018). The historical evolution of vocational education in the Nordic countries. In S. Michelsen & M.-L. Stenström (Eds.), *Vocational education in the Nordic countries: The historical evolution* (pp. 1–23). Routledge.

NHO. (2025). *Om Næringslivets Hovedorganisasjon (NHO)* [About the Confederation of Norwegian Enterprise (NHO)]. Webpage https://www.nho.no/om-nho/

Norwegian Government. (2021a). *Meld. St. 14 (2020–2021). Perspektivmeldingen 2021.* Ministry of Finance. https://www.regjeringen.no/no/dokumenter/meld.-st.-14-20202021/id2834218/?ch=3

Norwegian Government. (2021b). *NOU 2021: 2 Kompetanse, aktivitet og inntektssikring – Tiltak for økt sysselsetting*. Ministry of Labour and Social Inclusion. https://www.regjeringen.no/no/dokumenter/nou-2021-2/id2832582/?ch=4

Norwegian Government. (2021c). *Meld. St. 21 (2020–2021). Fullføringsreformen – med åpne dører til verden og fremtiden*. Ministry of Education and Research. https://www.regjeringen.no/no/dokumenter/meld.-st.-21-20202021/id2840771/

Norwegian Government. (2023). *Meld. St. 33 (2023–2024). En forsterket arbeidslinje – flere i jobb og færre på trygd*. Ministry of Labour and Social Inclusion https://www.regjeringen.no/no/dokumenter/meld.-st.-33-20232024/id3052167/?ch=2

Nyen, T., & Tønder, A. H. (2014). Yrkesfagene under press. Universitetsforlaget.

Nylund, M., & Rosvall, P. Å. (2019). Vocational education, transitions, marginalisation and social justice in the Nordic countries. *European Educational research journal, 18*(3), 271–277. https://doi.org/10.1177/1474904119838893

OECD. (2024). *Norway. Overview of the education system (EAG 2024)*. https://gpseducation.oecd.org/CountryProfile?primaryCountry=NOR&treshold=10&topic=EO

Olsen, O. J. (2008). *Institusjonelle endringsprosesser i norsk fag- og yrkesutdanning. Fornyelse eller gradvis omdannelse?* Stein Rokkan Centre for Social Studies https://hdl.handle.net/1956/4267

Øverbye, E., & Stjernø, S. (2012). *Arbeidslinja: arbeidsmotivasjonen og velferdsstaten*. [The work-first principle: Work motivation and the welfare state].Universitetsforlaget.

Persson Thunqvist, D., Hagen Tønder, A., & Reegård, K. (2019). A tale of two reforms: Institutional change in vocational education and training in Norway and Sweden in the 1990s. *European Educational Research Journal, 18*(3), 298–313. https://doi.org/10.1177/1474904118823104

Smeplass, E. (2024). *Yrkesskoler*. Store norske leksikon. https://snl.no/yrkesskole.

Smeplass, H., & Haugseth, J. F. (2025). A pragmatic exploration of Norwegian education: National curriculum and professional engagement. *Historical Social Research / Historische Sozialforschung, 50*(3), 127–156. https://doi.org/10.12759/hsr.50.2025.33

Smeplass, E., & Schmees, J. K. (2023). Stärken und Herausforderungen des sequenziellen Systems der Berufsausbildung in Norwegen. *Berufsbildung*. https://doi.org/10.3278/BB2303W016

Statistics Norway. (2024a). *Videregående opplæring og annen videregående utdanning. 12609: Utvalgte nøkkeltall for videregående opplæring, etter region*, statistikkvariabel og år https://www.ssb.no/statbank/table/12609/tableViewLayout1/

Statistics Norway. (2024b). *Så høy er lærlinglønnen* [These are the rates of the apprenticeship salaries]. https://www.ssb.no/arbeid-og-lonn/lonn-og-arbeidskraftkostnader/artikler/sa-hoy-er-laerlinglonnen

Statistics Norway (2024c). *Arbeidskraftundersøkelsen*. https://www.ssb.no/arbeid-og-lonn/sysselsetting/statistikk/arbeidskraftundersokelsen

Stengrundet, A. & Nygård, G. (2024). *«Alle» med yrkesfaglig utdanning i Norge er i jobb*. Statistics Norway. https://www.ssb.no/utdanning/videregaende-utdanning/artikler/alle-med-yrkesfaglig-utdanning-i-norge-er-i-jobb

Teige, B. K., & Stuart, M. (2013). Trade union involvement in lifelong learning in Norway. In Cooney, R. & Stuart, M. Trade unions and workplace training (pp. 126–143). Routledge.

Telhaug, A. (2006). *Skolen mellom stat og marked: norsk skoletenkning fra år til år 1990-2005*. Didakta norsk forlag.

Telhaug, A., Medias, O., & Aasen, P. (2006). The Nordic model in education: Education as part of the political system in the last 50 years. *Scandinavian Journal of Educational Research, 50*, 245–283.

Thinqvist, D., Hagen Tønder, A., & Reegård, K. (2019). A tale of two reforms: Institutional change in vocational education and training in Norway and Sweden in the 1990s. *European Educational Research Journal, 18*(3), 298–313. https://doi.org/10.1177/1474904118823104

Tønnessen, L. K. B. (2011). *Norsk utdanningshistorie: en innføring med fokus på grunnskolens utvikling*. Fagbokforlaget.

Udir. (2022). *The Norwegian Education Mirror 2022*. https://www.udir.no/in-english/the-education-mirror-2022/cost-of-kindergarten-provision-and-primary-and-secondary-education/higher-spending-per-pupil-in-norway-than-in-most-other-countries/

Udir. (2024a). *Flere søker yrkesfag* [More applicants to vocational education and training]. https://www.udir.no/tall-og-forskning/statistikk/statistikk-videregaende-skole/analyser/2024/sokertall-vgo-2024/

Udir. (2024b). *Satser lærlinger, praksisbrevkandidater, lærekandidater og kandidater for fagbrev på jobb 2024* [Rates for apprentices, practice certificate candidates, training candidates, and candidates for trade certificates at work 2024]. https://www.udir.no/om-udir/tilskudd-og-prosjektmidler/tilskuddssatser/2024/satser-larlinger-praksisbrevkandidater-larekandidater-og-kandidater-for-fagbrev-pa-jobb-2024/

United Nations. (2024). *Human development report 2023/2024. Reimagining cooperation in a polarized world*. United Nations Development Programme. https://hdr.undp.org/system/files/documents/global-report-document/hdr2023-24reporten.pdf.

Volckmar, N. (2008). Knowledge and solidarity: The Norwegian social-democratic school project in a period of change, 1945–2000. *Scandinavian Journal of Educational Research, 52*, 1–15.

CHAPTER 13

Localised Capacity to Realise the Potential of Vocational Education

Stephen Billett

Meeting and Advancing Localised Needs Through VET

It has become orthodox for the purposes and processes of vocational education and training (VET) to be developed in and mandated by centralised national agencies (e.g., governments, employers, unions, professional bodies) in countries such as the United Kingdom, Australia, New Zealand, Singapore and those in Scandinavia, for examples. This often occurs under the aegis of meeting the demands of key national interests and agencies (Edwards, 2002). To a degree, there is good reason for this emphasis. That is, these stakeholders can indicate the kinds of demand and skills required for occupations nationally, which then informs the educational purposes for courses responding that demand and means of recognising and certifying occupational capacities nationally. Yet, part of the responding to the demand side is also being able to meet how

S. Billett (✉)
Griffith University, Brisbane, QLD, Australia
e-mail: s.billett@griffith.edu.au

E. Smeplass and C. Zirkle (eds.), *Opportunity Structures in Vocational Education and Training*, Palgrave Studies in Technical and Vocational Education and Training,
https://doi.org/10.1007/978-3-032-04410-5_13

such requirements are manifested at the local level. This is because how occupations are practiced is not standardised and uniform and enterprises have particular needs, and it is they that make decisions on employment and training of their staff, not centralised agencies or abstracted industry or professional institutions (Billett, 2000). So, whilst necessary to a degree being responsive to such agencies, alone it is an insufficient basis for VET systems to be directed, organised, and their efforts enacted, and their success judged. Not often privileged, but important at the local level, beyond being responsive is to bring about change by advancing the capacities and potential of local communities in which the enterprises are located. Indeed, perhaps there has never been more important time when VET provisions play this role, as local enterprises and communities face the challenges of responding to social, environmental and economic changes. This includes working with public and private sector enterprises and encouraging and supporting innovations, extending their economic activities, through skill formation, supporting innovations and building capacities locally. All of this is consistent with the overall focus of this volume, the concept of 'opportunity structures', viewed here as organisational frameworks within education, training and the workforce that open pathways, fostering accessibility and progression for individuals through VET. These structures comprise policies, practices and collaborative networks that can together create a positive social and economic environment in which individuals can be informed about and navigate their educational and career trajectories. Here these are proposed as needing to occur at the local level. A central premise, therefore, is the need to professionalise education and training capacities and routes at the local level and in ways that lead not only to formal certification, open new career and employment opportunities for both young people and older working age adults and of all kinds of backgrounds and educational achievement.

Whilst more than any other educational sector, VET is well placed to perform these roles and it needs to collaborate at a local level and through engagement with public and private sector enterprises, schools and social partners based within their communities. Only at the local level can the needs and circumstances of local enterprises, schools, community needs be understood, partnerships formed and outcomes realised and responded to (Billett, 1998b; Skilbeck et al., 1994). This localised requirement seems to be the case regardless of whether the nation state has a developed or developing economy (UNESCO-UNEVOC, 2018a;

UNESCO, 2018). Moreover, what constitutes the 'local' needs to be understood situationally. It is usually much more than being purely about being close physically or geographically, as it has social and personal dimensions. Geographically, it can variously comprise a small sized location, such as a suburb of a city, or much larger such as an entire region in a rural or remote location, defined either by distance or geographical factors, or even the kind of economic activity that might occur there. That is, it could primarily be associated with a particular set of socio-economic, ethnic or industry factors (e.g., one in which migrant communities might inhabit, or a farming or mining community). Alternatively, it may comprise a geographical region that shares similar dialects, societal or religious sentiments. However, how individuals view, identify or dis-identify with those communities and engage with them is central to how they come to associate or disassociate with them (Hodges, 1998). Therefore, it is these relational factors that might construe what constitutes 'local'. This complex of factors is also instructive as it indicates how these are likely not to be fully understood and be responded to by central agencies. Hence, there are considerations about how VET can respond to local needs as it encompasses issues of proximity or distance, homogeneous or heterogeneous populations and individual subjectivities.

Yet, in the current era, VET has been often positioned as an educational system whose key role is to be responsive and reactive to the needs of central agencies (national/regional government, key industry, employer and union bodies) (Stevenson, 2001; Wheelahan et al., 2022). This often leads to top-down intents, implementations and evaluations that fail to understand, be responsive to and meet the needs of its communities that have their own imperatives, aspirations, institutions, capacities and potentials, as these remain unknown to and and unacknowledged by such agencies (Billett, 2011). Such an orientation is quite remote from what Dewey (1916) proposed about VET needing to balance the needs of employers against those of broader educational goals and interests. Moreover, exercising centralised mandates is often accompanied by overtly instrumental processes that are antithetical to meeting localised needs. This has been shown to have deleterious outcomes, such as focusing on 'training' aligned to meet specific goals rather than adaptable outcomes and not through educational experiences that are shaped to meet the needs of and engage local communities thus ending up being unattractive to young people (Aarkrog, 2020; Billett, 2014; Rintala & Nokelainen,

2020). Such mandates that favour measurable outcomes may also be inappropriate or unhelpful in the development of higher-order capacities and adaptable outcomes, such as generating problem-solving skills required for sustaining employability (Organisation for Economic Co-operational and Development [OECD], 2013; Stevenson, 2001). A consequence here, globally, is for VET to be viewed by young people, their parents and enterprises as having a lower status and narrower outcomes than a university education (Clement, 2014; UNESCO-UNEVOC, 2018b). All of this does not position VET to respond to the range of global challenges being faced contemporaneously, many of which need to be addressed locally. However, the orthodoxy is for VET to be organised in quite top-down ways, perhaps largely because the social purposes it serves are seen to be beyond the understandings and capacities of those who might otherwise organise educational experiences.

The case is made through, firstly, discussing the positioning of VET in the contemporary era and its focus on responding to the entire demand side, including that generated locally, what constitutes that demand and how it can be addressed. Prior to proposing approaches informed by that program of research, how VET might be positioned, organised, and enacted to be more locally engaged and relevant that program of research is briefly overviewed. In concluding, the worth of proposing a localised approach to organising and enacting VET provisions locally is rehearsed.

A Top Down and Centralised Provision Vocational Education

It is not perhaps surprising that many, if not most, VET systems have their purposes and practices dictated in top-down ways. Indeed, most education systems were founded by state and/or religious institutions and VET systems are no exception here. There were particular purposes for the formation and development of such systems. Indeed, there are only a few instances of educational provisions being founded by individuals or local institutions. Some are associated with particular approaches to children's education (e.g., Montessori). For provisions of adult education, there are also examples being founded by individuals for specific provisions (e.g., Birkbeck), and through locally driven movements such as in Canada (Nesbit, 2011), and also the Scandinavian folk education movement. But these are the exceptions. Far more generally, VET systems have primarily been developed to address specific nationally derived goals or imperatives and their continuing progress has been based upon such imperatives (Hanf, 2002). The formation of the German VET system occurred at the time of the formation of the modern German state. Whilst its purposes were aligned with developing the skills required by that state, there were elements of it aligned with engaging young people (i.e., males) with the values of the new nation state, including avoiding the attractive rhetoric of an emerging social movement (i.e., socialism) (Gonon, 2009). Similarly, in Australia, the need for a national system of VET arose to provide material for prosecuting two world wars and then to meet the occupational education needs of returning service men and women (Dymock & Billett, 2010).

These purposes of VET have also been shaped through national debates and imperatives. Most noteworthy debate was between Dewey and Snedden (Dewey, 1915/1979; Drost, 1967) about what would constitute the American VET system given the need to provide for the hundreds of thousands of young Americans returning from the battlefields of Europe, yet who were ill-equipped for employment in civilian life. That debate was focused on the degree by which the American VET system would be either about broader educational goals (Dewey's perspective) (Dewey, 1915/1979) or being highly responsive to industry needs (Snedden)(Drost, 1967): i.e., about either social or economic efficiency (Labaree, 2011). The French VET system was shaped by a societal sentiment in the form of French republicanism that sought to separate education from work (Remery & Merele, 2014; Troger, 2002) that was quite distinct from the sentiment in Germany, which was to embed (vocational) education with work (Deissinger, 2000). But the general point here is that it has been societal sentiments and imperatives that have emphasised a top-down approach as being normative for this education sector. Moreover, this centrally driven approach is most exercised during times of economic and social crises when education systems of all kinds are subject to greater centralised control and mandates (Stevenson, 1991). Yet, it is often at these times that mobilising the interests and talents of entire working populations is perhaps most needed. Take for example the incredible outcomes of mobilising entire nations in the reconstruction of Germany and Japan after World War II (Sennett, 2008).

Yet, despite such orthodoxies, there is no tradition or justification for the belief that government and industry views do hold a monopoly on what constitutes VET and how it and its purposes are defined. The development of skilful capacities has occurred across human history and most of it in the absence of governmental efforts and industry mandates. Even the guilds were largely regulating bodies rather than being actively involved in the process of skill formation (Epstein, 2008; Rosser, 1997). They were concerned with regulating the ability to practice occupations, not their development. However, there are other voices and perspectives needing to be included when deciding about and enacting education provisions, albeit in schooling or tertiary education. For instance, the school-based curriculum movement (Skilbeck, 1984) emphasised the importance of educational purposes and provisions being shaped by local advice and imperatives, including accounting for the kinds and calibres of students, and providing relevant experiences for them. Whilst it is acknowledged that teachers are usually employees of state or religiously shaped institutions, they have no inherent right to make educational decisions (Skilbeck, 1984; Skilbeck et al., 1994). However, their ability to respond to student needs in shaping appropriate and relevant experiences is central to effective education provisions. This can be viewed as responding to the demand-side equation. So, in terms of both educational and social efficacy, there is a fundamental case for VET provisions to be shaped, enacted and evaluated in terms of localised imperatives, albeit informed by national social and economic goals. This is what is exercised, to some degree, through the German VET system that allows discretion at the local level as enacted through chambers of commerce (Deissinger, 1997; Deissinger & Ott, 2016).

Yet, more fundamentally, beyond practical considerations of educational effectiveness aside, there is something quite illiberal about such top-down and centralised organisation, governance and mandates of VET systems. This is not to suggest an anarchic 'anything goes' approach to VET provisions based on ad hoc localised needs. Instead, it is having an approach that is respectful of and responsive to localised circumstances and imperatives and able to be understood or responded to at a central level which co-ordinates nationally consistent goals and processes. However, to realise such an outcome, it is necessary to have local social, educational and administrative infrastructure. This then raises the question of what constitutes the demand side of the education equation, and the balance between centralised and localised demands.

Global Challenges and Localised Responses

Nation states, their regions and communities now face significant and sometimes existential challenges from the need to respond to climate change, enhanced global competition, accommodating mass migration, managing geopolitical tensions and addressing the need for greater self-reliance and self-sufficiency arising from those tensions. These imperatives are now shaping the key purposes of VET (Billett, 2024). So, there has perhaps never been an era when it is more important to review the purposes to which VET is directed to and how these can be realised locally, as defined above. That is, in ways accommodating the requirements of geographic place, community needs and engaging individuals. Yet, advancing communities' social and economic goals requires more than VET being highly responsive and reactive to national needs, but exercising the capacity to bring about change as well. These can include contributing to developing the community's capacities to extend its social and economic activities, identifying and realising localised potentials and generating the capacities to do so. These kinds of changes are becoming important goals for VET provisions. Indeed, when added to the constant changes in those requirements and workplace practices, the situated nature of occupational competence and workplace needs demands working age adults' continuing development across working life. This necessitates workplaces to provide support and guidance for all kinds and classifications of workers and, wherever possible, combining workplace innovations and workers' learning (Leow et al., 2023). Hence, this includes providing opportunities for workers to engage in identifying, initiating and appraising innovations (Hoyrup et al., 2012; Lemmetty & Billet, 2023).

Responsiveness and the Demand Side

Meeting these kinds of personal and societal goals is central to the project of VET and stands as premises through which its full promise and potential might be realised. However, to realise these outcomes means going beyond national mandates, ordering of occupational standards, regulation and legislation. There is a need for the development and enactment of local administrative, social (i.e., social partnerships) and educational (i.e., educational expertise) infrastructure (Billett, 2024). This includes having local infrastructure and expertise to develop effective partnerships, identifying educational needs and the organisation and enactment of experiences to achieve required outcomes. *Social infrastructure* includes partnerships supporting work placements, work experiences, employment opportunities and expressions of enterprises' requirements. *Administrative infrastructure* includes the intentional organisation and enactment of educational provisions and their certification. *Educational infrastructure* includes the provision and alignment of experiences to achieve learning outcomes, teachers' guidance and support and extending those efforts beyond the educational institution. These forms of infrastructure are required as means to meet communities' needs and optimise achieving VET's purposes. Much of what is proposed here can be optimised by engagements, interactions and partnerships outside of VET institutions, comprising the social infrastructure that brings together partners with shared interests and abilities to make distinct contributions in securing mutually understood goals.

Consequently, whilst national mandates for and the top-down organisation of VET are important, often and perhaps always, achieving these outcomes requires effective social, administrative and educational infrastructure at the local level. Certainly, advancing communities' social and economic goals requires VET to engage with industry and government needs. It also requires encouraging and supporting innovations in workplaces, extending existing social and economic activities locally and building local capacities, as occurs in and through what might be described as mature VET systems (Billett et al., 2023). Contrary to this, often in countries with both developed and developing modern economies, the orthodoxy is quite different. That is for VET to be organised, enacted and judged on its reactiveness to industry needs (i.e., the corporate demand side) and in very top-down ways. The common motif here is for VET systems to be responsive to central mandates, usually with little room for negotiation at the local level. Both global agencies (OECD, 2019) and national governments (Hanf, 2002) have sought to secure this responsiveness using detailed and prescriptive syllabus, and stipulated means of instruction and assessment to meet the demands of VET (Billett, 2011). Responsiveness to the needs of professional and industry bodies is a necessary element of educational intents (i.e., the knowledge to be learnt for working in that sector of employment). However, on its own, it is an insufficient means by which VET systems should be directed, enacted and judged. Not the least here because how occupations are practiced, and what constitutes performance requirements of work tasks differ across situations and circumstances (Billett, 2001). Moreover, such requirements cannot be confidently and effectively predicted at the central level, and instead need to be understood in the circumstances where they are manifested (e.g., locally), that is, by the local demand side. Consequently, perhaps more than other educational sector, being able to understand what constitutes the demand side of VET locally is necessary. It follows that having functional partnerships within local communities is necessary to respond to the demand side.

So, when considering what constitutes the privileging of the demand side, it is now timely to account more fully the local demand side as these are central to meeting the needs of communities, individuals who reside and enterprises that are located within them, and then collectively the needs and well-being of nation states. The importance of being locally responsive to the demand side and encourage innovations has been a consistent finding across a program of research into the purposes and practices of VET conducted over the last three decades, mainly, but not wholly in Australia.

A Program of Research

What is being advanced here is largely drawn from a three-decade long program of research aiming to understand what should constitute the purposes for VET, how it should be enacted and experienced to meet personal, social and economic needs of the nations, communities and individuals that it seeks to serve. As foreshadowed above, an outcome of this research program has been to delineate five contemporary purposes of VET: (i) engaging young people with VET; (ii) assisting them identify occupations to which they are suited; (iii) preparing them for occupations; (iv) continuing education across working life to meet changing needs and goals; and (v) aligning workplace innovations with workers' learning (Billett, 2024). However, the focus here is on how VET should be organised and enacted to realise those purposes. Not all, but some of the projects comprising that program of research are listed in Table 13.1. An early project comprised an evaluation of models of entry-level VET that found the combination of work and educational experiences was important for developing the kinds of capacities required for effective initial occupational preparation. This involved appraising the kinds of experiences able to be provided and how these need to be supported educationally, and that these can only be understood, enacted and exercised through local engagement, and not by remote mandates (Billett, 1994). An initial set of findings about having highly situated approaches to decision-making in VET arose from a study that investigated what constitutes demand side of VET in metropolitan, regional and remote communities for a national research agency (Billett, 1998a). It highlighted the different needs, resources and circumstances across these communities and identified the importance of commitment to and the enactment of local collaborations in addressing goals of young people's employment and tertiary education (Billett & Seddon, 2004; Seddon & Billett, 2004).

It was found in another study that provisions of continuing education and training (CET) were most favoured by working age adults when they were accessible locally and, in their workplaces (Billett et al., 2016). Given that there is no compunction for these adults to participate in CET, issues of accessibility, relevance of educational experiences and outcomes and quality of provisions become salient. This earlier finding was rehearsed in a recent study in Singapore that involved interviews with graduates from CET programs and a survey of working age Singaporeans

Table 13.1 Projects informing the case for localised decision-making and capacities

Projects	*Description*	*References*
1	Models of entry-level VET	Billett (1994)
2	Demand side of VET	Billett (1998a)
3	Social partnerships in VET for community capacity building	Billett et al. (2007)
4	CET models and approaches	Billett et al. (2016)
5	Standing and status of VET—Australia, and European, Asia, Africa & America	Billett et al. (2020, 2022a, b); UNESCO-UNEVOC (2018a, b)
6	Workplace innovations in small to medium size enterprises	Billett et al. (2011, 2022b)
7	Engaging indigenous and non-indigenous young people in allied health roles	Billett et al. (2022a)
8	School–industry partnerships to realise post-school pathways	Billett et al. (2023)
9	Effective CET provisions—Singapore	Leow and Billett (2022); Leow et al. (2023)

(Leow et al., 2023). It was found that working age adults participated in CET for diverse reasons, some personal and some professional, and they exhibited diverse patterns of readiness to participate. Consequently, it was concluded to understand their needs, and before participating in those programs, counselling should be available to ensure the most appropriate fit between their needs and aspirations and available CET programs. Having completed those programs, they would be given guidance about where they might seek new employment opportunities and abilities to practice and further develop their newly acquired knowledge. Such provisions can only be understood, organised and enacted locally. This is because it requires direct engagement with working age adults to understand and respond to their needs.

Also, when addressing what stands as the most significant problem VET faces—its low standing against other forms of schooling and tertiary education—it has been found globally (UNESCO-UNEVOC, 2018a) that local actions of engaging and informing young people, and their parents and familiars about VET and the occupations it serves are necessary. It is locally that these young people and their parents/caregivers make decisions about post-school pathways (Billett et al., 2020; Clement, 2014). Whilst there is a role to be played by central agencies and professional bodies, ultimately, it is how the information provided by these agencies is interpreted and considered in terms of young people's circumstances and aspirations. These can only be engaged with and understood at the local level. Again, when trying to advance enterprises' capacities to respond to emerging challenges, such as climate change, geopolitical tensions and those brought about by new technologies and ways of working, their abilities to respond to their often hybrid circumstances (i.e., locally unique) were central to their development (Billett et al., 2022a, b). Then, when addressing the needs of specific groups of learners (e.g., young Indigenous) within communities to engage in allied healthcare work, it was again found that VET provisions be based and informed locally (Billett et al., 2022a, b). Similarly, a recent study on school–industry partnerships to support post-school transitions found that factors shaping the success of those partnerships were local and had to be negotiated at the local level and engaged with by local enterprises (Billett et al., 2023). All of these studies point to the need to account for and be responsive at the local level, thereby emphasising this dimension of what constitutes the demand side.

Hence, in terms of the five purposes for VET that were identified earlier (see Chap. 11): (i) engaging young people with VET; (ii) assisting them identify occupations to which they are suited; (iii) preparing them for occupations; (iv) continuing education across working life to meet changing needs and goals; and (v) aligning workplace innovations with workers' learning, all of these need to be addressed at the local level and through VET systems that are supported by social, educational and administrative infrastructure. It follows then that there is a need to focus on how these imperatives can be addressed.

More Locally Engaged and Relevant VET Provisions

As noted, each of these studies referred to above has emphasised the importance of understanding and responding to educational purposes and processes at the local level. Whether it is working to: (i) elevate the standing of VET, (ii) meet the needs of local students and communities (i.e., 'the situated' demand side), (iii) include how Indigenous and migrant communities can best engage, (iv) promote the formation and continuity of social partnerships, (v) meet employers' needs, (vi) identify what constitutes innovations in enterprises, and (vii) advance communities' social and economic goals, all of these need to be understood and enacted locally. Consequently, there is a need to consider how more locally engaged and relevant VET provisions can be identified, organised and enacted.

As foreshadowed, much of what constitutes achieving the purposes of VET proposed here can be optimised by engagements, interactions and partnerships with interlocutors located outside of VET institutions. This includes engagements with workplaces, communities and industry bodies beyond the 'school fence' through local partnerships. Hence, rather than relying solely upon those institutions' capacities and the experiences they can provide, there is a need to engage outside of those institutions to understand fully the needs of the local communities (including private and public sector enterprises) and community members (e.g., parents, young people, local employers, community workers) to organise and enact experiences to support pertinent and effective educational provisions. It is posited from the projects reported above that when those ranges of factors are understood, they can have a profound impact upon how the education provision is designed, organised and implemented, and this includes the purposes those programs are meant to achieve. In the early project on understanding localised demand for VET, a group of mature age women participating in business courses to negotiate their ways back into the workforce reported wanting to have work experiences and opportunities that allow them to utilise their new skills (e.g., word processing, use of spreadsheets) (Billett, 1998b), and actively resisted work opportunities that were based upon earlier technologies and less advanced business practices. Yet, their educators were constrained by the need to adhere to nationally standard syllabus. Consequently, this builds on what has been proposed above about the design and enactment of VET provisions needs to be based upon the actions of educators, administrators, local employers as well as students at the local level.

It follows that meeting these imperatives cannot be realised by central agencies, who are remote from these circumstances and, instead, their principal role should be to organise arrangements that make it possible to adapt national curriculum outcomes, content and educational provisions to meet these needs. In addition, the central agencies might instead work towards a more decentralised approach that provides resources and supports the localised administrative, educational and social infrastructure required to realise these goals. Therefore, it is essential that considerations are now given to, firstly, the kinds of discretion that can be afforded educators and social partners at the local level, and then how these forms of infrastructure can be deployed to achieve these kinds of outcomes. Consequently, rather than imposing greater levels of mandates, prescription and regulation, greater discretion is required for VET systems, administrators and educators to realise both a level of responsiveness to national and local imperatives and change within the communities in terms of economic and social processes and outcomes.

Administrative, Educational and Social Infrastructure

As proposed, rather than a reliance upon highly centralised governance, curriculum design and mandated experiences and content being able to address the local demand side (i.e., needs and requirements) and to bring about change and innovations in communities, this requires a level of local social, administrative and educational infrastructure.

Social infrastructure includes local partnerships amongst public and private sector enterprises, tertiary and vocational institutions and industry representatives that can assist in identifying specific needs for the knowledge to be learnt and the readiness of students to engage and learn that knowledge as well as the available resources to provide effective educational experiences. That infrastructure extends to organising and supporting work placements, work experiences, employment opportunities and expressions of enterprises' requirements. As has been found in research on social partnerships within education, there needs to be commitments from the social partners that include them exercising their interests, but not to the degree that threatens the overall goals of supporting economic and social initiatives locally. It was found that a key quality of social partners was when they were willing to give up something which reflected their interests to realise a commonly identified goal within their communities (Billett & Seddon, 2004). Whilst such willingness might seem ideal and fanciful, such commitments are perhaps most likely to arise within local communities where these needs are visible and, in some ways, personal.

Administrative infrastructure includes intentional organisation and enactment of educational provisions and their certification at the local level. The model from Germany utilising chambers of commerce is an instance of how such administrative infrastructure is part of the local social infrastructure and can work in bipartite ways to achieve the goals of individuals, enterprises and educational institutions. Their role in organising and certification of the knowledge that has been learnt opens up the range of educational provisions from just being offered through standalone VET institutions. Instead, the provision and recognition of learning through work and in work settings can be encouraged when the certification of that learning is more accessible than when held to be a monopoly of VET institutions' processes and practices. So, there is a facilitative, rather than a regularity bureaucratic, role that needs to be played, opening up possibilities for educational provisions, learning outcomes and means by which individuals can learn and innovations and change be promoted through VET. Yet, all of this requires a level of educational infrastructure that, in many instances, has been the monopoly of centralised agencies and institutions.

Educational infrastructure refers to the expertise to adapt and support effective educational provisions and alignment of experiences to achieve learning outcomes and innovative practices. A key element of this infrastructure will be the expertise of vocational educators as curriculum makers and ability to engage with others (e.g., public and private sector enterprises, schools, community workers) to provide guidance and support for learners regardless of whether they are enrolled in VET courses, in workplaces or other community settings.

Importantly, the guidance and support provided by vocational educators and their ability to extend the capacities beyond the educational institution. Envisaged here is a far broader role for vocational teachers and administrators to empower them more broadly adapt their professional practice and exercise their agency in enacting these professional roles. Instead of viewing these educators as mere implementers of curriculum decided elsewhere and by others (Blachford, 1986), they need to play a key role in those processes. In short, they will become curriculum makers. So, whereas there has always been a need for teachers to understand their students' readiness and align educational experiences with them, the potential roles and imperatives to be addressed here are far greater. These include selecting and enacting educational activities that align students' readiness with achieving the kind of goals that their potential workplaces and communities need from them. Moreover, they will need to consider what comprises engaging experiences for the students so they will engage and participate in ways required to learn demanding knowledge. And, because part of the educational project here is to develop students' occupational adaptability to promote their employability and being innovative at work, educational processes must seek to develop that adaptability and acting, generating new and innovative responses to tasks and activities directed towards their learning and promoting higher order thinking. This can be achieved through problem-solving and projects that engage and require students to be generative of higher order forms of knowledge. Such approaches to VET, being responsive nationally, locally and focused on bringing about innovations, require administrative, educational and social infrastructure at the local level of the kind required for the contemporary era.

Building Communities Through Vocational Education

What has been proposed here is the need to address the five current purposes of VET at the local level. It is only in the specific circumstances that VET provisions are being designed and enacted, learning needs be understood, enterprise requirements responded to, broader goals of learning and career development achieved and communities' reach of social and economic activities extended. This is not a rejection of the role of national agencies, industry advisory processes and administrative arrangements and the establishment and delineating national occupation requirements and standards. Instead, it is about permitting more space for educational purposes and processes to be negotiated locally, thereby revisiting and revising earlier concepts such as the School-Based Curriculum Development movement (Skilbeck, 1984) in the context of needing to address imperatives at the community level. This includes addressing the concept of 'opportunity structures' that is the key premise of this volume. This includes opening pathways, fostering accessibility and progression for all kinds and classifications of individuals through VET provisions. It has been proposed here that localised approaches have the potential to open up the prospects for, roles and potential of local networks, partnerships and engagements to be enacted and in ways that would not be constrained by highly centralised descriptions, regulations and mandates. Part of this is about positioning of VET system not just in terms of standalone institutions, but those embedded in, responsive to and, potentially, providing leadership in change and innovations within local communities. Such a concept heightens roles and expectations of VET to realise both personal changes (i.e., individual learning, career development and employability) and local transformations. It also comprises broadening and deepening the role of VET as perhaps articulated through those five purposes for contemporary VET. Part of this is broadening and expanding the role of vocational educators to going beyond being mere implementers of curriculum developed elsewhere by others to adapting national prescriptions and requirements to the needs of local communities. Added here is an important role of encouraging and engaging students to be agentic by tailoring experiences within the communities that likely bring about such adjustments. All of this necessitates a greater engagement with the local communities that VET institutions and educators serve. Finally, given the almost global orthodoxy of centralised and top-down provisions of VET, what is proposed here in terms of local discretion, engagement and decision-making stands to make VET less illiberal. That is more participatory, inclusive and more broadly responsive engaged. This may well serve wider goals, such as elevating its standing and status within both local and national community. It is ironic that the very qualities that are objected by many western nations, (i.e., illiberal democracies, autocracies, tight governance and societal control) are being exercised, albeit qualitatively in softer forms within many VET systems through top-down systems, that favour certain interests over others. For instance, what is taken as the 'demand side' is adopted quite selectively. It usually refers to the corporate demand side, not that of the individual needs and aspirations and those of communities that comprise the local demand side. The orientation proposed through this chapter about local decision-making seeks to redress that imbalance.

References

Aarkrog, V. (2020). The standing and status of vocational education and training in Denmark. *Journal of Vocational Education and Training, 72*(2), 170–188. https://doi.org/10.1080/13636820.2020.1717586

Billett, S. (1994). *Evaluating modes of skill acquisition.*

Billett, S. (1998a). Changing relationships between the vocational education system and their clients. *Unicorn, 24*(3), 60–72.

Billett, S. (1998b). Situation, social systems and learning. *Journal of Education and Work, 11*(3), 255–274.

Billett, S. (2000). Defining the demand side of VET: Industry, enterprises, individuals and regions. *Journal of Vocational Education and Training, 50*(1), 5–30. https://doi.org/10.1080/13636820000200104

Billett, S. (2001). Knowing in practice: Re-conceptualising vocational expertise. *Learning and Instruction, 11*(6), 431–452. https://doi.org/10.1016/S0959-4752(00)00040-2

Billett, S. (2011). *Vocational education: Purposes, traditions and prospects.* Springer.

Billett, S. (2014). The standing of vocational education: Sources of its societal esteem and implications for its enactment. *Journal of Vocational Education and Training, 66*(1), 1–21. https://doi.org/10.1080/13636820.2013.867525

Billett, S., Dymock, D., & Choy, S. (2016). *Supporting learning across working life.* Springer.

Billett, S., Hay, S., Choy, S., Gibbs, K., McKay, L., & Le, A. H. (2023). Initiating, building and sustaining Industry-school partnerships to realise diverse post—school pathways. Journal of Education and Work, 37(1–4), 149–164. https://doi.org/10.1080/13639080.2024.2383564

Billett, S. (2024). Building VET systems to advance communities: beyond responsiveness. *Ammattikasvatuksen aikakauskirja, 26*(1), 98–104. https://doi.org/10.54329/akakk.143479

Billett, S., & Seddon, T. (2004). Building community through social partnerships around vocational education and training. *Journal of Vocational Education & Training, 56*(1), 51–68. https://doi.org/10.1080/13636820400200245

Billett, S., Choy, S., & Hodge, S. (2020). Enhancing the standing of vocational education and the occupations it serves: Australia. *Journal of Vocational Education and Training, 72*(2), 270–296. https://doi.org/10.1080/13636820.2020.1751247

Billett, S., Le, A.H., Millichap, D., & Meyer, V. (2022a). *Health education to employment pathways program.* Queensland Health.

Billett, S., Tan, J., Chan, C., Chong, W. H., & Keat, J. S. C. (2022b). Employee-driven innovations: Zones of initiation, enactment and learning. In W. O. Lee, P. Brown, A. L. Goodwin, & A. Green (Eds.), *International handbook on education development in Asia-Pacific* (pp. 1–19). Springer Nature. https://doi.org/10.1007/978-981-16-2327-1_67-1

Billett, S., Choy, S., Gibbs, K., Le, A. H., McKay, L., & Hay, S. (2024). Initiating, enacting and sustaining partnerships to inform post-school pathways. *Journal of Education and Work, 37*(1–4), 149–164. https://doi.org/10.1080/13639080.2024.2383564

Blachford, K. (1986). *Orientations to curriculum in TAFE.* Melbourne, Australia: Hawthorn Institute of Education.

Clement, U. (2014). *Improving the image of technical and vocational education and training.* Bonn Germany: Federal Ministry for Economic Co-operation and Development. https://www.dcdualvet.org/wp-content/uploads/2014_GIZ_Improving-the-Image-of-Technical-and-Vocational-Education-and-Training_A-synthesis.pdf

Deissinger, T. (1997). The German dual system—A model for Europe? *Education and Training, 39*(8), 297–302.

Deissinger, T. (2000). The German 'philosophy' of linking academic and work-based learning in higher education: The case for vocational academies. *Journal of Vocational Education and Training, 52*(4), 605–625. https://doi.org/10.1080/13636820000200134

Deissinger, T., & Ott, M. (2016). Tertiarisation of vocational education and training and its implications – Problems and issues in Germany and France. In S. Bohlinger, T. K. A. Dang, & M. Glatt (Eds.), *education policy: Mapping the landscape of scope* (pp. 268–296). Peter Laing.

Dewey, J. (1915/1979). Education vs. trade-training: Reply to David Snedden. In J. A. Boydston (Ed.), *The middle works of John Dewey, 1899–1924* (pp. 36–48). Southern Illinois University Press.

Dewey, J. (1916). *Democracy and education.* The Free Press.

Drost, W. H. (1967). David Snedden and education for social efficiency. University of Wisconson Press.

Dymock, D., & Billett, S. (2010). Skilling Australians: Lessons from World War II national workforce development programs. *Australian Journal of Adult Learning, 50*(3), 468–496. https://search.informit.org/doi/10.3316/informit.839155474732654

Edwards, R. (2002). Mobilizing lifelong learning: governmentality in educational practices. *Journal of Educational Policy, 17*(3), 353–365. https://doi.org/10.1080/02680930210127603

Epstein, S. R. (2008). Craft guilds in the pre-modern economy: A discussion. *Economic History Review, 61*(1), 155–174. https://doi.org/10.1111/j.1468-0289.2007.00411.x

Gonon, P. (2009). The quest for modern vocational education: Georg Kerschensteiner between Dewey, Weber and Simmel (Vol. 9). Peter Lang.

Hanf, G. (2002). Introduction. In *Towards a history of vocational education and training (VET) in Europe in a comparative perspective, Florence*. Office for Official Publications of the European Communities.

Hodges, D. C. (1998). Participation as dis-identification with/in a community of practice. *Mind, Culture and Activity, 5*(4), 272–290.

Hoyrup, S., Bonnafous-Boucher, M., Hasse, C., Lotz, M., & Molller, K. (Eds.). (2012). *Employee-driven innovation: A new approach*. Palgrave Macmillan.

Labaree, D. F. (2011). How Dewey lost: The victory of David Snedden and social efficiency in the reform of American education. In D. Tröhler, T. Schlag, & F. Ostervalder (Eds.), *Pragmatism and modernities* (pp. 163–188). Sense Publishers.

Lemmetty, S., & Billet, S. (2023). Employee-driven learning and innovation (EDLI) as a phenomenon of continuous learning at work. *Journal of Workplace Learning, 35*(9), 162–176. https://doi.org/10.1108/JWL-12-2022-0175

Leow, A., Billett, S., & Le, A. H. (2023). Towards a continuing education and training eco system: A case study of Singapore. *International Journal of Training Research*, 1–17. https://doi.org/10.1080/14480220.2023.2203944

Nesbit, T. (2011). Canadian adult education: Still a movement. *Canadian Journal of University Continuing Education, 37*, 1–13. https://doi.org/10.21225/D52S3S

Organisation for Economic Co-operational and Development. (2013). *OECD skills outlook 2013: first results from the Survey of Adult Skills*. OECD.

Organisation for Economic Co-operational and Development. (2019). *Skills matter*. https://doi.org/10.1787/1f029d8f-en

Remery, V., & Merele, V. (2014). French approaches to accreditation of prior learning: Practices and research. In T. Halttunen, M. Koivisto, & S. Billett (Eds.), *Promoting, assessing, recognizing and certifying lifelong learning: International perspectives and practices* (pp. 265–280). Springer.

Rintala, H., & Nokelainen, P. (2020). Standing and attractiveness of vocational education and training in Finland: Focus on learning environments. *Journal of Vocational Education and Training, 72*(2), 250–269. https://doi.org/10.1080/13636820.2020.1744696

Rosser, G. (1997). Crafts, guilds and the negotiation of work in the medieval town. *Past and Present*, (154), 3–31. https://doi.org/10.1093/past/154.1.3

Seddon, T., & Billett, S. (2004). *Social partnerships in vocational education: Building community capacity*. National Centre for Vocational Education Research

Sennett, R. (2008). *The craftsman*. Yale University Press.

Skilbeck, M. (1984). *School based curriculum development*. Harper and Row.

Stevenson, J. C. (1991). Cognitive structures for the teaching of adaptability in vocational education. In G. Evans (Ed.), *Learning and teaching cognitive skills* (pp. 144–163). ACER.

Skilbeck, M., Connel, H., Lowe, N., & Tait, K. (1994). *The vocational quest: New directions in education and training*. Routledge.

Stevenson, J. (2001). Vocational knowledge and its specification. *Journal of Vocational Education and Training, 53*(4), 647–662. https://doi.org/10.1080/13636820100200182

Troger, V. (2002). Vocational training in French schools: the fragile state-employer alliance. Towards a history of vocational education and training (VET). In *Towards a history of vocational education and training (VET) in Europe in a comparative perspective, Florence*. Office for Official Publications of the European Communities.

UNESCO. (2018). *Work-based learning programmes for young people in the Arab region: A comparative and longitudinal analysis - full report*. UBESCO.

UNESCO-UNEVOC. (2018a). *Improving the image of TVET: Making TVET attractive to youth.*

UNESCO-UNEVOC. (2018b). *Short summary of the Virtual Conference on improving the image of TVET: Making TVET attractive to youth.*

Wheelahan, L., Moodie, G., & Doughney, J. (2022). Challenging the skills fetish. *British Journal of Sociology of Education, 43*(3), 475–494. https://doi.org/10.1080/01425692.2022.2045186

CHAPTER 14

Quality Management in Vocational Education and Training: A Reflection on Educational Governance in Austria, Germany, and Spain

Junmin Li, Ekaterina Schlüter, Hannes Hautz, Julián Bell Sebastián, and Fernando Marhuenda-Fluixá

Introduction

In the period up to the early 1990s, the school system's organisational structure, encompassing school-based vocational education, was characterised by hierarchical arrangements and centralised administration

Fernando Marhuenda-Fluixá received a research grant from the Spanish Ministry of Universities (PRX22/00214) to undertake a research stay on VET at EHB-Zollikofen, where he contributed to this chapter between March and July 2024.

J. Li (✉)
TU Dortmund University, Dortmund, Germany
e-mail: junmin.li@tu-dortmund.de

E. Smeplass and C. Zirkle (eds.), *Opportunity Structures in Vocational Education and Training*, Palgrave Studies in Technical and Vocational Education and Training,
https://doi.org/10.1007/978-3-032-04410-5_14

in many regions of Europe (Amelsvoort & Scheerens, 1997; Powell et al., 2012). Schools operated under external governance and were required to comply with administrative directives issued by the respective education authorities. Criticisms of the inefficiencies associated with this bureaucratic model of control, coupled with the advent of New Public Management, prompted a shift towards decentralised school development in many countries. This shift introduced a separation between strategic and operational school management (Dubs, 2003), marking a transition of schools from state enforcement entities to service-oriented organisations with increased decision-making independence (Schedler, 1995). One of the rationales behind the introduction of this initiative was to provide schools with greater flexibility, enabling them to respond more promptly to the evolving needs of their learners. The new level of autonomy within school organisation gave rise to concerns about possible undesirable developments at the level of individual schools, thereby heightening the need for system-wide coordination and school-wide management mechanisms (Altrichter, 2010; Altrichter & Heinrich, 2007). Subsequently, in the late 1990s and early 2000s, the establishment of comprehensive Quality Management (QM) systems and the regular monitoring of school performance became pivotally important in educational governance across Europe (Altrichter & Heinrich, 2007).

Zaunstöck (2020) has explored how European (and more extensive international) processes are impacting national systems, arguing that this is not a top-down process, but acknowledging the governing capacities of nation states which are therefore capable of shaping European policies, partly in their own interests. In this regard, following the Lisbon-Copenhagen-Bruges-Osnabrück process, facilitating the use of tools to assure quality has been one of the ways in which European governance has tried to manage actions at the national level, in the form of soft-framing

---

E. Schlüter
University of Cologne, Cologne, Germany

H. Hautz
University of Innsbruck, Innsbruck, Austria

J. Bell Sebastián · F. Marhuenda-Fluixá
Universitat de València, Valencia, Spain

mechanisms (Zaunstöck et al., 2021). Concurrently, the European reference framework for quality assurance within vocational education and training (VET) aimed at assisting member states, enhancing the quality of their VET systems, leading to the adoption of different QM systems reflecting the distinct approaches to governance prevalent in individual countries (Martínez-Izquierdo & Torres Sánchez, 2022).

More recently, governance research has been applied to the analysis of VET systems, recognising their intricate coordination mechanisms and governance structures across various levels (state, private providers, social partners, etc.). Educational governance research focuses on the formation, upkeep, and transformation of social orders and services within the education sector, emphasising the coordination of actions across diverse stakeholders within a multi-tiered system (Altrichter, 2010). While the governance approach and QM application may differ in VET when compared to general education, the coordination of stakeholders and actions shows considerable variation. Individual VET systems are hallmarked by diverse organisational structures to facilitate the coordination and direction of actions, leading to a spectrum of applied approaches across Europe. Through an educational governance lens, this study aims to investigate how QM practices are conceptualised in selected European countries. Embracing a comprehensive understanding of QM, which encompasses systematic procedures for evaluating, enhancing, and ensuring quality within VET institutions covering both instructional and organisational areas, the research question pursued is: which understanding of educational governance is reflected by the different concepts of QM in selected European VET systems, and how is the coordination of actions in these complex multi-level systems shaped by QM?

This research not only identifies country differences and similarities, but also critically examines whether one of the claims of New Public Management, namely to enable a stronger opportunity structure for educational institutions, is fulfilled by QM systems of the countries analysed. In international and comparative VET research, this question has not been analysed and discussed in detail up until now. The countries of Austria, Germany, and Spain were selected because they belong to different facets of educational governance regimes, and different VET systems in Europe. Although Austria and Germany have a similar VET system, they differ in terms of their educational governance and QM approach. Germany has a decentralised approach, which is organised differently in the various states as a result of federal organisation (Köhler,

1996). Austria pursues a centrally controlled approach to educational governance and QM (Donovan & Hautz, 2024). Spain, in turn, is hallmarked by two different approaches that treat state and private VET providers differently, while each autonomous region is also responsible for applying and adapting state-wide norms, in a way analogous to Germany (Marhuenda-Fluixá et al., 2019; Merino Pareja, 2023). As a result, the selection of countries in this study reflects a most-different approach (Anckar, 2008). At the same time, the educational governance approaches of the countries under examination here mirror those of many other European countries. These comparative results can be used by an international audience to better understand the connection between educational governance and QM strategies in VET institutions on the global stage, to reflect the influence of educational governance approaches on vocational QM system of their own countries and receive inspirations for potential modification of opportunity structures within the QM system for their vocational organisational frameworks.

Theoretical Framework

To examine the current levers of steering in the VET systems, we use the perspective of educational governance and connect this outlook with key QM elements.

Educational Governance

The field of governance research is a multifaceted and interdisciplinary domain that explores the coordination of actions and the production of collectively binding decisions in various societal contexts. The governance concept was developed in response to the failure of political planning in the 1980s and has been used to analyse and understand social structures (Kussau & Brüsemeister, 2007). More precisely, the governance perspective assumes that the state does not control social systems and steering from above—hierarchically—but that there are various stakeholders, including the state, whose multifaceted, sometimes contradictory actions lead to the emergence and structuring of social systems (Altrichter & Maag Merki, 2010). Therefore, governance research focuses on political regulatory outcomes and the evolution of control mechanisms, as well as the potential for extended participation in non-hierarchical social structures (Dose, 2005; Mayntz, 2009). The term "governance" encompasses all forms of action coordination through behaviour-steering norms, and it differs from purely market-based coordination mechanisms. It encompasses justifications, rationalisations, deliberations, and discourses as essential elements, particularly relevant when considering the (trans-)intentional provision of public goods or the production of collectively binding decisions (Zürn, 2008). Educational governance research specifically explores the analysis of the formation, maintenance, and transformation of social orders and services within the education sector. The focus lies on the coordination of actions among various educational stakeholders across different levels within a multi-tiered system (Altrichter & Maag Merki, 2010). According to Altrichter (2010), and Kussau and Brüsemeister (2007), educational governance rejects the assumption of direct controllability by the state and follows an extended understanding of steering. At the centre of this perspective is the problem of coordinating the actions of different stakeholders, who are all co-producers of system performance characteristics (Benz, 2004). The partly contradictory transactions, the inherent logics and dynamics of social systems, and indirect influences from external stakeholders may lead to unintended and "transintentional" consequences (Altrichter & Maag Merki, 2010). The concept of "action coordination" serves as a tool of analysis and explanation of how actors at different points in a complex system participate in the regulation and performance of that very system (Altrichter, 2010; Altrichter & Heinrich, 2007). Against the background of our research topic, we have analysed the educational governance of QM in the VET sector of the selected countries.

A key dimension within governance research is the distinction between governance content, governance structures, and governance processes. Research on governance content concentrates on specific regulations and tools. Governance structures pertain to the institutional configuration framing the creation of governance content. Governance processes involve negotiation procedures leading to governance content within a governance structure (Zürn, 2008). The focus of our research lies on the examination of governance structures within the VET sector, and less on their actual impact on VET practice.

Altrichter and Heinrich (2007) propose several categories for analysing governance structures and constellations: (a) "guiding values and impact claims" for the analysis of the intentions and effect of governance constellations, (b) "stakeholders" comprising a distinction between actors of control and actors of accountability, and the distribution of influence between the stakeholders, (c) "structures and rights of implementation" for the distribution of formal, non-formal, and informal rights of implementation among stakeholders, (d) "multi-level systems" for the examination of the delineation of the boundaries between the actors, and how these boundaries are crossed. Based on the analysis of these categories, a conclusion is drawn as regards the last category, (e) "coordination of action", dealing with the question of how various stakeholders with different rights of implementation coordinate their actions in a multi-level system. For the analysis and description of the coordination of action of a specific governance regime, an orientation towards "classical models of societal coordination" (Altrichter, 2010, p. 148) such as network, market, or hierarchy is possible. Moreover, Altrichter (2010) suggests, as a heuristic framework for analysing the differences between education systems, five dimensions that de Boer et al. (2007) have identified in a comparative study of the governance structure of university systems in specific countries. The primary dimension, "state regulation", signifies the conventional supervision of public systems by the state's authoritative control. This approach aims at intricately prescribing the conduct of subsystems. The second dimension, "external guidance by state or other stakeholders", comprises regulatory endeavours that guide other systems and institutions by establishing goals, offering advice, and conducting assessments. The third dimension, "competitive pressure and quasi-markets", pertains to the coordination of systems through market or "quasi-market" processes. The fourth dimension, "managerial self-governance", denotes the regulatory authority of internal hierarchies within organisations, including the leadership's influence in internal goal setting, fund allocation, and decision-making. The fifth and final dimension, "professional self-governance", underscores the influence of professionals in decision-making, institutionalised through collegial decision-making within universities and the peer review-based self-steering of academic communities.

These models and dimensions are ideal, characteristic descriptions, and in reality, they usually do not occur in pure form, but only as a combination, or in a more differentiated form. The heuristic categories can thus help develop an initial orientation about the specific educational governance structure (Altrichter & Heinrich, 2007). In the present study, this theoretical framework is used to analyse and juxtapose the principal characteristics of educational governance in the context of QM.

Quality Management in Education

The idea of QM was originally developed in the manufacturing industry, starting from the control of product quality. QM includes quality control, quality assurance, and the regulation of responsibilities, competencies, and resources to implement control and assurance procedures (Martin et al., 2020). Total quality management represents a comprehensive form of QM in companies, integrating all organisational levels and emphasising continuous improvement and participative management (Ott & Scheib, 2002). With the rise of New Public Management, the decentralisation of school systems, and the resulting separation between strategic and operational school management, QM concepts reached the education sector (e.g., Altrichter & Heinrich, 2007; Dubs, 2003), introducing accountability in educational policy and practice. Since then, work has been performed at the level of both educational administrations and schools, to record and assess quality, and to use the results to improve structures and processes, in part by adapting QM models from the manufacturing industry. In addition, the European Quality Initiative initiated by the European Council in the early 2000s has heightened the importance of quality and QM in VET across Europe. In 2002, the Copenhagen Process was launched (European Commission, 2002) with the aim of implementing the Lisbon Strategy "to become the most dynamic and competitive knowledge-based economy in the world capable of sustainable economic growth with more and better jobs and greater social cohesion" (European Commission, 2000). To achieve this goal, it was considered necessary to improve the quality of VET systems through the development of shared quality criteria and the mutual exchange of satisfactory quality assurance models (European Council, 2003). Therefore, the European Quality Assurance Reference Framework for Vocational Education and Training (EQAVET)[1] has been developed to support quality assurance in all Member States. It provides a set of guiding principles (for example, the four-stage quality cycle), criteria, and indicators that can contribute to the assessment and improvement of the quality of VET systems, but it does not prescribe a specific QM approach (European

[1] EQAVET builds on previous European quality assurance frameworks. The first European Common Quality Assurance Framework (CQAF) was developed back in 2004. In 2009, it was replaced by the enhanced European Quality Assurance Reference Framework (EQARF), which has since been improved and is now termed EQAVET.

Commission, n.d.). The adoption of EQAVET is voluntary and it serves as a kind of "toolbox" (European Commission, n.d.) from which countries can choose specific elements individually, according to the needs of their respective QM and VET systems.

But quality can be understood in different ways, as it depends on the perspective adopted and the standards by which it is measured (Harvey & Green, 2000). The definition of quality is subject to the judgement of various stakeholders and may vary depending on the views, interests, and experience of learners, teachers, the training company, the school management, parents, or future employers (Martin et al., 2020). There is also the problem that learners themselves are significantly involved in the "production" of education through the development of their knowledge and skills. For these reasons, quality in the school context is generally understood here in a broad sense, and it may vary according to context, perspective, and country. The term "assessment" is closely linked to the concept of quality. In school practice, assessment is described as a process of systematically gathering and interpreting information. Countries therefore use different instruments under the umbrella of QM, principally based on a combination of quality indicators, quality assurance approaches, external and internal assessment, and certification, varying along a continuum between the degree of regulation and the degree of autonomy (Cedefop, 2015; Dubs, 2003).

From the perspective of educational governance, quality management involves different areas of interest at the school level and the educational administration level, which pursue various and sometimes contradictory objectives. The integration of these different needs requires the stakeholders' interaction, a challenging task that calls for transparency, openness, and communication. The educational governance regime of a country influences which actors are empowered to make which decisions, at which stage of the QM process, and in which way. Therefore, the way in which QM is organised in each country may reflect the form of educational governance in that nation.

Methodology

The study is based on the analysis of documents in the three selected countries, including government reports, recommendations, regulations, and requirements, but also research studies, including theoretical and empirical publications as well as reports from schools and school associations. The documental analysis involves three steps: (1) selection, (2) sampling, and (3) thematic analysing (Morgan, 2022). The factors of authenticity, credibility, representativeness, and meaning are used for a qualitative validation of the documents (Flick, 2018).

In total, 83 documents were subjected to a comprehensive analysis, comprising 29 documents for the Austrian case, 30 documents for the German case, and 24 documents for the Spanish case, all of which were carefully selected for intensive document scrutiny. The resulting thematic analysis adhered to the dimensions outlined in the respective theoretical framework for each country, as illustrated in the subsequent section. The analysis was underpinned by a reflexive approach, as outlined by Braun et al. (2019). To minimise interpretative bias in the documental analysis, four researchers discussed the country results together. Subsequently, the research team convened to compare and compile the findings, leading to a holistic perspective in response to the overall research topic.

Based on the documental analysis, a structured programme theory of QM concepts has been developed for each country. Programme theory identifies the causal chain of a programme, in our case QM, to examine the relationship between input activities and intended or observed outcomes (Rogers et al., 2000). Programme theory is the theory that illustrates the assumed effects of specific operations performed within a programme. A predefined structure of the programme theory enables a systematic comparison between the countries. So, the *tertium comparationis* are driven by programme theory (see Chen, 2005, for a detailed analysis) and the categories of the educational governance perspective (Altrichter & Heinrich, 2007). For all three countries (see Table 14.1), a table of categories and guiding questions for documental analysis was compiled, based on the analytical categories by Altrichter and Heinrich (2007) and Altrichter(2010).

Findings

In the following sections, we provide an overview of the findings of the governance analysis for individual countries. The findings are structured nation by nation, and they present the results for the four analytical categories of Table 14.1. In order to underline the interdependence between different categories, the presentation of the results does not strictly follow

Table 14.1 Analytical categories and guiding questions for the analysis of educational QM governance structure

Category	*Guiding questions, possible perspectives*
Guiding values	Which overall (educational) values/ effects justify the QM approaches?
Structures and actors	Which individual and institutional actors participate in the coordination of QM approaches?
Rights of disposal and structures of regulation	Which responsibilities and rights of disposal do the actors have, at which governance structural level of the approaches to QM?
Governance regimes	Which kind of governance regimes or combination of regimes are reflected in the approaches to QM?

the order of the table, but rather the governance logic of each country's QM approach. A comparative summary is provided in the discussion section.

Austria

The VET system in Austria encompasses an "extended educational mandate" (Schlögl et al., 2019, p. 276), which brings together general and vocational education. It's distinctive architecture, comprising both apprenticeships and full-time VET schools, facilitates "the opening up of social opportunities" (Lassnigg, 2011, p. 423). By offering individuals a diverse range of occupational qualifications at varying levels, along with potential pathways for progressing to higher education, VET in Austria plays a pivotal role in broadening perspectives and promoting "upward mobility" (Lassnigg, 2011, p. 423).

The implementation of QM has been mandatory for all Austrian schools since 2012 according to § 18 of the Federal School Supervision Law. In the VET sector, a comprehensive nationwide QM system for vocational schools called QIBB (VET Quality Initiative) was introduced by the Federal Ministry of Education (FME) in the school year 2006/07 (FMEW, 2015). QIBB has been part of the national strategy for the implementation of the European Quality Initiative, and it was developed according to the principles of EQAVET (Timischl, 2006). Initiated by the Austrian Education Reform Act of 2017, the FME developed and introduced, in cooperation with the Austrian Reference Point for Quality Assurance in General and Vocational Education and Training (RQB), a revised QM system for all school types, in the 2021/22 academic year. The new QM system, named QMS (Quality Management System for Schools), is based on QIBB and the QM system for the general education sector (SQA), and it combines both of them into a single overall framework (FMESR, 2023). Therefore, the QMS builds on existing structures and tools and adds new elements to them.

A fundamental *principle and guiding value* of QMS is the four-phase Deming-quality cycle according to the concept of systematic, evidence-oriented, and continuous improvement (Gramlinger & Jonach, 2022). In addition, QMS is inspired by the Q2E function model theorised by Landwehr & Steiner (2020), and it underpins both the data-based development of school and teaching processes and results, in addition to accountability provisions, for all stakeholders of a school (FMESR, 2021). Compared to other QM approaches, QMS is intended more to enhance teaching-related quality development, and it focuses on teacher teams as one main group of actors (Gramlinger & Jonach, 2022). It is broadly defined as an "educational quality management system that places learning and learners at the centre of quality work" (FMESR, 2021, p. 3). Key objectives comprise the guarantee of high-quality education, the encouragement of learners' individual potential, and the promotion of teachers' professionalism and cooperation (FMESR, 2021). In addition, the implementation of QMS should increase the transparency of workflows in the school towards external stakeholders, systematising internal processes on the basis of evidence, and improving the education system's permeability (Gramlinger & Jonach, 2022).

In order to achieve these objectives, pivotal tenets comprise a clear *structure* of the organisation, the processes and the responsibilities of the QM system at each school, a quality framework with a set of quality criteria that characterise "good schools and good teaching in Austria" (FMESR, 2021, p. 8), and various steering tools to shape QMS activities, such as a school development plan, a quality manual, and task profiles at team and individual teacher levels. Strong emphasis is also placed on generating data regarding the quality and effectiveness of the school's processes and teachers' actions through regular evaluations and systematic educational monitoring by the school inspectorate. Three central *actors*—the school management, teacher teams, and individual teachers—are designated as being responsible for the implementation of the measures, and the attainment of targeted QMS results (FMESR, 2021). In this way, a school's entire pedagogical staff should play an active role in quality development and assurance (Jonach & Pitzer, 2022). This implies an imperative change in teachers' attitudes and behaviour, who should align their thinking and actions according to the QM logics.

QMS is designed as a multi-level system with various *actors* (see in detail FMESR, 2022). It is coordinated and further developed conceptually at a centrally organised national level by the FME together with the RQB. The school inspectorate, which is comprised within the education boards of the nine federal states and the departments of the education regions, is responsible for encouraging the implementation of QMS in schools, for ongoing quality control, and for conducting meetings with school management to review and recalibrate targets (FMESR, 2019). As a result of this change in focus, the title of the school inspectors was amended to "school quality managers." At school level, in addition to the three main actors mentioned above, school quality coordinators (Q-SK) also play an important role in the QMS system. This special function is assigned to teachers selected by the school management to support the introduction and implementation of QM and to mediate between school management and the teaching staff (for a detailed analysis, see Hautz, 2022a). These players are tasked with "convincing as many teachers as possible of the added value of quality work and bringing them on board" (Freynschlag et al., 2022, p. 7; translated by H.H.). Regional quality coordinators (Q-RK) provide assistance to these teachers in their new role, and guide them in the implementation of the QMS (FMESR, 2021). Although the whole QMS activities should be tailored to the needs of the learners, in the design of quality measures, the opinions of students are not taken into direct consideration.

Overall, it is assumed that QMS makes school processes more efficient, improves the quality of teaching-learning arrangements, professionalises teachers, and even reduces their workload, since problems can be dealt with as a team (FMESR, 2021). However, there is no empirical evidence to date that the proclaimed goals have been achieved. Despite the considerable experience with QM in Austria, it remains unclear whether these programmes have a positive effect on teaching, learning outcomes, and the distribution of opportunities in VET (see e.g., Hopmann, 2021). Rather, recent research findings have yielded contradictory conclusions. In a qualitative study on the views of VET teachers in Austria on quality and QM, Hautz and Thoma (2021) have shown that although teachers consider themselves as central bearers for quality development and assurance, and are in principle motivated to play an active part in these functions, the level of teaching may even drop in quality as a result of increased standardisation, additional administrative tasks, and mandatory documentation efforts within the QM framework. The teachers interviewed feel that their professional independence is limited by top-down QM measures, narrowing their focus to core educational tasks. They perceive QIBB principally as a way of justifying school quality for stakeholders, applied at an organisational level, with little practical impact on teaching and learning (Hautz, 2022b). Donovan and Hautz (2024) have further argued that QM initiatives in Austria constitute "a crisis of trust" in VET teachers, leading to an erosion of pedagogical autonomy and the sacrifice of core educational values. Moreover, Landwehr (2021, p. 60) describes a "QM crisis in schools", noting that QM systems often do not extend to teaching activities, they are not fully accepted by teachers, and are questionable in their effectiveness.

The new nationwide QM system aims at focusing more on the teaching level by defining school and teaching quality and providing specific structures of regulation and standards. More specifically, changes with respect to the previous QM structure within the VET school system include the increase of central *structures of regulation* and requirements by the FME with regard to the criteria, processes, tools, and responsibilities for quality, and mandatory evaluations. Furthermore, a school's entire pedagogical staff is expected to take part in quality development and assurance, thus taking responsibility for the attainment of the school's and policy objectives. To achieve this, the school is granted certain areas of freedom and *rights of disposal*: for example, pedagogical guiding principles, team projects, and teachers' individual goals can be set independently within the QMS framework (FMESR, 2021). The provisions for independence as part of the 2017 Education Reform Act also give schools the chance to flexibly organise school timetables, group sizes, and teaching units. According to the FME, to ensure the best possible education for all learners, these areas of freedom require "systematic underlying quality control" (FMESR, 2018). This should be guaranteed by means of systematic monitoring by the school inspectorate, and by the ongoing internal and external evaluation of schools for VET according to a standardised quality framework (FMESR, 2018).

QM in Austria is characterised by a *governance regime* that is shaped by centralised external steering through output orientation, standardised targets, and pre-determined forms of monitoring and control. In addition, VET professionals in schools are expected to be motivated and actively engaged in autonomous school development, for which they are granted increased organisational and pedagogical freedom of design (FMESR, 2018). However, the purported level of self-governance is set within a limited, pre-structured realm of possibilities (Hautz, 2022a). On one hand, the basic criteria of quality are already determined by the QMS quality framework. On the other hand, teachers are influenced in their professional behaviour through "indirect forms of governance" (Hautz, 2022a, p. 211), such as the expertise of Q-SKs, the activation of responsibility for the attainment of standards, and the ongoing documentation of professional conduct. The introduction of a new level—teacher teams—in a school's QM hierarchy increases the influence of school leaders and Q-SKs on individual teachers. Within the QMS framework, the intention is to align the behaviour of teachers to policy objectives by making them more accountable for learners' outcomes and aligning their decisions with common standards of comparison. This potentially increases the pressure of competition within the teaching staff and between schools (Hautz, 2022b). The coordination of action of the Austrian QMS can therefore be considered as regulated self-regulation (Zlatkin-Troitschanskaia, 2007). External steering based on overall targets, processes, and quality instruments is coupled with pre-structured self-governance, which is largely hierarchical in nature, evidence-based, and oriented towards potential comparisons with other schools (Table 14.2).

Germany

VET in Germany stands for equal opportunities and offers prospects (DIHK 2024). Its specific elements support participants on their way into the world of work. For example, training preparation helps with orientation and with the aim of introducing participants to vocational training in a recognised occupation by providing the basics for acquiring vocational skills (BIBB, 2024). In addition, the possibility of dual qualification (the simultaneous acquisition of a vocational qualification and a higher education entrance qualification) promotes the transition and permeability of VET in Germany (Matanovic & Sacks, 1976; Frommberger, 2012). Furthermore, the close link between education and the

Table 14.2 Overview of the main characteristics of QM in Austria

Category	*Key points for Austria*
Guiding values	QMS aims to put learners at the heart of all quality efforts and is centred on ensuring that all learners receive the best possible conditions for their education. Following principles are considered to be essential for achieving this goal: 1. A systematic and evidence-based design of school and teaching processes. 2. The promotion of professional practice among and cooperation between teachers in order to foster a school-wide quality culture. 3. Increased transparency, accountability, and permeability throughout the entire school system.
Structures and actors	**State level:** Federal Ministry of education (FME)—responsible for development and coordination of QMS Austrian reference point for quality Assurance in General and Vocational Education and Training (RQB)—support in development of QMS **Regional level:** Regional school quality managers—support of schools in the implementation of QMS, ongoing quality controlling, review, and target setting meetings with the school management Regional quality coordinators (Q-RK)—support of school quality coordinators in the realisation of QMS **Institutional level:** School management—strategic planning and coordination of QM at school level School quality coordinators (Q-SK)—responsible for operative tasks of QM at school level, QM contact partner for teachers, and teacher teams Teachers and teacher teams—continuous development of school and teaching quality
Rights of disposal and structures of regulation	QMS contains regulatory requirements regarding the objectives, criteria, processes, tools, and responsibilities for quality work. The school is granted organisational and pedagogical freedom within the QMS framework, for example in setting academic principles, team projects, and individual teacher goals. School activities undergo systematic monitoring by internal and external evaluation according to a standardised quality framework.

(continued)

Table 14.2 (continued)

Category	*Key points for Austria*
Governance regimes	"Regulated self-regulation" which combines: Centralised external steering based on standardised targets, a common quality framework and pre-determined forms of monitoring and control. A pre-structured form of self-governance that is output-oriented, evidence-based, and characterised by continuous self-evaluation and self-responsibility.

labour market makes the transition from school to the world of work easier for young people (Ebner 2013, p. 37 f.), which has a positive impact on youth unemployment.

Since the mid-1990s, reforms have been introduced in Germany's public school system, with a focus on quality and QM (Wagner & Rückmann, 2017). A specific feature of the situation in Germany is that the framework conditions for the implementation of QM in VET schools are determined by the federal states (KMK, 2004). German VET shows several lines of QM development, based either on the Q2E, EFQM, or ISO models, or on minimum standards. As an example, Q2E has been adopted and further developed in some federal states. From 2003 to 2006, Baden-Württemberg developed the OES (Operationally independent school) concept on the basis of Q2E, and implemented it starting from 2007. In Bavaria, QM at VET schools was developed from 2006 on, according to these two concepts (Q2E and OES), followed by the development of its own concept—QM at VET schools in Bavaria (QmbS, Wilbers, 2024). While the QM systems in the federal states that are comprised within the above-specified line of development are predetermined and are mostly based on the Q2E model, other models, in particular the ISO model, are not as widespread (Mai, 2020). In a second line of development, VET schools are not organised by means of a fixed quality management system, but through the definition of minimum QM standards in VET schools. These standards envisage quality specifications in the form of orientation frameworks, and they require VET schools to conduct regular internal and external evaluations regularly. In this context, to date, efforts have been targeted at transforming the classic model of bureaucratic input control towards an evaluation-based output control (Feldhoff, 2011). In the German VET system, however, accountability in connection with the new distribution of power should be linked to liability, it should be practicable, and it should likewise be based on qualitative feedback systems (Becker et al., 2006).

QM approaches are justified primarily by the *guiding value* of providing impetus for the school's further development (Schulministerium NRW, 2023). This should encourage the authoritative link between self-evaluation and the schools' external evaluation of the school and focus on the fulfilment of the supra-regional educational mission (Becker et al., 2006, p.119). An analysis of the claims regarding intentions and impact shows that the overall goal proclaimed by the German government is to create equal opportunities in vocational education. This objective goes hand in hand with the need to incorporate sustainability aspects (BMBF, 2023). The implementation of these goals places notable demands on all stakeholders and requires a high level of school participation in the process (MSB NRW 2023, pp. 3–4). Therefore, fundamental principles, such as *guiding values* like the implementation of further education and training, the development of cooperation with stakeholders within society and science, an active participation in regional networks, and the promotion of democracy and tolerance among all stakeholders, are anchored within the regional action frameworks (MK, 2021) and could be strategically assessed within the QM process.

The *structure* of quality management in Germany is characterised by the coordinated efforts of both individual and institutional *actors on the* federal, regional, and local *level*. First, at the federal level, it comprises the German state, in the form of the federal ministries of education as a source of normative documents and regulations. Second, at the regional level, school inspectors, recruited from the schools' personnel and working in teams, operate as external assessors, in accordance with regulations and frameworks of reference. The regional school district departments act as consultants. Third, at the school level, the schools' own inspectors develop internal evaluations within the schools. The principals and teachers implement the evaluated measures. It could be argued that in the course of the transformation of Germany's QM structure, there has been a rebalancing of the influence of the various *actors*. Thus, responsibility for school development should be increasingly shifted from ministries toVET schools. At the same time, many measures also aim at transferring decision-making powers to schools, while strengthening their accountability to higher levels of authority and society as a whole, and making them more transparent (Becker et al., 2006, p. 10). By way of example, to attain a status of "extended autonomy", a school can experiment with modifications to the schooling system's structure and organisation, as well as changing or augmenting educational content and organisation, along with forms of school constitution and management that are limited in time and space (§25, Schulgesetz NRW—SchulG, 2022). In addition, external evaluation experts from private organisations could be given a greater importance as stakeholders. Despite individual approaches adopted at the federal state level, all actors are being addressed within the context of the transnational DEQA-VET point of reference. On one hand, it networks the stakeholders of VET (this includes societal partners, companies, VET schools, vocational colleges, ongoing education providers and certification organisations, VET researchers, and education policymakers and administrators at federal and state levels). On the other hand, it performs the function of intermediation between national and European levels (DEQA-VET, 2023). There is a lack of national evidence on the impact of quality management in vocational education. For example, a study on the use or non-use of evidence-based knowledge in schools in Baden-Württemberg found that both principals and teachers use process-related sources of information with reference to the specific teaching level, while they see little benefit in the new control instruments (school performance comparisons, learning status surveys, school inspections) and therefore use these sources only to a limited extent. The use of evidence tends to be more pronounced in family and innovative schools than in schools based on competition and conformity (Demski, 2019).

Fields of *power and the distribution of rights of disposal* vary among different actors. For example, the NRW quality analysis teams solely perform tasks of quality analysis, and they are not assigned additional roles and powers pertaining to school supervision. The results of the quality analysis should be communicated to the VET school, to the school board, and to the school supervisory authority in the form of a written quality report. They should also be made available to the departments of the district governments and to the respective ministry, in accordance with a standardised state wide procedure, for further evaluation (QA-VO 2007, § 1). Which new, changed, or redistributed rights of *disposal* the *actors* acquire in the course of transformative processes is still open to debate. For instance, in the evaluation process, data is collected from school records, classroom observations, and interviews with groups of students. This data is then provided to schools as a basis for planning and as a way of encouraging further development of their school and classroom environments. It could be argued that increasing the degree of data transparency for all stakeholders and reducing the number of documents that VET schools provide to the evaluation teams should strengthen the process (MSB NRW, 2023, pp. 4–5.). Educational standards and tests, as well as internal QM in schools that comply with said standards and the quality frameworks, correspond to the tools used to coordinate QM actions in Germany. These tools should enable VET schools to develop into regional competence centres (MK, 2021). However, VET schools should avoid situations of competition among themselves, as these would be undesirable in a federal state system, and they would also counteract the overall goal of equal opportunities in vocational education (Becker et al., 2006, pp. 96–97).

In our analysis, we suggest that the QM approaches in VET in Germany reflect a *governance regime* that is characterised by centralised input regulation by the state, and by external control. Key features of this approach include output orientation (for example, comparative tests) and overall substantial goals, mandatory methods of supervision, and evaluation procedures. Hybrid managerial and professional (namely teachers) practices of self-governance can be observed in regard to school independence, the external control of essential goals, and hierarchical school self-control. Mandatory target agreements on school development are compiled by school supervision organisations (ministry, inspectorate) along with the inspected VET schools, and these are based on the quality evaluation report. It could be argued that since companies are free to choose specific VET schools as a source of their apprentices, this may indirectly create competition between such VET schools. Furthermore, the quasi-markets of supply and demand in the field of apprenticeship and training in the dual system must also be taken into consideration (Becker et al., 2006) (Table 14.3).

Spain

QM in Spain is determined and affected by the relationship between school and labour markets (Olazaran et al., 2023; Souto-Otero & Ure, 2012). In this context, while the Ministry of Education and Vocational Training is responsible for the coordination of QM, its management is underpinned by a threefold approach regulated by the 2006 Education

Table 14.3 Overview of the main characteristics of QM in Germany

Category	*Key points for Germany*
Guiding values	Encourage a strong link between self-evaluation and external evaluation of schools and focus on the fulfilment of the supra-regional educational mission. To create equal opportunities in vocational education and active participation in regional networks and to promote democracy and tolerance.
Structures and actors	**Federal state level:** The ministries of education of the Federal States as a source of normative documents and regulations **Regional level:** The school inspectors act as external evaluators, while the departments of the regional school districts act as consultants **Institutional level:** Schools use internal staff as their own evaluators to carry out internal evaluations
Rights of disposal and structures of regulation	Rights of disposal and structures of regulation vary from one federal state to another. In the evaluation process, the data is made available to the schools as a basis for planning and as a stimulus for the development of the school. There could be differences in the authority of the teams of evaluators and the school inspectorate to the schools.
Governance regimes	Central input regulation by the state and by external control Hybrid managerial and professional (namely teachers) practices of self-governance in relation to school autonomy

Law, and modified by the 2020 Law: state, regional, and local levels (Cedefop, 2022).

The *guiding values* of the Spanish VET system are inclusivity, responsiveness to market and societal needs, and the pursuit of excellence. These justify the comprehensive QM strategies across the VET landscape. By highlighting equity, the system aims to provide all individuals with access to quality vocational education, fostering personal and professional growth. This commitment to adaptability ensures that VET programmes remain responsive to the dynamic demands of the labour market, fostering a workforce that is both skilled and versatile. Furthermore, the pursuit of excellence is crucial for maintaining competitive standards and for the personal fulfilment of each learner. In Spain, VET serves as a critical structure of opportunities by enabling individuals to acquire qualifications that are directly aligned with labour market demands. This alignment fosters employability and social mobility, particularly for those who may not pursue traditional academic routes. Moreover, the decentralised governance model allows regional and local adaptations of VET programmes, ensuring their responsiveness to diverse regional labour market needs while promoting equity and access across varied socioeconomic contexts. By integrating private and public actors, the system enhances pathways for learners to transition seamlessly between education and employment, thus strengthening their long-term prospects in a competitive market (Martínez-Morales & Marhuenda-Fluixá, 2020).

The *actors* of the VET system operate within a highly decentralised *structure* that spans national, regional, and local levels, rather than a vertical structure descending from the regional to the local level or to VET schools; this structure is significantly characterised by decentralisation (Dobbins & Christ, 2019). This three-level governance, regulated by the 2006 Law and amended in 2020, facilitates customised governance and QM practices in different regions. In recent years, QM in the VET system has been considered a critical factor for the country's progress (Rupérez, 2022). At the national level, the Ministry of Education and Vocational Training oversees the strategic coordination of QM, with the support of the National Education Evaluation Institute (INEE) and the National Institute of Professional Qualifications (INCUAL).

In addition, some Regional Qualifications Institutes and the University Quality Assessment Agencies, such as that of the Canaries, perform a central role in regulating QM, because CFGS (*Ciclos Formativos de Grado Superior*) are considered as tertiary education. These institutions maintain and improve educational standards. At the regional level, agencies enforce QM standards that reflect regional priorities and labour market needs, emphasising a three-fold approach to QM. Local educational institutions are instrumental in adapting regional guidelines to their contexts, effectively implementing QM processes.

Private quality assessment companies also have a significant impact on the governance and continuous improvement of the VET system by establishing direct links with the labour market and promoting public–private partnerships, both in the private and public sectors, underlining the role of QM in shaping VET (Olazaran et al., 2023; Souto-Otero & Ure, 2012).

Regardless of the type of governance assessment, a large number of actors at national and regional levels, both public and private (private companies and public administrations), contribute to the functioning of the VET system, coordinating activities among different institutions and actors to develop QA (Sanz De Miguel, 2017). This collaboration extends to tools for verification and regulated certification systems for VET providers. Additionally, VET governance requires the cooperation of different actors, including private institutions, which must meet the same requirements established by the Ministry of Education and Vocational Training (MEFP) in collaboration with the Regional Education Administrations. These ensure that all educational providers, regardless of their source of funding, adhere to pre-determined standards of quality. These requirements have been defined by the MEFP together with the Regional Education Administrations.

In terms of rights of *disposal* and structures *of regulation*, the governance of the Spanish VET system is encapsulated by Law 2/2006, modified in 2020, and further redefined by Law 3/2022.

This law defines VET governance and QM systems: the evaluation of VET quality should be performed according to the indicators established within EQAVET. This can be better understood if we consider that in Spain, VET consists of three parallel subsystems which have developed since the mid-1960s (non-formal, formal, and ongoing VET). The State Higher Inspectorate and regional inspection services are key components of this framework, along with regulations on Integrated and Reference VET schools.

QM tools are important as distinctive hallmarks for private providers, non-profit organisations, and communities, demonstrating their adherence to quality standards. This use of QM extends beyond traditional sectors of education, because employers' associations and union training foundations have ventured into the realm of QM. Training providers that do not seek ISO or similar accreditation of their quality standards may find themselves at a disadvantage in the competitive training landscape.

Regarding *governance regimes* in the Spanish VET system, they reveal a combination of centralised state control and a regionalised, market-influenced approach, which Martínez-Morales & Marhuenda-Fluixá (2020) describe as having undergone significant changes since the mid-1960s. This system includes traditional methods of bureaucratic governance through the State High Inspectorate and regional inspection services. It also highlights the way in which some regions are challenging the Spanish government's proposals based on partisan or regional interests.

Traditionally, educational inspections have been considered as evaluation and control mechanisms. However, according to De Lel and European Commission (2021), inspectors are increasingly tasked with promoting more participatory and collaborative approaches among different stakeholders, including schools, school leaders, teachers, and other educational staff. This shift is intended to modify the role of inspectors from that of primarily monitoring the situation, to that of significantly improving educational practices, thereby fostering supportive interactions that help improve the performance of the entire education system.

Samanes and Clares (2021) underline the need for governance that promotes educational support and inclusiveness, ensuring that all parts of the system work together towards common goals. It can be concluded that each ecosystem requires specific governance, affecting all components of the VET system.

Over the years, Spain has benefited from VET structures and dynamics that have served the needs of the country. This does not mean, however, that all actors are equally involved in increasing accountability and decentralisation among VET providers; the Ministry retains responsibility for VET policy, which is managed and delineated by the regions according to local needs. In this way, the VET subsystems have been encouraged to attain higher quality and to respond more effectively to the needs of society, in line with international standards. The deployment of the potential of the operative elements within the GSC framework must move towards the attainment of the Spanish VET's objective of guaranteeing quality education and vocational training from a perspective of equity that promotes equal access to vocational qualifications for all students, for the official recognition and homologation of the professional qualifications acquired through any of the three subsystems. This is of fundamental importance, because up until now, the Ministry of Education has retained its status of granting the only valid qualifications. Therefore, for formal VET, the results demonstrate state-led governance, along with managerial self-governance at VET schools. As for the non-formal market-regulated VET, competitive pressure leads to quasi-market governance.

Current state of the art of quality evaluation and management in VET has been perceived as an opportunity to improve the attractiveness, prestige, and worth of VET provision, and there are good examples of these in cases like the Center for Innovation of VET in Aragón (CIFPA), the Center for Applied Innovation in VET in the Basque country (Tknika), or the Spanish professional association of public VET schools (FPempresa). All three examples aim to increase the excellence of VET, particularly at tertiary level but also at upper secondary. At the other side of the ladder, there is also an excellent example among the Spanish Association of Second Chance Schools (See also Chap. 9), which gathers almost 50 schools throughout the country that engage young people in Basic vocational education as well as in different kinds of vocational training provision, often at the lower levels of qualification: These students have a record of poor previous educational experience and they face obstacles to enter formal upper secondary education; and the Association has developed and improved a quality framework model for the sake of better serving the needs of these young people at risk (Table 14.4).

Discussion and Conclusion

If we compare the QM systems for VET in the three selected countries, it becomes clear that the decentralisation of quality development in all countries on the dimension of "*guiding values and impact claims*" has led to evidence-based governance and a culture of accountability (FMESR, 2021; Becker et al., 2006). The overall educational goal of QM in all countries considered is high quality of vocational education. In order to achieve this objective, different adjustments are made. In all three countries, the QMS reflects the development of cooperation with societal stakeholders, with active participation in regional networks and the inclusion of demand from the labour market (BMBF, 2023). In Austria, the QMS focuses more strongly on fostering learners' individual potential and promoting teacher professionalism and cooperation through the establishment of teacher teams (Gramlinger & Jonach, 2022). In Germany, the approach to high-quality VET is strongly related to achieving equal opportunities in the educational system, in order to increase permeability between different educational pathways (Cedefop, 2023). In the case of Spain, one notes the pursuit of excellence within the development of QM, in order to increase the value and recognition of VET, thus facilitating the transition from VET into higher education (Rupérez, 2022).

In all three countries, a multi-level QM system has been established, with a clear division of tasks between the "*actor* of control" and the "*actor*

Table 14.4 Overview of the main characteristics of QM in Spain

Category	*Key points for Spain*
Guiding values	Enhance inclusivity, responsiveness to market and societal needs, and the pursuit of excellence Comprehensive QM strategies across the VET landscape
Structures and actors	**State level:** Ministry of Education and vocational Training (MEFP)—establishes the national framework laws The educational inspectorate managed by the state and operating across all regions **Regional level:** Educational inspection services managed by regional administrations Regional administrations of education (autonomous communities (ACs)) responsible for the control and evaluation of QMS **Institutional level:** School management: staff elaborate the internal evaluations within the school, school managers and teachers implement the measures evaluated. Private evaluation businesses offer their services with their own quality Assurance model (*eg*. Eduqatia working and charging the Spanish Association of Second Chance Schools for the external evaluation they conduct).[2]
Rights of disposal and structures of regulation	The evaluation of VET quality should be performed according to the indicators established within EQAVET. The state higher inspectorate and regional inspection services are key components of this framework, along with regulations on integrated and reference VET schools. QM tools are important as distinctive hallmarks for private providers, non-profit organisations, and communities, demonstrating their adherence to quality standards.
Governance regimes	Combination of centralised state control and a regionalised, market-influenced approach

of accountability and distribution". In this context, regional administrations (state/private agencies) are becoming increasingly important in monitoring schools and quality control (FMESR, 2021; Dobbins &

[2] https://www.e2oespana.org/modelo-e2o/acreditacion/

Christ, 2019; Frais et al., 2017; Maritzen, 2015). In all three countries, the school actors within the QM framework have a limited scope for action and the structures in the countries differ. To illustrate, the structure and organisation of QM in Austria are subject to centralised regulation, extending even to internal school processes and responsibilities. In Germany, there is a variation in structures and actors between the federal states: for example, the separation between general education and vocational schools in the QM system varies according to the federal state. This can lead to certain quality-related aspects of vocational training being overlooked or neglected, or becoming dependent on subjective interpretation by controlling organisations. In Spain, state requirements also apply to private providers, which have established their own QM mechanisms in order to compete in the educational market (Olazaran et al., 2023; Souto-Otero & Ure, 2012). However, this situation has significant limitations, because the presence of the private sector and its influence can lead to a fragmentation of quality standards, resulting in a lack of coherence and uniformity in applying QM policies at the national level. This can cause variations in educational quality and make it difficult to ensure a consistent level of vocational training across the country.

In Austria, the rights of *disposal* and regulations are focused more on the work of the pedagogical staff and their participation in quality development and assurance. QM governance gives pedagogical personnel certain obligations, while also granting a degree of freedom and rights of disposal for their work (FMESR, 2021). In this way, the QM system aims to promote a form of "professional self-governance" (De Boer et al., 2007) within the school context. In Germany, the focus is more on the powers of the quality inspectors and the extent to which schools are required to facilitate the evaluation operations performed by school inspectors, or QM functions are lifted (MSB NRW, 2023). In Spain, a more differentiated picture can be identified as a result of the three official VET subsystems: non-regulated, regulated, and continuing VET. In this case, there are different rights of disposal and regulations here. The new Organic Law 3/2022 was intended as a mode of guaranteeing complete recognition across all three subsystems, and it requires the establishment of a more market-oriented joint quality management mechanism. It is interesting to note the importance of QM tools as a distinctive feature for private providers, non-profit organisations, and local authorities in order to demonstrate compliance with quality standards.

In Austria, compared to the other two countries, an overall move towards a centralised governance *regime* tends to be pursued. In addition, recommendations and guidelines in Austria impact the organisational structure within schools, for example through the formation of teacher teams, and the comprehensive quality framework that is specified by the Ministry of Education. This is not the case in Spain and Germany, where there are no guidelines for schools' internal organisational structure. Instead, in Germany, there is a separation between the structures performing assessment and those that hold decision-making power for further development in schools. In Spain, public schools have virtually no independence in issues such as hiring staff (for which the regional administration holds responsibility), nor in determining new VET qualifications. The latter is another function that regional administrations retain for themselves, and it is sometimes applied in accordance with subjective considerations, rather than for the needs of society or the labour market. Furthermore, in Spain, QM aspects are important in terms of visibility for private VET providers, which have established their own QM concept in areas of both formal and non-formal VET. This, in the case of formal VET, has led to the independent quest for QM labels on the part of public VET schools. It is important to mention that, as regards QM governance in Spain, it is clear that in this country, the administration and the market often follow different approaches. More specifically, non-formal VET and continuing vocational education and training (CVET) institutions tend to operate to a greater degree like market-influenced organisations rather than structured systems. Consequently, QM is not only used for legislation, but also to some extent as a form of competition within the training market. This is manifestly obvious in Spain. The same effect is not explicitly reported in Germany, but such considerations definitely play a role in terms of communications on the part of training companies.

With reference to Altrichter's (2010) heuristic framework of educational governance, it can be seen that Austria and Germany both receive "external guidance from the state" in relation to the determination of quality indicators and targets. This form of educational governance is combined with "managerial and professional self-governance" in order to determine how quality indicators and targets can be achieved. Similarly, in Spain, this type of educational governance can also be observed in the state or regionally regulated sector of the VET system. In the market-regulated part of the non-formal VET system, the characteristic of "competitive quasi-market pressure" can be seen, but this is also combined with "managerial self-governance" in terms of implementation.

From the perspective of educational governance, a comparative analysis of the findings reveals the dominance of two primary QM functions within the countries considered, determined by the prevailing governance structures. Within the formal VET system, there appears to be a growing emphasis on functions of accountability and performance. Conversely, in the non-formal VET system, the marketing function takes precedence. One notable feature is that the aspect of quality development or enhancement at the teaching and learning level is often relegated to a secondary position. This factor raises concerns about the potential emergence of "non-intended results" in VET systems caused by the coordination of actions, which risks diverting attention and significance away from the "pedagogical core business" (Coffield et al., 2007, p. 736).

Since the rise of New Public Management in the 1990s, in all three countries, there has been a growing expectation for increasingly participatory governance among VET educational stakeholders. This expectation particularly involves school management, teachers, and training companies. Although there appears to be a trend towards a more decentralised approach to educational governance, this does not necessarily imply a de facto shift towards a participatory QM model (Dubs, 2003; Schedler, 1995). Our analysis indicates that standards, quality requirements, and educational goals are still predominantly established top-down by the ministries of education. Consequently, stakeholders within VET institutions are managerially responsible for meeting these requirements. The analysis of policy documents reveals a notable absence of emphasis on the autonomous development of schools' opportunity structures, manifested in the form of modification to the standards and performance targets at the level of individual schools (Becker et al., 2006).

The findings of this study suggest that the central focus of current QM systems is the decentralisation of responsibility as regards pre-determined outcomes and the practical implementation of governmental requirements. There is a noticeable shift of operational tasks from the ministries of education to the stakeholders within VET institutions, resulting in increased workload and modifications in the responsibilities of school-system stakeholders (Hautz, 2022a). At the same time, there is a narrow window of opportunity for involvement in QM strategy and conceptual direction, which raises questions about the extent of autonomy within the current governance framework of formal VET systems in all three countries. In light of the aforementioned considerations, it can be posited that the function of QM is to direct the quality development of VET schools in a manner that aligns with the specified understanding of quality. The creation of additional opportunity structures is a challenging undertaking.

Furthermore, a market-driven demand for QM systems has emerged (see also Locke & Maton, 2019; Pasura, 2014; Rönnberg, 2011). Here, QM is less about ensuring equal participation of relevant stakeholders and fostering pedagogical values, and more about representing a tool for communicating with customers and attaining additional visibility with respect to competitors in the marketplace.

Against this background, after decades of statements on decentralisation in educational governance, it is clear that the stronger integration of management tasks into schools' activity profile requires a reconsideration and redefinition of the school organisation. This implies the creation of new structures for support and professional development that can reduce teachers' responsibilities with respect to administrative and management tasks. This conclusion is based on the analysis of the educational governance structures in the three countries on the basis of documental research. Where possible, the governance structure was enriched with empirical secondary data. The actor constellation can be interpreted and implemented differently by the actor groups, depending on the framework conditions of governance implementation. It would also be interesting and important to examine how QM potentially has improved the quality in VET institutions and how quality of VET is related to VET success. Thus far, no empirical evidence has been presented in any of the three studied countries to demonstrate the efficacy of QM measures on learning outcomes and the advancement of social equity. This may be attributed to the fact that individual performance is contingent upon factors that are largely beyond the purview of any conceivable QM system (Hopmann, 2021). Furthermore, it is important to consider the inherent limitations of educational governance. These arise from the complex interplay of multiple actors with their distinct interests, the intrinsic logics, and dynamics of educational subsystems and school organisations, as well as the influence of a diverse range of stakeholders (Altrichter & Maag Merki, 2010). As a result, QM can lead to unintended consequences and adverse effects on the quality of VET (see e.g., Donovan & Hautz, 2024; Hautz & Thoma, 2021). In light of the findings of this study, it would be particularly important to conduct further research examining the specific effects of QM and educational governance on the opportunity structures of VET in diverse national contexts.

References

Altrichter, H. (2010). Theory and evidence on governance: Conceptual and empirical strategies of research on governance in education. *European Educational Research Journal, 9*(2), 147–158. https://doi.org/10.2304/eerj.2010.9.2.147

Altrichter, H., & Heinrich, M. (2007). Kategorien der Governance-Analyse und Transformationen der Systemsteuerung in Österreich [Categories of governance analysis and transformations of system governance in Austria]. In H. Altrichter, T. Brüsemeister, & J. Wissinger (Eds.), *Educational governance: Handlungskoordination und Steuerung im Bildungssystem [Educational governance: Action coordination and steering in educational system]* (pp. 55–103). VS Verlag für Sozialwissenschaften. https://doi.org/10.1007/978-3-531-90498-6_3

Altrichter, H., & Maag Merki, K. (2010). Steuerung der Entwicklung des Schulwesens [Managing the development of the school system]. In H. Altrichter & K. Maag Merki (Eds.), *Handbuch neue Steuerung im Schulsystem [Manual for new control in the school system]* (pp. 15–39). VS Verlag für Sozialwissenschaften.

Amelsvoort, H. G. V., & Scheerens, J. (1997). Policy issues surrounding processes of centralization and decentralization in European education systems. *Educational Research and Evaluation, 3*(4), 340–363. https://doi.org/10.1080/1380361970030404

Anckar, C. (2008). On the applicability of the Most similar systems design and the most different systems design in comparative research. *International Journal of Social Research Methodology, 11*(5), 389–401. https://doi.org/10.1080/13645570701401552

Becker, M., Spöttl, G., & Dreher, R. (2006). Berufsbildende Schulen als eigenständig agierende lernende Organisationen. Stand der Weiterentwicklung berufsbildender Schulen zu eigenständig agierenden lernenden Organisationen als Partner der regionalen Berufsbildung (BEAGLE) [Vocational schools as autonomously acting learning organisations. Status of the further development of vocational schools into autonomously acting learning organisations as partners in regional vocational education and training (BEAGLE); research report]. In M. Becker, G. Spöttl, & R. Dreher (Eds.), *Materialien zur Bildungsplanung und zur Forschungsförderung [Materials on education planning and research funding]*. BLK. http://www.pedocs.de/volltexte/2008/225/pdf/heft135.pdf

Benz, A. (2004). Multilevel Governance—Governance in Mehrebenensystemen [Multilevel Governance—Governance in Mehrebenensystemen]. In A. Benz (Ed.), *Governance—Regieren in komplexen Regelsystemen. Eine Einführung [Governance—Governing in complex regulatory systems. An introduction]* (pp. 125–146). VS Verlag für Sozialwissenschaften.

BMBF [Federal Ministry of Education and Research]. (2023). *Richtlinie zur Förderung von Projekten im Rahmen des Programms "Nachhaltig im Beruf—zukunftsorientiert ausbilden" (NIB). [Guideline for the funding of projects under the programme "Sustainable at work—future-oriented training" (NIB)].* Bundesanzeiger vom 27.01.2023. https://www.bmbf.de/bmbf/de/bildung/berufliche-bildung/foerderinitiativen-und-programme/nachhaltigkeit-in-der-beruflichen-bildung/nachhaltigkeit-in-der-beruflichen-bildung_node.html

Braun, V., Clarke, V., Hayfield, N., & Terry, G.. (2019). *Answers to frequently asked questions about thematic analysis*. Retrieved from https://cdn.auckland.ac.nz/assets/psych/about/our-research/documents/Answers%20to%20frequently%20asked%20questions%20about%20thematic%20analysis%20April%202019.pdf

Cedefop; Fundae. (2022). *Vocational education and training in Europe - Spain: system description* [From Cedefop; ReferNet. Vocational education and training in Europe database]. https://www.cedefop.europa.eu/en/tools/vet-in-europe/systems/spain-u2

Cedefop. (2023). *Vocational education and training policy briefs 2023: Monitoring and analysis of VET policies - Germany*. Cedefop. Retrieved from https://www.cedefop.europa.eu/en

Cedefop [European Centre for the Development of Vocational Training]. (2015). *Handbook for VET providers: Supporting internal quality management and quality culture* (Cedefop reference series 99). Publications Office of the European Union. https://doi.org/10.2801/82638

Chen, H. T. (2005). *Practical program evaluation: Assessing and improving planning, implementation, and effectiveness*. Sage.

Coffield, F., Edward, S., Finlay, I., Hodgson, A., Spours, K., Steer, R., & Gregson, M. (2007). How policy impacts on practice and how practice does not impact on policy. *British Educational Research Journal, 33*(5), 723–741. https://doi.org/10.1080/01411920701582363

De Boer, H., Ender, J., & Schimank, U. (2007). On the way towards new public management? The governance of university systems in England, The Netherlands, Austria, and Germany. In D. Jansen (Ed.), *New forms of governance in research organisations* (pp. 137–152). Springer.

De Lel, G., & European Commission: European Education and Culture Executive Agency. (2021). *The organisation of school time in Europe: Primary and general secondary education: 2021/22*. Publications Office of the European Union. https://data.europa.eu/doi/10.2797/63021

Demski, D. (2019). Nutzung von internen und externen Evaluationen in der Schulpraxis [Use of internal and external evaluations in school practice]. In T. Stricker (Ed.), *Zehn Jahre Fremdevaluation in Baden-Württemberg [Ten years of external evaluation in Baden-Württemberg]*. Springer VS. https://doi.org/10.1007/978-3-658-25778-1_4

DEQA-VET. (2023). *Deutsche Referenzstelle für Qualitätssicherung in der beruflichen Bildung [German Reference Point for Quality Assurance in Vocational Education and Training]*. https://www.deqa-vet.de/de/daqavet_88364.php

Dobbins, M., & Christ, C. (2019). Do they matter in education politics? The influence of political parties and teacher unions on school governance reforms in Spain. *Journal of Education Policy, 34*(1), 61–82. https://doi.org/10.1080/02680939.2017.1406153

Donovan, C., & Hautz, H. (2024). Thresholds for dis/trust: exploring the impact of standards-based reforms on the process of being and becoming a VET teacher in England and Austria. *Education + Training, 66*(5), 541–556. https://doi.org/10.1108/ET-09-2023-0386

Dose, N. (2005). Governance als problemorientierte Steuerung. Steuerung angesichts alter Probleme und neuer Herausforderungen [Governance as problem-orientated control. Governance in the face of old problems and new challenges]. In G. F. Schuppert (Ed.), *Governance Forschung. Vergewisserung über stand und Entwicklungslinien [Governance research. Verification of status and lines of development]* (pp. 77–94). Nomos.

Dubs, R. (2003). *Qualitätsmanagement für Schulen [Quality management for schools]*. IWP Institut für Wirtschaftspädagogik.

Ebner, C. (2013). *Erfolgreich in den Arbeitsmarkt? Die duale Berufsbildung im internationalen Vergleich [Successfully entering the labour market? An international comparison of dual vocational education and training]*. Campus Verlag.

European Commission. (2000). *Lisbon European council 23 and 24 March 2000: Presidency conclusions*. http://www.europarl.europa.eu/summits/lis1_en.htm

European Commission. (2002). *The Copenhagen declaration*. European Commission.

European Commission. (n.d.). *EQAVET - European quality assurance in vocational education and training*. https://ec.europa.eu/social/main.jsp?langId=en&catId=1536

European Council. (2003). Council resolution of 19 December 2002 on the promotion of enhanced European cooperation in vocational education and training. *Official Journal of the European Communities*, 2–4.

Feldhoff, T. (2011). *Schule organisieren [Organise school]*. VS Verlag für Sozialwissenschaften.

Flick, U. (2018). *An introduction to qualitative research*. Sage.

FMESR [Feder Ministry for Education, Science and Research]. (2022). *Akteurinnen und Akteure [Actors]*. https://www.qms.at/ueber-qms/akteurinnen-und-akteure

FMESR [Feder Ministry for Education, Science and Research]. (2023). *General information on QMS in English*. https://www.qms.at/english

FMESR [Federal Ministry for Education, Science and Research]. (2018). *2017 education reform: Freedom for schools, transparency for the system, model regions for education as a whole*. https://www.bmbwf.gv.at/en/Topics/school/krp/2017_ed_ref.html

FMESR [Federal Ministry for Education, Science and Research]. (2019). *Verordnung des Bundesministers für Bildung, Wissenschaft und Forschung betreffend das Schulqualitätsmanagement [Regulation of the Federal Minister of Education, Science and Research concerning the School Quality Management]*, BGBl. II Nr. 158/2019.

FMESR [Federal Ministry for Education, Science and Research]. (2021). QMS at a glance. Overall presentation of the quality management *system for schools*. Version 1.0.

FMEW [Federal Ministry of Education and Women's Affairs]. (2015). *VET schools and colleges in Austria*. Information brochure of the General Directorate for Vocational Education and Training, Adult Education and School Sports. FMEW.

Frais, M., Schoppmann, U., & Renz, M. (2017). Evaluation in Schulen. Rückblick und Ausblick [Evaluation in schools. Review and outlook]. *Zeitschrift Für Evaluation, 16*(2), 288–292.

Freynschlag, D., Göbl, B., Egger, A., Granzner, A., Grubmüller, S., Hessmann, D., & Pass, B. (2022). Q-SK + Q-RK = wunderbar! QMS bringt einen MehrWERT an Schulqualität [Q-SK + Q-RK = wonderful! QMS brings added VALUE to school quality]. *Magazin der pädagogischen Hochschule OÖ, 6*(2), 7.

Frommberger, D. (2012). Von der Berufsbildung in die Hochschulbildung (Dritter Bildungsweg): Eine berufs- und wirtschaftspädagogische Einordnung unter besonderer Berücksichtigung aktueller Rahmenwerke zur Förderung von Übergängen und Durchlässigkeit [From vocational training to higher education (third educational pathway): A vocational and business education categorisation with particular reference to current frameworks for promoting transitions and permeability]. *Zeitschrift Für Berufs- Und Wirtschaftspädagogik*, 169–193. Franz Steiner Verlag.

Gramlinger, F., & Jonach, M. (2022). QMS. Das neue Qualitätsmanagementsystem für Schulen wird implementiert [QMS. The new quality management system for schools is being implemented]. *Schulverwaltung: Fachzeitschrift Für Schulentwicklung Und Schulmanagement, 10*(4), 98–101.

Harvey, L., & Green, D. (2000). Qualität definieren. Fünf unterschiedliche Ansätze [Define quality. Five different approaches]. In A. Helmke, W. Hornstein, & E. Terhart (Eds.), *Qualität und Qualitätssicherung im Bildungsbereich; Schule, Sozialpädagogik, Hochschule [Quality and quality assurance in education; school, social pedagogy, university]* (Vol. 41, pp. 17–39). Zeitschrift für Pädagogik. https://doi.org/10.25656/01:8483

Hautz, H. (2022a). The 'conduct of conduct' of VET teachers: Governmentality and teacher professionalism. *Journal of Vocational Education & Training, 74*(2), 210–227. https://doi.org/10.1080/13636820.2020.1754278

Hautz, H. (2022b). Qualität und Qualitätsmanagement aus der Perspektive von Lehrpersonen. Eine kritische Bestandsaufnahme [Quality and quality management from the perspective of teachers. A critical review]. *SchulVerwaltung aktuell. Fachzeitschrift für Schulentwicklung und Schulmanagement, 10*(4), 115–117.

Hautz, H., & Thoma, M. (2021). Teacher subjectivation in the quality dispositive: The example of VET in Austria. *British Journal of Sociology of Education, 42*(5–6), 791–811. https://doi.org/10.1080/01425692.2021.1908117

Hopmann, S. T. (2021). Qualitätsmanagement als Perpetuum Mobile. Ein kritischer Blick [Quality management as a perpetual mobile. A critical view]. *SchulVerwaltung aktuell. Fachzeitschrift für Schulentwicklung und Schulmanagement, 10*(4), 118–121.

Jonach, M., & Pitzer, B. (2022). QMS—das Qualitätsmanagementsystem für Schulen oder: Warum die Qualität von Schule und Unterricht kein Zufall sein darf! Was ist eine gute Schule? [QMS—the quality management system for schools or: Why the quality of school and teaching should not be a matter of chance! What is a good school?]. *Magazin der pädagogischen Hochschule OÖ, 2*(6), 4–5.

Köhler, H. (1996). Bildung ist Ländersache - Zur Entwicklung des Schulwesens im föderalistischen Staat [Education is a matter for the federal states - On the development of the school system in a federalist state]. In A. Leschinsky (Ed.), *Die Institutionalisierung von Lehren und Lernen. Beiträge zu einer Theorie der Schule (Zeitschrift für Pädagogik, Beiheft) [The institutionalisation of teaching and learning. Contributions to a theory of the school (Zeitschrift für Pädagogik, supplement)]*. https://doi.org/10.25656/01:9788

Konferenz der Kultusminister in der Bundesrepublik Deutschland (KMK). (2004). *Qualität Beruflicher Schulen: Empfehlung für die Analyse und Bewertung der Arbeit beruflicher Schulen als Beitrag für die kontinuierliche Qualitätsentwicklung. Beschluss der Kultusministerkonferenz vom 14.03.2004 [Quality of Vocational Schools: Recommendation for analysing and evaluating the work of vocational schools as a contribution to continuous quality development. Resolution of the Standing Conference of the Ministers of Education and Cultural Affairs of 14 March 2004]*. KMK.

Kussau, J., & Brüsemeister, T. (2007). Educational governance: Zur analyse der Handlungskoordination im Mehrebenensystem der Schule [Educational governance: To analyse the action coordination in a multi-layer-system of school]. In H. Altrichter, T. Brüsemeister, & J. Wissinger (Eds.), *Educational governance: Handlungskoordination und Steuerung im Bildungssystem [Educational governance: Action coordination and steering in educational system]* (pp. 15–54). VS Verlag für Sozialwissenschaften.

Landwehr, N. (2021). Das unterrichtsbezogene Qualitätsmanagement im Spannungsfeld von Oberflächen- und Tiefenstrukturen [Teaching-related quality management in the tension between surface and deep structures]. *Erziehung & Unterricht, 171*(9–10), 60–74.

Landwehr, N., & Steiner, P. (2020). *Schulinternes Qualitätsmanagement nach dem Modell Q2E: Funktionenmodell. [School internal quality management according to the Q2E model: Function model]*. https://www.q2e.ch/wp-content/uploads/sites/162/2020/05/q2e-das-funktionenmodell-ausf-hrliche-fassung-blau.pdf

Lassnigg, L. (2011). The 'duality' of VET in Austria: Institutional competition between school and apprenticeship. *Journal of Vocational Education & Training, 63*(3), 417–438. https://doi.org/10.1080/13636820.2011.590220

Locke, P., & Maton, K. (2019). Serving two masters: How vocational educators experience marketisation reforms. *Journal of Vocational Education & Training, 71*(1), 1–20. https://doi.org/10.1080/13636820.2018.1480521

Mai, F. (2020). Qualitätsmanagement in der Bildungsbranche: Ein Leitfaden für Bildungseinrichtungen und Lerndienstleiter [Quality management in the education sector: A guide for educational institutions and learning service providers]. Springer Gabler.

Marhuenda-Fluixá, F., Chisvert-Tarazona, M. J., & Palomares-Montero, D. (2019). *The school-based vocational education and training system in Spain*. Springer.

Maritzen, N. (2015). Schulinspektion—Aspekte der transformation von governance-Strukturen [School inspection - aspects of the transformation of governance structures]. In M. Pietsch, B. Scholand, & K. Schulte (Eds.), *Schulinspektion in Hamburg. Der erste Zyklus 2007–2013: Grundlagen, Befunde, Perspektiven [School inspection in Hamburg. The first cycle 2007–2013: basics, findings, perspectives]* (pp. 13–36). Waxmann.

Martin, J., Elg, M., & Gremyr, I. (2020). The many meanings of quality: Towards a definition in support of sustainable operations. *Total Quality Management & Business Excellence*, 1–14. https://doi.org/10.1080/14783363.2020.1844564

Martínez-Izquierdo, L., & Torres Sánchez, M. (2022). Dual vocational education and training systems' governance model and policy transfer: The role of the European Union in its diffusion. *Social Sciences, 11*(9), 403. https://doi.org/10.3390/socsci11090403

Martínez-Morales, I., & Marhuenda-Fluixá, F. (2020). Vocational education and training in Spain: Steady improvement and increasing value. *Journal of Vocational Education & Training, 72*(2), 209–227. https://doi.org/10.1080/13636820.2020.1729840

Matanovic, W., & Sacks, P. (1976). Doppelqualifikation—unter besonderer Berücksichtigung berufsqualifizierender Bildungsgänge: Ein Diskussionsbeitrag [Dual qualification - with special consideration of vocational qualification programmes: A contribution to the discussion]. *Berufsbildung in Wissenschaft und Praxis - BWP, 3*(5), 8–10.

Mayntz, R. (2009). Von politischer Steuerung zu Governance? Überlegungen zur Architektur von Innovationspolitik [From political control to governance? Reflections on the architecture of innovation policy]. In R. Mayntz (Ed.), *Über Governance. Institutionen und Prozesse politischer Regelung [About governance. Institutions and processes of political regulation]* (pp. 105–120). Campus.

Merino Pareja, R. (2023). La formación profesional reglada [Formal vocational training]. In A. Blanco, S. Mora, & J. A. López-Ruiz (Eds.), *Informe España 2023: Cátedra José María Martín Patino de la Cultura del Encuentro* (pp. 129–208).

MK. (2021). *Strategischer Handlungsrahmen für berufsbildende Schulen in Niedersachsen. Regionale Kompetenzzentren weiterentwickeln [Strategic framework for vocational schools in Lower Saxony. Further developing regional competence centers]*. Ministry of Education and Cultural Affairs of Lower Saxony.

Morgan, H. (2022). Conducting a qualitative document analysis. *The Qualitative Report, 27*(1), 64–77. https://doi.org/10.46743/2160-3715/2022.5044

MSB NRW. (2023). *Konzept zur Weiterentwicklung der Qualitätsanalyse [Concept for the further development of quality analysis]*. Ministry for School and Education of the State of North Rhine-Westphalia.

Olazaran, M., Otero, B., Lavia, C., & Albizu, E. (2023). Implementation of dual vocational education and training in Spain: The case of the Basque Country. *Journal of Vocational Education & Training*, 1–20. https://doi.org/10.1080/13636820.2023.2180422

Ott, B., Scheib, T. (2002). Qualitäts- und Projektmanagement in der beruflichen Bildung: Einführung und Leitfaden für die Aus- und Fortbildung [Quality and project management in vocational education and training: introduction and guidelines for initial and continuing training].: Cornelsen Verlag.

Pasura, R. (2014). Neoliberal economic markets in vocational education and training: shifts in perceptions and practices in private vocational education and training in Melbourne, Australia. *Globalisation, Societies And Education, 12*(4), 564–582. https://doi.org/10.1080/14767724.2014.906300

Powell, J. J., Bernhard, N., & Graf, L. (2012). The emergent European model in skill formation: Comparing higher education and vocational training in the Bologna and Copenhagen processes. *Sociology of Education, 85*(3), 240–258. https://doi.org/10.1177/0038040711427313

QA-VO. (2007). *Verordnung über die Qualitätsanalyse [Ordinance on quality analysis]*. Online verfügbar unter https://www.schulministerium.nrw/sites/default/files/documents/Qualitaetsanalyse-Verordnung.pdf

Rogers, P. J., Petrosino, A. J., Hacsi, T., & Huebner, T. A. (2000). *Program theory evaluation: Challenges and opportunities, new directions for evaluation 87*. Jossey-Bass.

Rönnberg, L. (2011). Education governance by marketisation and quality assurance—Introductory notes. *Education Inquiry, 2*(4), 555–561. https://doi.org/10.3402/edui.v2i4.21999

Rupérez, F. L. (2022). The quality of governance of the education system. The case of the LOMLOE. *Revista Española De Pedagogía, 80*(281), 14. https://doi.org/10.22550/REP80-1-2022-02

Samanes, B. E., & Clares, P. M. (2021). Hacia un ecosistema de investigación sobre formación profesional en España. *Revista de Investigación Educativa [Towards a research ecosystem on vocational education and training in Spain. Journal of Educational Research], 39*(1), 249. https://doi.org/10.6018/RIE.424901

Sanz De Miguel, P. (2017). Gobernanza de la formación profesional dual Española: entre la descoordinación y la falta de objetivos [Governance of Spanish dual vocational training: between lack of coordination and lack of objectives]. *Revista Española De Educación Comparada*, (30), 77–98. https://doi.org/10.5944/reec.30.2017.18705

Schedler, K. (1995). Ansätze einer wirkungsorientierten Verwaltungsführung [Approaches of outcome-oriented administration leadership]. Haupt.

Schlögl, P., Stock, M., & Mayerl, M. (2019). Berufliche Erstausbildung: Herausforderungen und Entwicklungsaufgaben in einem bedeutsamen Bildungssegment Österreichs [Initial vocational education and training: Challenges and development issues in a key education segment in Austria]. In S. Breit, F. Eder, K. Krainer, C. Schreiner, A. Seel, & C. Spiel (Eds.), *Nationaler Bildungsbericht Österreich 2018, Band 2, Fokussierte Analysen und Zukunftsperspektiven für das Bildungswesen [National Education Report Austria 2018, Volume 2, Focused analyses and future perspectives for the education system]* (pp. 269–305). Leykam.

Schulministerium NRW [Ministry of Education NRW]. (2023). *Qualitätsanalyse [Quality analysis]*. https://www.schulministerium.nrw/schule-bildung/schulorganisation/qualitaetsanalyse

Souto-Otero, M., & Ure, O. B. (2012). The coherence of vocational education and training in Norway and Spain: National traditions and the reshaping of VET governance in hybrid VET systems. *Compare: A Journal of Comparative and International Education, 42*(1), 91–111. https://doi.org/10.1080/03057925.2011.628836

Timischl, W. (2006). *QIBB—The Austrian VET initiative to assure and further develop quality in the school-based education system.* Contribution to the Directors General for Vocational Education & Training Conference within the framework of Austrian EU presidency (27–28 March 2006). Federal Ministry for Education, Science and Culture.

Wagner, C., & Rückmann, J. (2017). Qualitätsmanagement in der beruflichen Bildung in Deutschland: Ein aktueller Überblick mit dem Schwerpunkt der schulinternen Evaluation [Quality management in vocational education and training in Germany: A current overview focussing on internal school evaluation]. In P. Schlögl, M. Stock, & D. Moser (Eds.), *Berufsbildung, eine Renaissance? [Vocational training, a renaissance?]* (pp. 253–264). https://doi.org/10.3278/6004552w253

Wilbers, K. (2024). Qualitätsmanagement für berufliche Schulen: Wissenschaftliche Bezüge und Perspektiven [Quality management for vocational schools: Scientific references and perspectives]. In Kultusministerkonferenz (Eds.), *Qualität Beruflicher Schulen. Empfehlung für die Analyse und Bewertung der Arbeit beruflicher Schulen als Beitrag für die kontinuierliche Qualitätsentwicklung. Beschluss der Kultusministerkonferenz vom 14.03.2024 [Quality of Vocational Schools. Recommendation for analysing and evaluating the work of vocational schools as a contribution to continuous quality development. Resolution of the Standing Conference of the Ministers of Education and Cultural Affairs of 14 March 2024].* (pp. 15–31). Kultusministerkonferenz.

Zaunstöck, T. (2020). *Europeanisation of VET—The Spanish vocational education and training system under the influence of European education policy.* [Master-thesis., Karlsruhe Institute of Technology and University of Valencia].

Zaunstöck, T., Marhuenda-Fluixá, F., Ros-Garrido, A., & Fischer, M. (2021). Europeanisation of VET–the Spanish vocational education and training system and the influence of European education policy. *Journal of Vocational Education & Training, 73*(2), 316–335. https://doi.org/10.1080/13636820.2021.1889645

Zlatkin-Troitschanskaia, O. (2007). Steuerungsfähigkeit des öffentlichen Schulwesens versus Steuerbarkeit der Schule—Paradigmenwechsel? [Governability of the public school system versus manageability of the school - paradigm shift?]. In J. van Buer & C. Wagner (Eds.), *Qualität von Schule—Ein kritisches Handbuch [School quality - a critical handbook]* (pp. 67–81). Lang.

Zürn, M. (2008). Governance in einer sich wandelnden Welt—eine Zwischenbilanz [Governance in a changing world - an interim assessment]. In G. F. Schuppert & M. Zürn (Eds.), *Governance in einer sich wandelnden Welt [Governance in a changing world]* (pp. 553–580). VS Verlag für Sozialwissenschaften.

CHAPTER 15

The Vocation—A Core Quality of Vocational Education and Training

Christian Helms Jørgensen

Introduction

VET is organised around vocations like skilled carpenters, clerks and health care workers. This chapter examines the qualities of vocations for the education system and for working life. The concept 'vocation' is here used to include both the institutional dimensions, vocational institutions and occupational organisations, as well as the personal and social dimensions. One reason for examining vocations is that research has questioned their future role and position (Baethge et al., 1998; Fuller, 2019). Many highly cited studies of working life (Braverman, 1974; Sennett, 1998) have written the epitaph on the vocations due to first Taylorism and later flexible capitalism. Vocations were seen as outdated and inflexible institutions that do not match the new requirements for flexible and interdisciplinary work (Dostal, 2001). The manual skills associated with traditional craft-like vocations have little relevance for the 'knowledge

C. H. Jørgensen (✉)
Roskilde University, Roskilde, Denmark
e-mail: chrhj@outlook.dk

E. Smeplass and C. Zirkle (eds.), *Opportunity Structures in Vocational Education and Training*, Palgrave Studies in Technical and Vocational Education and Training,
https://doi.org/10.1007/978-3-032-04410-5_15

economy' (Fuller, 2019). In addition, neoliberal policies in education and employment policy have for decades criticised vocations and professions for monopolising specific types of work and for limiting flexibility (Thelen, 2014). This has involved policies for deregulation and flexibilisation, for example by modularising VET programmes and tailoring VET to the requirement of individual firms and students. More recently, the influx of immigrant labour tends to undermine the role of VET, for example in the Construction industries of Norway and Denmark (Haakestad & Friberg, 2020; Arnholtz & Østhus, 2024). In light of these and other challenges for vocations, this chapter will examine the diverse qualities of vocations and the advantages of an institutional regime that is based on vocations (Deissinger, 1998; Thelen, 2004; Kurtz, 2005).

Empirically, the chapter is based on studies of the Danish and Nordic VET systems (Jørgensen et al., 2018) as well as on international comparative VET research (Greinert, 1999; Thelen, 2004; Busemeyer & Trampusch, 2019). Using a comparative methodology, it contrasts vocations with the professions that are also based on specialised skills and expert knowledge. It uses a historical approach to explore important junctures in the historical development of vocations and VET systems. Founded on this analysis, the chapter identifies three ideal-typical regimes of skill formation, one of them being collective skill formation, where vocations are essential institutions. Vocations are characterised as institutions and didactical principles that bridge between the world of education and the world of work (Deissinger, 1998). Thereby, they support young people's learning from both worlds and their transition from education to working life (Marczuk, 2024). Vocations enhance flexibility in the labour market by providing broad skills with sectoral and industry-wide portability. For employees, vocations are associated with high levels of self-management at work and participation in social communities beyond the individual firm (Deissinger, 1998; Brockmann et al., 2008).

Based on this analysis, vocations are described as complex and composite institutions. The chapter points at the internal tensions in the vocations—between theoretical knowledge, practical knowledge, organisational interests and vocational cultures. However, a key quality of vocations is that they establish an institutional framework that coordinates and negotiates between different interests, which in other regimes are more disconnected, and therefore separates education from training.

What Is the Difference between Vocations and Professions?

Being a professional is generally associated with high quality and status. Vocations and occupations related to the VET system have many similarities with the traditional professions. When trying to qualify the concept of vocations in relation to professions, this section draws on theories of professionalism that have identified common features of professions (Larson, 2012; Freidson, 2001). The sociology of professions has traditionally concentrated on medics, priests, lawyers and engineers, and has more recently extended its interest to the new 'semi-professions', such as nurses and teachers. But it has not included the vocations and occupations of the skilled workers. The English term 'vocational' is sometimes associated with *'manual or practical activities'* and contrasted this to the activities of highly qualified and academic professions. In an international perspective, this is however misleading. The English VET system has a narrow, practical and fragmented understanding of skills compared to VET other countries, where the VET curriculum includes broader forms of knowledge and competencies (Brockmann et al., 2008; Clarke & Winch, 2006).

The conceptual separation of vocational and academic knowledge is also refuted by research. The German 'Berufssoziologie' includes in the category 'Berufe' both the skilled trades and the academic professions (Heidenreich, 1999). Something similar applies to the Norwegian and Swedish term *'yrke'* that includes carpenters, nurses and doctors. When vocations are not included in Anglo-Saxon research on professions, it is attributable to the weak position of VET and vocations in these countries. After Liberalism's dissolution of the pre-industrial guilds in the USA and England, no strong modern institutions were established to set standards and ensure the quality of vocational training, which then deteriorated (Greinert, 1999). In the USA, the parties on both sides were too weakly organised to develop a consensus-oriented collaboration on VET similar to what happened in the German speaking and the Nordic countries (Thelen, 2004).

In the labour market, the vocations and skilled trades define themselves both upwards in relation to the professions and downwards in relation to the un- and semiskilled workers. These groups have typically been organised in different unions and confederations (Hyman, 2001). In Denmark, a strong tradition of craft unions rather than industrial unions has predominated, and here the division between skilled and unskilled workers has historically been deeper than in the other Nordic countries (Christensen, 1978; Jørgensen et al., 2018). In keeping with technological progress, the skilled workers' organisations have on the one hand accepted that unskilled workers took over parts of their more simple work areas, and on the other hand, they have tried to upgrade their skills and take over tasks from higher educated groups, for example engineers and nurses.

Upwards, the skilled trades differ from both the traditional and the new professions. There are, however, many similarities between the professions and vocations, as both are based on a combination of experiential, applied and codified knowledge (Saks, 2012). Vocational education and training, like many professions, are organised as alternating training, where a significant part of the skills are transmitted during internships or apprenticeships at a workplace. Both groups often have strong trade unions or professional organisations, where membership is conditional on a degree or an educational certificate. But vocations and professions also differ in several dimensions.

The first dimension is the *knowledge base*. Vocational practice relies more on practical, tacit and experiential knowledge, whereas professions rely more on codified, scientific knowledge. An indication of this can be found in the extent of work-based learning in the education programmes. The Danish VET programmes include around 70% work-based training, nurses have 40% and the academic professions have little or no work-based learning in their education programmes. Some academic professions, like doctors, have however formalised internship as part of their career paths after the completion of the degree.

The second dimension relates to the length of the *education* programme and the status of the profession. Professions most often require a higher education degree and enjoy higher esteem than skilled workers and craftspersons, who have completed an upper secondary VET programme. However, in some areas, the differences between vocations and professions are small. In Denmark, the Bachelor programme for early childhood teachers lasts three and a half years, of which one year is a paid internship. The VET programme for computer technicians lasts five years, of which three years are apprenticeship training in a company. The length of school-based education is roughly the same in the two education programmes. Measured in terms of salary after completion, many vocations in VET have higher value than professional degrees at Bachelor and Master levels in Denmark.

A third dimension concerns the vocations' and the professions' *control* over the education and their specialist field of work. Many professions (like doctors and nurses) have formal protection of their field by means of licenses, and they can enforce mandatory membership of a trade organisation. Licensing is a requirement for some vocations (electricians and plumbers), but is rarer.

Licensing is justified, among other things, by a third dimension, the vocational and professional *ethics* and the role of internal bodies of the profession to enforce the ethical guidelines via collegiate self-control. Such formal bodies are seldom for vocations, where professional standards for quality are primarily determined by the employers or disseminated informally via practice.

A fourth dimension concerns the *autonomy* or self-governance of vocations and professions. Autonomy relates both to a formal governmental delegation of responsibility to the profession via authorisation and licensing, and to citizens entrusting responsibility to professionals. For the vocations, autonomy relates to the discretionary self-determination at work that skilled tradespersons have gained. Some vocations have achieved high levels of self-management in planning their work, choosing working methods and assessing the quality of their own work.

Fifthly, there is a difference with regard to the nature and aims of the occupational or professional *organisations*. Skilled workers are more often employed and subject to an employer's control compared to professionals, who more often are independent and self-employed, especially in the UK and USA (Freidson, 2001). Consequently, skilled workers' trade unions have given higher priority to the pursuit of better payment and working conditions, while the professions have given priority to the cultivation of the expert knowledge and practice of their professional communities.

The professional and vocational *identity* is a sixth dimension, which is important for young peoples' choice of education and working life career—and their sense of being and belonging (Chan, 2013). Vocations as well as professions afford opportunities for experiencing sense of meaning and purpose, self-esteem and pride and of belonging to a distinct community of practice.

This sketchy comparison of vocations and professions in different dimensions suggests that they share many basic qualities. They typically include considerable discretion in work, social esteem, belonging to a specialist community, a specific work identity and a sense of shared meaning. The clearest difference between the two is based on the lower hierarchical position of upper secondary VET compared to the intermediate and higher education degrees of the professions. However, VET is not just lower in academic terms, but has different qualities. It involves direct interaction with and transformation of the material world and the production of useful, tangible goods and services. It involves a broader repertoire of skills and knowledge beyond the codified knowledge of the academic professions. If we look at salary levels, autonomy at work and indications of social status, the differences between vocations and professions are less clear. Vocations can thus be seen as professions at a lower level in the hierarchies of work and education. It is therefore useful to consider professionalism as a potential element in any form of skilled work by virtue of its expertise and autonomy.

Professions can be understood as special forms of vocations, which have achieved a greater degree of formal control over their expert field and more autonomy at work. Their jobs require higher levels of scientific and codified knowledge including explicit ethical standards. Due to these differences as well as the different historical trajectories of vocations and professions, it is relevant to maintain a conceptual distinction between the two. However, vocations exhibit many of the same qualities as the professions with regard to prestige, pay, work identity, social belonging and self-determination at work. However, vocations as well as professions differ regarding these qualities. Many previously prestigious crafts have been degraded due to the growth of industrial mass production, and they have lost many of these qualities. Other vocational educations have been upgraded to become more advanced theoretically, and have attained levels of expertise at the level of professionals with a Bachelor degree.

Conceptual Challenges of Studying Vocations

In sociology, the concept 'vocation' is commonly used to denote a recognised field of specialist work and the skills required to perform it. It can also include the institutions and the organisations of the specialists that are also called 'occupations' (Billett, 2011). In Norway and Sweden, the term *'yrke'* includes the traditional crafts and trades as well as academic professions. In Denmark, the term *'erhvervsfag'* denotes vocations associated with VET, while the term *'profession'* is associated with a higher education degree. In international research, it is a challenge to define vocations conceptually, since there are no direct equivalence between the Nordic term *'yrke'*, the German *'Beruf'* and the English *'vocation'* or *'occupation'*. The differences concern not only the terms, but also reflect differences in the role and organisation of VET in these countries (Greinert, 1999; Thelen, 2004; Clarke & Winch, 2006).

A common understanding across these differences is that vocations have two dimensions, an institutional and a personal (Billett, 2011). On the one hand, vocations reflect the social division of labour in society that has resulted in the occupational organisation of specialised forms of work. On the other hand, vocations have shared cultural meanings and express personal interest and identities. This is indicated by the common etymological core of the terms *'Beruf'* and *'vocation'* as a *'calling'*. In his comprehensive examination of this issue, Billett (2011) proposes that these two dimensions are expressed by the concepts of *'occupation'* and *'vocation'* respectively. Occupations are social institutions that reflect social interests and requirements and vocations reflect personal goals and trajectories.

It is useful to make this analytical distinction between the social functions of occupations and the personal meanings of vocations, though they are two sides of the same reality. In this chapter, I will however generally use the concept vocation and reserve the term 'occupation' for the institution in the labour market, e.g. an 'occupational labour market' (Marsden, 1999). With a systemic perspective, Kurtz (2005) argues that the main function of vocations (*'Berufe'*) is to connect the education system with the employment system. Yet others have studied vocations with respect to the meaning they represent for the people who have a vocation or identify with a vocation (e.g. to be a clerk)—what it means to them (Chan, 2013). A social institution like a vocation cannot survive just by virtue of its functions; it must also be experienced as meaningful by the people who make up the institution. This meaning is created by the individual, but also by the community in the form of an occupational culture (Heinz, 1995). These different dimensions indicate that vocations are shaped and constituted by multiple social forces.

In trying to identify the role and meaning of vocations, it is helpful to trace the historical roots of vocations. These roots can vary considerably between sectors and countries, because some vocations have a long history with roots in the former guild system. Others have emerged more recently. However, some general lines of development can be identified (Deissinger, 1998; Thelen, 2004; Epstein, 2008).

In the early apprenticeship system, vocational training and productive work were fully integrated and separate from the formal school system. Employers (masters, artisans) and employees (journeymen) jointly belonged to the guild (Epstein, 2008). The development of modern vocations took place through a differentiation of respectively the *employment system* where skills are applied at work, the *labour market* where labour is exchanged and the *education system* where teaching and training takes place (Kurtz, 2005). From Durkheim and Marx to Luhmann, such differentiation processes have primarily been explained from a structural and functional perspective (Schimank, 2007). In other words, specialisation was a prerequisite for effectiveness and professionalism. The weakness of such structural approaches is that they neglect the social subjects and the strategies that have shaped the vocations. This can be included in two ways. First, by an analysis of the particular historical struggles and learning processes that have shaped the vocations and assigned them different roles across countries and industries (Thelen, 2004). Secondly, by a study of the development of personal and cultural meanings of vocations (Heinz, 1995). Both dimensions will be analysed in this chapter. First, I will briefly present and compare three different types of VET systems (VET regimes) and explain the specific qualities of the vocation in the collective skill formation regime.

The Position of Vocations in Three VET Regimes

The role and meaning of vocations in society depend on the organisation of the national VET system. Comparative research often operates with three ideal-typical VET regimes. In summary form, Table 15.1. contrasts the collective skill formation regime based on vocational self-governance with the market-led and the state-led regime (Deissinger, 1998; Greinert, 1999; Busemeyer & Trampusch, 2019).

In a *market* regime, state regulation of VET is weak and vocational training is driven by the skills requirements of individual companies, for example in training schemes of large firms. The skills are mostly firm specific with limited portability across companies, because industry-wide or sectoral standards for vocational training are weak. The main work identity encouraged here is identification with the employing company. The acquisition of specific skills frequently takes place in connection with career advancement in internal labour markets of large firms. A central dividing line of the social structure is between the core workforce, the

Table 15.1 Three ideal types of VET regimes

Form of regulation	*Market*	*State*	*Collective*
Key logic	Logic of employment and production	Meritocratic logic of educational credentials	Occupational-vocational logic (*'Beruf'*)
Political culture	Liberalist	Centralist conservative (*étatisme*)	Corporatist - collective skill formation
Main institutional framework	The business/the firm	The school system	The vocation, the trade
Key criteria of valid knowledge	Is it useful for employability?	Is it correct according to theory?	Is it relevant for the purpose of the vocation?
Dominant type of labour market	Internal labour market / secondary labour market	Professional and internal labour markets	Occupational labour markets—craft unions
Dominant type of social differentiation	Insiders versus outsiders in labour market	Higher versus lower educational credentials	Skilled versus unskilled versus academics
Main social identity	Organisational identity	Citizen and educational title	Vocational identity

insiders, in larger companies that offer training and internal career paths, and the outsiders, the peripheral workforce with low skills without these opportunities.

In a *state-led* regime, the state took on the task of organising VET in the period after the dissolution of the guilds and the introduction of the freedom of trade. VET was gradually subordinated to the aims of state building, general education and the ideals of upper secondary school (*'the primacy of speech, theory and abstraction'* Greinert, 1999, 70). It was mainly organised as school-based vocational training possibly with shorter internship periods—seldom as apprenticeships. The social structure was shaped by credentialism, as social status primarily depends on the degree achieved in the state education system, (Hadjar & Becker, 2016). The VET system emphasises the formation of citizens over the usefulness of skills for business or the relevance for vocations.

In the *collective skill formation* regime (Busemeyer & Trampusch, 2019), the governance of VET lies not with the companies or the state, but with the labour market organisations in a system of vocational self-governance. This self-governance is supported by the state and legally institutionalised to reduce the problem of 'free riders'—i.e. the risk that individual companies evade collective regulation. VET in this regime is mainly organised on the model of apprenticeships, which requires a high level of mutual trust between the labour market organisations. While VET in the two other regimes tends to separate education from training, and the execution from the conception of work, the collective skills regime connects these elements closely. VET includes both education in vocational schools and prolonged training in workplaces, and the governance of VET is partly left to the labour market partners in collaboration.

In this regime, workers identify more with their vocation than with the company. Their employment security is less dependent on the individual firm, as they can move easily between firms in occupational labour markets (Marsden, 1999; Hyman, 2001). For society, it is an advantage that the broad vocational skills support mobility between firms, where the organisation of work matches the vocational skills profiles. This is sustained by general recognition of these skills by employers, collective agreements with wide coverage and high levels of employment protection (Hyman, 2001; Martin, 2012). For the individual, the 'professionalisation' of non-academic work in this regime is an advantage, as it offers higher levels of autonomy at work and job satisfaction (Gallie, 2007). In addition, the vocational culture includes a well-defined sense of meaning and purpose, and opportunities for self-esteem and of belonging to a vocational community (Chan, 2013). For society, the broad and high skills associated with this VET regime strengthen the economic competitiveness by supporting the development of a diversified quality production (Sorge & Streeck, 2018).

Note that this is a simplified presentation of ideal-typical regimes, which do not exist in pure form. The comparative literature mentions in connection with these regimes, France (state-led), England (market-led) and Germany (collective skill formation) (Greinert, 1999). The point of this presentation of VET regimes is to demonstrate that vocations are embedded in a larger social order, which includes a special form of educational governance, labour market structure, political culture and dominant work identity (Thelen, 2004; Hillmert, 2002).

What Constitutes Vocations?

Vocations are constituted by a combination of an exclusive field of expert work, an association of professional practitioners in this field (e.g. trade union) and the related education and training programme. Access to the specialist work is frequently restricted by some forms of employment protection. The vocation is an overarching institution that extends over and connects three social sub-systems: the employment system, the labour market and the education system. In addition to these systemic dimensions, the vocation is constituted by shared cultural meanings and vocational identities. The vocation must be experienced as meaningful by its members through opportunities for '*being and belonging*'—for acquiring a vocational identity and participate in a vocational community. Vocations are thus complex, composite institutions that represent something different in each of these areas. They are constituted by different social agents and interests in each of them. As this is important for assessing the future significance of vocations, I will elaborate further on this regarding the labour process, the labour market and the VET system.

First, the autonomy at work of skilled labour involves skills that are both broad and high. Their broad skills integrate a variety of tasks in the horizontal division of labour, (drilling, milling, turning, etc.). Their high skills include a variety of tasks in the vertical, decision-making dimension (plan and define the job, negotiate with users, choose working methods, execute the job and assess its quality). These qualities include opportunities for exercising professional judgement during the planning, execution and evaluation of the work process and common standards for quality as an integral part of the vocational skills.

The broad skills and workers' self-management have historically been under attack from Taylorism and managerial efforts to gain direct control over the work process, as this has been studied in the tradition of Labour Process Theory (Braverman, 1974). The spread of industrial mass production weakened many vocations as un- and semiskilled workers replaced skilled workers in traditional crafts such as bakers and butchers (Jørgensen, 2006). However, in countries and industries where skilled workers had a strong position, industry could take advantage of this and upgrade production, by shifting for example into forms of '*Diversified Quality Production*' (Sorge & Streeck, 2018). In addition, the model of vocational work could extend into many previously low skilled jobs as they were covered by vocational education and converted into skilled work.

Secondly, with the development of organised labour markets, the role of vocations came to depend on the unions' organisational power, collective agreements and employment protection to protect the value of skills (Busemeyer & Trampusch, 2019). A key issue here was the capacity of trade unions for protecting skilled jobs from the competition from unorganised and un-skilled workers (Marsden, 1999). For example in the Danish Apprenticeship Act, the labour market organisations in cooperation early gained the right to authorise companies to train apprentices and to supervise the training. By means of collective action, the craft unions gained control over the employment practice in skilled fields of work (Christensen, 1978). In addition, a comprehensive public further training system protected the value of their skills through opportunities for continuous upgrading of skills. In Denmark, the craft unions have been extensively involved in the regulation of education and labour market policy. The close cooperation between unions and employers' organisations on skill formation has been institutionalised in vocational self-governance and in relation to the state in tripartite cooperation. The strong involvement of the labour unions in the governance of VET has given unions a stronger professional character compared to the 'trade unionism' and militancy, which dominates in countries like England and the USA (Hyman, 2001). Nationally, vocations developed along different lines because of particular historical dynamics and alliances that gave them different positions in working life.

Thirdly, the position and status of VET are crucial for the strength of the vocations (Billett, 2011). Historically, vocational skills and cultures were transferred as an integral part of apprentices' training and participation in work. It continues to do so in internships and modern forms apprenticeships. Here, the apprentices learn not only the skills of the trade, but also the customs and culture of the vocational community. With the growing specialisation of production and the expansion of the state education system, the work-based training was partially transferred to vocational schools. In most Nordic countries, VET was integrated into a united upper secondary school for all young people from the 1970s. Hereby, the dissemination of implicit knowledge and practical skills was supplemented by, and replaced by, school-based teaching in explicit and theoretical knowledge. The vocational subjects from working life were partially substituted by the general subjects that structure the school system (Greinert, 1999; Dostal, 2001). These school subjects derive their justification from scientific disciplines, in contrast to vocational subjects, which are based mainly on experiential learning from work practice. The driving forces behind this 'schoolification' are diverse. Technological changes raised the requirements for theoretical knowledge acquired in vocational schools. The labour movement called for an expansion of the general subjects in VET as part of the struggle for democratisation and social equality. Young people from the working class should be prepared to participate in society as democratic citizens—not just as skilled labour. In addition, the expansion of the school-based part of VET has been justified by the need to keep young people in school in periods of high youth unemployment and a shortage of apprenticeships (Jørgensen, 2015). More recently, VET is drawn closer to general education to avoid VET appearing as a 'dead end' with weak connection to higher education. In addition, the education boom in the post-war period had a self-reinforcing effect, as an educational degree increasingly became a positional good with a value in itself (Archer, 1979). In sum, VET was historically integrated more closely into the state education system and subjected to the specific logics of this system. This has challenged the specific vocational nature of VET and the role of the vocations.

Vocational Identities

A fourth constitutive field for the vocations is the vocational identities and cultures. This has gained more attention as the attractiveness of VET is declining in many countries. An advantage of VET systems, especially apprenticeship systems, is that they offer clear and distinctive vocational identities (Heinz, 1995; Billett, 2011). This is a contrast to the 'student' identity that predominates in general education—and sometimes in school-based VET. Educational pathways based on vocations appear transparent to young people, because even the school-based courses are clearly connected to well-known occupations. While general education primarily offers access to codified knowledge, VET offers opportunities for *becoming* (for example a hairdresser) for *belonging* to an occupational community and for bodily *interactions* with the social and material world of the occupation (Colley et al., 2003). VET with internships or apprenticeships is known to smoothen young peoples' transition from education to work, because this transition is included and supported in the work-based part of the programmes (Gangl, 2001; Marczuk, 2024).

For the students/apprentices, VET has additional advantages when compared to general education. Due to its vocational nature, VET offers experiences of authenticity, meaning and engagement in work that has value to others. Especially the learning environment of the workplace appears meaningful to the students/apprentices (Billett, 2011). This can be the meaning of doing care work for patients or the lasting value of building a house. Students emphasise the authenticity of learning in the workplace, where tasks have a tangible purpose and the results of their efforts can be directly observed. In the workplace, they meet the current technologies, materials and methods used in the vocational field. And not least, here they can meet experienced professionals, who represent the vocational identity that they aspire to attain (Jørgensen, 2015). Alongside the formal trainer (instructor/supervisor), the colleagues can demonstrate the applied know-how and situated judgement of vocational practice. Communities of practice are an important resource for learning professional language, culture, traditions, dress codes and appropriate behaviour. However, these communities can also include exclusionary practices and abusive behaviours, which can limit the opportunities for learning.

VET students/apprentices frequently emphasise the diverse, joyful and bodily engaging nature of workshops and workplaces. This is often contrasted with schools' preference for symbolic, linguistic and written practice—and an inactive body. In the workplace, they 'learn by doing' when they engage with tools, technologies, social relationships and transform materials into products and services. The learning process takes place through the bodily, sensual and emotionally engaging work with people, tools and materials (Klope, 2020). The participation in authentic work process has a physical and aesthetic dimension. Craftsmanship requires a sense of taste, design, style, appearance and composition, similar to artistic work. The carpenter must develop a sense for the properties of the wood, and the elder care worker must develop a sense for old peoples' moods and needs.

Regardless of these qualities of vocations, the esteem of VET appears to decline in many countries. One reason is that the traditional vocational identities are challenged by the individualisation of young people's lives. Their entry into working life has been postponed by the extension of the compulsory school years. The expansion of welfare has opened greater opportunities for life before and outside of working life, and new youth cultures have emerged (Giddens, 1990). Youth has been extended as a separate phase in the life course, where individualised youth identities can develop (Beck & Beck-Gernsheim, 2002). As a result of this differentiation of civil life apart from working life, the cultural meaning and status of vocations are disseminated through youth cultures and social media. This questions the established status of the vocations. In addition, the liberalisation of the education system requires that schools deliver flexible programmes tailored to the individual student. In the Danish VET system, all students must draw up a 'personal education plan' at the beginning of their programme. This weakens the common skills profile of the vocation in favour of more individualised profiles and reduces the opportunities for a shared identification with the vocation.

Discussion and Conclusion

Vocations are key institutions in vocational education and training, and especially in the collective skills formation regime, which has several advantages compared to the state-led and marked-led regimes. Vocations provide framework for the standardisation of skills with high portability across firms in the labour market. Vocations guide the matching of VET programmes to the job profiles in occupational labour markets. This is especially the case for apprenticeships and other forms of VET with alternating training and high vocational specificity (Gangl, 2001). They bridge the worlds of education and work by integrating students into working life during the programme and by integrating skills and knowledge from both worlds.

These VET systems provide valuable alternatives to academic education, and offer smooth transitions from education to work (Blommaert et al., 2020). In Norway and Denmark, more than half of the apprentices from VET continue as regular employees in the company, where they are trained during their VET programme (Jørgensen, 2019). In these countries, the apprenticeship system has been modernised and expanded into new sectors. Previously un- and semiskilled forms of work have been upgraded to become vocational and covered by VET programmes. This is supported by comprehensive public further training systems that sustain the value of skills through continuous upgrading.

Broad and high vocational skills have promoted 'flat' work organisations with high levels of job discretion and self-management. The flat organisation of work has supported collaboration between employers and employees on innovation, upgrading of skills and workplace learning

(Gallie, 2007; Brockmann, et al., 2008). This has encouraged trade-union strategies to support technological modernisation of work and upgrading of their members' skills (Hyman, 2001; Thelen, 2014; Sorge & Streeck, 2018). The low skill differential and low power distance of work organisations improves the sharing of knowledge between the managerial-planning levels and the executing levels of work (Gallie, 2007). In this way, the vocational organisation of VET and working life can be described as components of a beneficial configurational causation (Mahoney & Thelen, 2009). This involves positive feedback cycles between broad and high skills, the extension of job discretion, the upgrading of high-quality production and the continual upgrading of skills (Kristensen & Lilja, 2012).

In addition to these benefits for society and the labour market, vocations include various benefits for VET students and workers. Vocational education provides opportunities for the development of clear vocational identities beyond the 'student' identity. This is because the programmes are aiming at well-known positions in society's division of labour, which young people can relate to and identify with (Heinz, 1995). Thereby, vocations define a special way of socialising young people into working life. This includes a specific pedagogical approach and learning culture that bridges the worlds of education and work (Deissinger, 1998). The objectives and responsibilities of the vocation configure the curricular content and guide the teaching in VET (Rauner, 2004). For students, the vocation offers purpose and meaning to their learning and connects learning in vocational schools to learning in the workplace.

In this chapter, the benefits of vocations have been emphasised, because they are often taken for granted. Vocations also have negative qualities, and some of them must be mentioned. Due to the numerous stakeholders involved in the governance structure, vocations can be slow to react to changes in working life. Strong vocational identities and communities can limit interdisciplinary collaboration. Vocational identities and cultures are often highly gendered and exclusive for minorities. When vocations define themselves in contrast to academic disciplines, they can reinforce the problematic vocational-academic divide. Vocations, like professions, are strongholds of power and particular interests that can impair the pursuit of the common good. These adverse qualities reinforce the analyses in this chapter of vocations as complex and composite institutions. This questions their stability and direct attention to the subverting forces that challenge their future role.

One is that the priority of vocational subjects in VET must yield to general and academic subjects, because VET is expected to give access to higher education. This is also driven by the academic drift of young peoples' educational choices. In addition, VET is expected to prepare for lifelong learning and mobility in the labour market and not just to qualify for a specific vocation. In working life, the high speed of change means that the employees are increasingly working with many diverse tasks that differ from what they learned in initial VET. The deregulation and marketisation of education required by neoliberal policies have since the 1980s weakened the vocations (Ball, 2012). Employers' pursuit of more flexible and 'lean' organisations tends to replace the established vocational division of labour with a functional division of labour (Baethge et al., 1998). In addition, it is a common assumption that specific vocational skills are less important than generic skills and personal competences in the future labour market (Dostal, 2001). For young people, VET appears to be less attractive and the traditional vocational identities seem to be less compatible with young peoples' individualised identities. These changes are weakening vocations. They raise the question if the diverse constituent parts of vocations are falling apart. Are vocations still relevant as a framework for connecting working life, education and young peoples' imagined futures?

The numerous qualities of vocations analysed in this chapter indicate that they will still be relevant in the future. Specific vocations will continue to wither away, as they have done previously, for example due to technological change. And new vocations will emerge to match the requirements of new types of expert work. But the vocation as a key institution in VET and working life and as a meaningful framework for work biographies will hardly be less relevant in the future. A decisive reason is the capacity of vocations to bridge education and work.

It is a general feature of modern societies that the worlds of education and work are drifting apart (Archer, 1979; Green, 1990). This presents these societies with the fundamental challenge of connecting education and work at two levels. They must coordinate the labour market's need for skills with the education system's production of skills. They must support young people's transition from education to working life and coordinate their educational choices with the opportunities for employment (Hillmert, 2002).

In a market-led regime, this is mediated via the market as educational providers seek to adapt to the changing skills demands of businesses.

However, market regulation is mostly reactive and fragmented, and lacks mechanisms that aggregate the diverse and short-term skills requirements of individual companies (Jørgensen, 2006; Sorge & Streeck, 2018). VET in a marked-led regime is generally more fragmented, because in-company training is not guided and standardised by an intermediary institution like the vocation. This regime provides little support for the transition from education to work of young people.

In a state-led regime, the education system has weak connections to the labour market and gives priority to general and academically oriented subjects. VET programmes are mainly school-based and the students' acquisition of specific work-process knowledge takes place after the completion of the formal education. This is for example the case in the Swedish VET programmes for the Construction industry and hairdresser (Jørgensen et al., 2018; Klope, 2020). Only recently, new institutions have been set up in Sweden to coordinate between vocational schools and businesses (Persson & Hermelin, 2018). In the state-led regime, education and training are weakly connected and young people are more their own, when they navigate the transition from education to working life.

In the collective skills regime, the vocation has a key role as intermediary institution for skill formation. It is composed of the labour market organisations, and the vocational self-governance by industry and sector has been quite adaptive to changes in working life. While employer organisations in other countries withdrew from corporatist types of collaboration following the neoliberal turn, they have generally retained their commitment to the institutions of the collective skills regimes (Thelen, 2014).

The vocation serves as a central opportunity structure for solving the coordination problem of modern societies and for supporting young peoples' transitions from education to work (Kurtz, 2005). Vocations provide standardised templates for synchronising the organisation of work with VET and for making the transition from education to work transparent for young people (Deissinger, 1998). As the skills requirements in working life are constantly changing, this coordination must take place as an ongoing process that mediates between many stakeholders and their conflicting interests. The institutional framework of vocations includes the main stakeholders and commits them to negotiate and resolve conflict of interest on skill formation. This makes it likely that the vocation will continue to be a key opportunity structure for skill formation and working life biographies in the future.

References

Archer, M. S. (1979). *Social origins of educational systems*. Sage Publications.

Arnholtz, J., & Østhus, S. (2024). Is labour migration disrupting dual vocational education and training systems? Empirical evidence from the Danish and Norwegian construction sectors. *European Journal of Industrial Relations.*, *30*, 245.

Baethge, M., Baethge-Kinsky, V. & Kupka, P. (1998). Facharbeit—Auslaufmodell oder neue Perspektive? i *SOFI Mitteilungen* Nr. 26/1998, 8199.

Ball, S. J. (2012). *Politics and policy making in education: Explorations in sociology*. Routledge.

Beck, U., & Beck-Gernsheim, E. (2002). *Individualization: Institutionalized individualism and its social and political consequences*. Sage.

Billett, S. (2011). *Vocational education: Purposes, traditions and prospects*. Springer.

Blommaert, L., Muja, A., Gesthuizen, M., & Wolbers, M. H. (2020). The vocational specificity of educational systems and youth labour market integration: A literature review and meta-analysis. *European Sociological Review*, *36*(5), 720–740. https://doi.org/10.1093/esr/jcaa017

Braverman, H. (1974). *Labor and monopoly capital: The degradation of work in the twentieth century*. NYU Press.

Brockmann, M., Clarke, L., & Winch, C. (2008). Knowledge, skills, competence: European divergences in vocational education and training (VET)—the English, German and Dutch cases. *Oxford Review of Education*, *34*(5), 547–567. https://doi.org/10.1080/03054980701782098

Busemeyer, M. R., & Trampusch, C. (2019). The politics of vocational training: Theories, typologies, and public policies. In *The Wiley handbook of vocational education and training* (pp. 137–164). Wiley.

Chan, S. (2013). Learning through apprenticeship: Belonging to a workplace, becoming and being. *Vocations and Learning*, *6*, 367–383. https://doi.org/10.1007/s12186-013-9100-x

Christensen, E. (1978). *Konflikter mellem faglærte og ufaglærte arbejdere*. Aalborg Universitetscenter.

Clarke, L., & Winch, C. (2006). A European skills framework?—but what are skills? Anglo-Saxon versus German concepts. *Journal of Education and Work*, *19*(3), 255–269. https://doi.org/10.1080/13639080600776870

Colley, H., James, J., Diment, K., & Tedder., M. (2003). Learning as becoming in vocational education and training: Class, gender and the role of vocational habitus. *Journal of Vocational Education & Training*, *55*(4), 471–498. https://doi.org/10.1080/13636820300200240

Deissinger, Th. (1998). Beruflichkeit als "or ganisierendes Prinzip" der deutschen Berufsaus bildung, Eusl Verlag.

Dostal, W. (2001). Beruflichkeit in der Wissensgesellschaft. In R. Sackmann & M. Wingens (Eds.), *Bildung und Beruf*. Juventa Verlag.

Epstein, S. R. (2008). Craft guilds in the pre-modern economy: A discussion. *The Economic History Review, 61*(1), 155–174. https://doi.org/10.1111/j.1468-0289.2007.00411.x

Freidson, E. (2001). *Professionalism. The third logic*. The University of Chicago Press.

Fuller, A. (2019). Does vocational education still need the concept of occupation? In *The Wiley handbook of vocational education and training* (pp. 81–96). Wiley.

Gallie, D. (2007). Production regimes and the quality of employment in Europe. *Annual Review of Sociology, 33*(1), 85–104. https://doi.org/10.1146/annurev.soc.33.040406.131724

Gangl, M. (2001). European patterns of labour market entry—A dichotomy of occupationalized vs. non-occupationalized systems? *European Societies, 3*(4), 471–494. https://doi.org/10.1080/14616690120112226

Giddens, A. (1990). *The consequences of modernity*. Stanford University Press.

Green, A. (1990). *Education and state formation*. Palgrave Macmillan.

Greinert, W. D. (1999). Berufsqualifizierung und dritte industrielle Revolution. *Nomos*. VerlagsGesellschaft.

Haakestad, H., & Friberg, J. H. (2020). Deskilling revisited: Labour migration, neo-Taylorism and the degradation of craft work in the Norwegian construction industry. *Economic and Industrial Democracy, 41*(3), 630–651. https://doi.org/10.1080/14616690120112226

Hadjar, A., & Becker, R. (2016). Education systems and meritocracy: Social origin, educational and status attainment. In Hadjarand, A. & C. Gross (Eds.) *Education systems and inequalities* (pp. 231–258). Policy Press.

Heidenreich, M. (1999). Berufskonstruktion und Professionalisierung. In H. J. Apel (Ed.), *Professionalisierung pädagogischer Berufe im historischen Prozess* (pp. 35–58). Julius Klinkhardt.

Heinz, W. R. (1995). *Arbeit, Beruf und Lebenslauf*. Juventa.

Hillmert, S. (2002). Labour market integration and institutions: an Anglo-German comparison. *Work, Employment & Society, 19*(4), 675–701. https://doi.org/10.1177/09500170232158742

Hyman, R. (2001). *Understanding European trade unionism: Between market, class and society*. Sage Publications.

Jørgensen, C. H., Olsen, O. J., & Persson, T. D. (Eds.). (2018). *Vocational education in the Nordic countries: Learning from diversity*. Routledge.

Jørgensen, C. H. (2006). *Utbildningsplanering: samspel mellan utbildning och arbete*. Studentlitteratur.

Jørgensen, C. H. (2015). Challenges for work-based learning in vocational education and training in the Nordic countries. In I. S. Bohlinger, U. Haake,

C. H. Jørgensen, H. Toiviainen, & A. Wallo (Eds.), *Work-based learning and professional development in times of globalisation* (pp. 159–171). Brill | Sense.

Jørgensen, C. H. (2019). Young peoples' access to working life in three Nordic countries: what is the role of vocational education and training? In I. H. Hvid & E. Falkum (Eds.), *Work and wellbeing in the Nordic countries: Critical perspectives on the world's best working life* (pp. 155–176). Routledge.

Klope, E. (2020). *Respektabla frisörer: Femininitet och (yrkes) identitet bland tjejer i gymnasieskolans frisörutbildning*. Linnaeus University Press.

Kristensen, P. H., & Lilja, K. (Eds.). (2012). *Nordic capitalisms and globalization: New forms of economic organization and welfare institutions*. Oxford University Press.

Kurtz, T. (2005). *Die Berufsform der Gesellschaft*. Velbrück Wissenschaft.

Larson, M. S. (2012). *The Rise of Professionalism. Monopolies of Competence and Sheltered Markets*. TransactionBooks.

Mahoney, J., & Thelen, K. (Eds.). (2009). *Explaining institutional change: Ambiguity, agency, and power*. Cambridge University Press.

Marczuk, A. (2024). Literature review of comparative school-to-work research: How institutional settings shape individual labour market outcomes. *Journal for Labour Market Research, 58*(1), 19–34. https://doi.org/10.1186/s12651-024-00375-w

Marsden, D. (1999). *A theory of employment systems: microfoundations of societal diversity*. Oxford University Press.

Martin, C. J. (2012). Political institutions and the origins of collective skill formation systems. In M. R. Busemeyer & C. Trampusch (Eds.), *The political economy of collective skill formation* (pp. 41–67). Oxford University Press.

Persson, B., & Hermelin, B. (2018). Mobilising for change in vocational education and training in Sweden–A case study of the 'Technical College'scheme. *Journal of Vocational Education & Training, 70*(3), 476–496. https://doi.org/10.1080/13636820.2018.1443971

Rauner, F. (2004). Work analysis and curriculum based on the Beruf concept. Fischer, M,. Boreham, N., & Nyhan, B., European perspectives on learning at work: the acquisition of work process knowledge. (pp237–257), Cedefop Reference Series.

Saks, M. (2012). Defining a profession: The role of knowledge and expertise. *Professions and Professionalism, 2*(1), 1–10. https://doi.org/10.7577/pp.v2i1.151

Schimank, U. (2007). *Theorien gesellschaftlicher Differenzierung*. VS Verlag.

Sennett, R. (1998). *The corrosion of character: The personal consequences of work in the new capitalism*. WW Norton & Company.

Sorge, A., & Streeck, W. (2018). Diversified quality production revisited: Its contribution to German socio-economic performance over time. *Socio-Economic Review, 16*(3), 587–612. https://doi.org/10.1093/ser/mwy022

Thelen, K. (2004). *How institutions evolve: the political economy of skills in Germany, Britain, the United States, and Japan*. Cambridge University Press.
Thelen, K. (2014). Varieties of liberalization and the new politics of social solidarity. Cambridge University Press.

CHAPTER 16

Looking Backward and Forward—Concluding Remarks

Eli Smeplass and Chris Zirkle

Identifying the Functions of VET

"Education is not preparation for life; education is life itself." (Dewey, 1916)

In the introduction chapter, we explained how the intention of this project was to contribute to a more positive discourse on a field that is often viewed through an overly critical lens. How might the research

E. Smeplass (✉) · C. Zirkle
Norwegian University of Science and Technology (NTNU), Trondheim, Norway
e-mail: Eli.smeplass@ntnu.no

C. Zirkle
e-mail: zirkle.6@osu.edu

C. Zirkle
The Ohio State University, Columbus, OH, USA

E. Smeplass and C. Zirkle (eds.), *Opportunity Structures in Vocational Education and Training*, Palgrave Studies in Technical and Vocational Education and Training,
https://doi.org/10.1007/978-3-032-04410-5_16

community shift its focus to ensure that strengths are not overshadowed by the pursuit of critical analysis? One of the longstanding traditions in social and educational research is a strong emphasis on identifying and analyzing problems rather than documenting successes. This orientation likely stems from the belief that problem-focused research is more likely to inform interventions, shape policy, and address pressing societal challenges. However, this approach also carries limitations. Biesta (2015) critiqued the modern tendency to reduce education to a system of measurable outputs and deficiencies, arguing that education must be understood as a force for democracy, social engagement, and personal development rather than simply an instrument for fixing societal problems. Abbott (2010) highlighted how social sciences often values critical diagnosis over celebratory description, prioritizing analyses that reveal structural constraints, inequities, or inefficiencies. We believe this inclination toward critique can limit the visibility of strengths within systems like VET, inadvertently narrowing the field's overall perspective. Without diminishing the importance of critical analysis, the research community must consider how to elevate constructive elements, and the positive contributions VET makes to both individual lives and society as a whole.

As this collection of chapters has shown, vocational education and training (VET) is far more than a series of localized systems for preparing individuals for the labor market. Instead, it is a global tapestry of interconnected practices, values, and ambitions—each shaped by cultural, historical, and economic contexts. Across diverse nations, from Norway to Zimbabwe, Costa Rica to Germany, VET has been shown to be not only a pathway to employment, but a key force in promoting social mobility, community, and economic development. This collection enters a longstanding international dialogue on the significance and potential of VET within global education systems and labor markets. For decades, researchers have explored how vocational education both reflects and shapes the socio-economic landscapes in which it operates. The work of Avis (2012), for example, has underscored the need to understand VET as part of a larger process of global reconstruction, where shifting economic demands and societal expectations require constant adaptation of VET policies and practices. Our project contributes to this conversation by offering a nuanced view of opportunity structures—providing both a theoretical framework and empirical evidence that highlight VET's capacity to foster equity, mobility, and innovation in diverse settings. Fraser's (2008) perspective on justice and political space offers a lens for examining some of the broader implications of VET. By considering how opportunity structures distribute educational and career pathways across different social groups, this book aligns with Fraser's call for rethinking equity and inclusion in increasingly globalized contexts. It extends her argument by showing how vocational pathways can be reimagined to address systemic inequities, not as isolated national projects but as part of a shared international endeavor to create more just educational landscapes. Our work also builds on the comparative traditions, for example outlined by Gonon and Deissinger (2021), who advocate for an international comparative history of vocational education. This volume brings together diverse national perspectives that illustrate how vocational education is shaped by varied historical and policy contexts. In doing so, we advance the field's understanding of how varied national histories and policy environments can inform more comprehensive frameworks for studying and improving VET systems globally. Furthermore, Pilz's (2012) methodological insights into comparative VET research have inspired this project. By incorporating cases from multiple countries, we contribute to the growing body of literature that uses comparative methods to discover common patterns, identify best practices, and suggest how different systems might learn from one another. However, comparative research must go beyond merely identifying similarities and differences; it must also engage in deeper, more analytical investigations into how and why vocational systems function as they do. Building on this, Pilz and Li (2020) emphasized the need to move beyond simple descriptive comparisons toward more theoretically driven approaches that explain the mechanisms shaping vocational education across different contexts. This volume continues that initiative by proposing opportunity structures as a conceptual tool that helps bridge diverse settings. Through this lens, we position VET as a vital component of contemporary education systems and invite the field to explore how different systems can engage in meaningful, reciprocal learning to address global challenges.

VET is more than just a bridge between school and the labor market; it plays a central role in shaping modern societies. From a functionalist perspective, VET is one of the essential components that sustains the social organism that constitutes society. Rather than being an isolated educational track, VET operates as a system designed to meet both individual needs and labor market demands while contributing to broader social cohesion and stability. Durkheim emphasized that various institutions—education among them—are integral to the cohesive functioning of society. In this sense, VET can be understood as a structural mechanism that enables individuals to find meaningful roles, acquire specialized skills, and contribute productively to economic and social life. However, while this alignment ensures that people have access to work and that work serves the broader needs of a well-functioning society, functionalist perspectives often emphasize stability over change. By integrating the concept of opportunity structures, we can move beyond a purely functionalist view to explore how VET systems can either reinforce existing social divisions or serve as a vehicle for social development.

The different authors in this volume have taken on the challenge of exploring both the positive aspects and the persistent issues within VET systems. This dual approach was not merely an editorial preference but a reflection of a critical methodological stance—the recognition that understanding VET requires embracing complexity. To highlight only the strengths of VET systems would be to risk overlooking structural inequalities, mismatches between training and market demands, or barriers to access. Conversely, focusing solely on problems would obscure the practical successes and innovations that make VET indispensable. The task, then, is not to choose between optimism and critique, but to hold both in view—to keep two thoughts in mind at once. We find that the authors in this book have skillfully navigated this balance, offering insights that highlight the strengths of VET while also probing the systemic challenges it faces. Their analyses demonstrate that it is possible—and indeed necessary—to appreciate the positive contributions of VET even as we critically engage with its limitations. This dual perspective enriches our understanding, fostering a constructive dialogue about the role of vocational education in modern societies.

Returning to Dewey's insight, we see how his perspective challenges the view of VET as a limited or secondary form of education. If education is life itself, then vocational learning must be understood as an integral part of that life, shaping professional competencies *and* personal and societal development. Dewey's philosophy of experiential learning suggests that VET should not be confined to narrow technical instruction but should cultivate adaptability, creativity, and the ability to engage critically with the evolving demands of work and society. This aligns with contemporary discussions on lifelong learning, emphasizing that vocational education must be dynamic and responsive rather than static and predetermined. By embracing this broader perspective, we can move beyond traditional dichotomies that separate vocational and academic pathways, instead recognizing their interconnected roles in fostering opportunity and participation. This approach reinforces the argument that VET is a continuous process that supports individuals throughout their lives, enabling them to contribute meaningfully to both their workplaces and their communities.

Examining the Features We Take for Granted

Throughout the book, our contributors have demonstrated how carefully chosen concepts can shed light on the complexities of VET. Drawing inspiration from theoretical and methodological pragmatism, we see how choosing relevant conceptual tools can describe social phenomena and open up new avenues for critical inquiry. Swedberg (2014) argued that the generation of fruitful concepts is at the heart of social science. A productive concept is one that generates interesting questions, sharpens the contours of existing problems, and prompts a deeper engagement with both empirical and theoretical puzzles. Concepts like opportunity structures, when thoughtfully applied, can help researchers move beyond surface descriptions and dig into the structures, relationships, and processes that underpin complex systems. This is a pragmatic understanding of social science theory. Peirce's pragmatism underscores the importance of practical implications in conceptual work. Peirce (1905) posited that the meaning of a concept lies in its observable effects and its ability to guide thought toward useful conclusions. In the context of this book, the application of opportunity structures exemplifies this principle. Rather than serving as a mere label, the concept becomes a lens through which the contributors analyze how VET systems can enable access, mobility, and inclusion—while also revealing where those systems fall short. It transforms the way we think about VET's role in society, by raising critical questions about how education and training systems function, for whom and why.

By leveraging the concept of opportunity structures, the authors have in various ways revealed patterns and connections that might otherwise remain hidden. Swedberg (2014) noted that fruitful concepts provoke further theoretical and empirical exploration. In this book, opportunity structures do exactly that—it pushes us to ask how systems create pathways for learners, why certain configurations succeed in promoting social mobility, and what conditions are necessary for equitable and sustainable outcomes. Meanwhile, Peirce's pragmatist orientation can be an encouragement to consider how these theoretical insights translate into actionable improvements—how policymakers, educators, and industry leaders might refine their approaches to strengthen VET systems. Thus, the *fruitfulness* of concepts lies both in their ability to describe the world as it is, as well as in their capacity to reorient our perspective, uncover new research questions, and guide practical efforts toward better educational and social outcomes. In this way, Swedberg's call for a more dynamic, concept-driven social science and its practical meaning converge, offering an understanding of VET in contemporary society. We were inspired by these conceptual insights to foster more meaningful dialogues across borders, enabling the global VET research community to connect more deeply and learn from one another's successes and challenges.

An example we can mention here is connected to some of the main points raised by Stephen Billett (2025a) in Chap. 11, discussing how vocational education must be understood as more than a set of training programs designed to meet immediate labor market demands. Instead, he argues that VET systems hold the potential to foster lifelong learning, cultivate innovation, and enable individuals to navigate evolving career trajectories. Drawing on the concept of opportunity structures, Billett demonstrates how well-designed vocational pathways empower people to find meaningful, stable work while also contributing to workplace creativity and productivity. His approach frames VET not merely as a supply-side mechanism for addressing skills shortages, but as an integral component of both personal and societal development. The contributions from Billett illustrate the fruitfulness of the opportunity structure concept by showing how it can help scholars move beyond a narrowly economic or technical understanding of VET. His analysis highlights the institutional and social frameworks that make vocational education successful—partnerships with industry, adaptable training pathways, and strong support for learners as they transition into the workforce. By using opportunity structures as a lens, Billett sheds light on what works within certain systems and invites a deeper inquiry into how these systems can be replicated or adapted in other contexts. Billett's work demonstrates that the concept of opportunity structures serves as a guide for imagining new approaches to VET policy and research. For instance, he points out that vocational education systems can be designed to help workers engage in innovation, reshaping their individual roles, as well as the organizations and industries they are part of. This, in turn, expands the role of VET from a pipeline for skilled labor into a platform for sustaining long-term economic and social resilience. In addition, in chap. 13 Billett (2025b) highlights the importance of considering local and contextual factors when applying the opportunity structures concept. His insights encourage scholars to think critically about how different governance models, cultural values, and economic conditions influence the ways opportunity structures operate. As a result, it serves as inspiration for other researchers who wish to use this concept to understand vocational systems, but also to suggest pathways for improvement.

Chapter 2 on Costa Rica by Daniel Láscarez Smith (2025) provides a compelling examination of how dual apprenticeships were introduced to tackle youth unemployment. Láscarez Smith identifies opportunity structures as a key analytical framework to understand the policy transfer process. His chapter outlines how these structures—consisting of pathways created by the interplay of education policies, labor market conditions, and stakeholder involvement—can shape access to training and employment. In Costa Rica's case, the implementation of dual apprenticeships revealed both the potential and limitations of these opportunity structures. While the program introduced new avenues for skill development and work placement, it also exposed systemic barriers such as insufficient employer engagement and a lack of alignment between training programs and the actual needs of the labor market. What makes this chapter particularly illuminating is how it applies the concept of opportunity structures to critically assess the real-world challenges of policy adaptation. By tracing the historical evolution of vocational education in Costa Rica, Láscarez Smith highlights the difficulty of transplanting systems from one national context to another. He demonstrates that even a seemingly successful model like the European dual apprenticeship system can falter when local conditions—such as political will, economic stability, and employer capacity—are not adequately addressed. This analysis underscores the importance of contextualizing opportunity structures, as it shows that their effectiveness relies heavily on how well they are integrated into the existing socio-economic and institutional fabric. Moreover, Láscarez Smith's work illustrates how the concept of opportunity structures can be used to identify and understand failures, as the gap between policy intentions and practical realities diverges. By highlighting both the strengths and weaknesses of Costa Rica's dual apprenticeship initiative, Láscarez Smith offers valuable lessons for the international VET research community, reminding us that specific

VET models are not universally applicable; rather, they must be carefully tailored to local conditions. Through this lens, the Costa Rican case becomes a source of broader insight into how VET systems can be reimagined and refined.

In chapter 12 on Norway's vocational education system (Smeplass, 2025), it is evident how VET as opportunity structures are intentionally designed to guide young people through educational and career pathways. Norway's tightly integrated apprenticeship model exemplifies how these structures foster both equity and adaptability. Central to the model is a tripartite collaboration among government, industry, and labor unions, ensuring vocational pathways remain relevant, accessible, and aligned with labor market demands. These structures provide routes into relatively stable employment while allowing students to transition between vocational and academic tracks if their goals change. The individual costs of lifelong learning remain lower than in other contexts and the government continues to invest heavily in labor market integration. While Norway's VET system has certain strengths, the chapter highlights ongoing challenges. Reforms over recent decades have aimed to broaden inclusivity and ensure equal access for all young people, but systemic barriers still remain. Disparities in apprenticeship availability and the integration of young immigrants into the system continue to pose difficulties. The case of Norway's VET system shows a need for continuous evaluation and policy adjustment. Identifying investment in youth as part of a nations opportunity structure, the chapter illustrates how the concept serves as a conceptual tool for analysis of inclusive and adaptive vocational pathways.

Beyond these cases, other chapters in this volume expand our understanding of opportunity structures by highlighting different dimensions of access, equity, and system responsiveness. Lyckander et al. (2025, Chap. 3) examine how adult education in Norway has evolved to provide flexible learning pathways, emphasizing the importance of adaptability in lifelong learning systems. Jeong (2025, Chap. 5) demonstrates how South Korea's Meister High Schools have elevated the status of vocational education through strong industry partnerships and policy-driven interventions. Seitamaa and Hansen (2025, Chap. 6) discuss how Finnish VET has undergone shifts between educational equality and labor market responsiveness, revealing the tensions between different policy objectives. Posch et al. (2025, Chap. 7) explore the role of VET in refugee integration in Germany, illustrating how opportunity structures are influenced by institutional quality measures and labor market participation. Mupinga and Chakamba (2025, Chap. 10) highlight the barriers and potential of vocational training in Zimbabwe, emphasizing the resilience of VET systems despite economic constraints. Palomares-Montero (2025, Chap. 9) focuses on Spain's Second Chance Schools as an alternative pathway for young people who have left education, showcasing how opportunity structures can be designed to reintegrate marginalized young people into training and employment.

These examples illustrate how opportunity structures are present across vastly different economic and institutional settings, yet they are often taken for granted within their own systems. By bringing these perspectives together, this book challenges us to critically examine what enables access to vocational education, what factors sustain these pathways, and what underlying structures make them effective. In doing so, it calls attention to the features of VET that are often overlooked—not because they are insignificant, but because they have become so deeply embedded in the fabric of education and labor market systems that they escape scrutiny. Revisiting these foundational structures with a fresh lens allows us to see how they shape individual opportunities and social mobility, reminding us that the success of vocational education depends on its ability to adapt to change but also on its capacity to sustain the structures that make those opportunities possible.

The Co-Existence of Overlapping Interests

The lesson from this book teaches us how opportunity structures are possible at the intersection of individual aspirations, institutional frameworks, and societal conditions. Opportunity structures emerge when these dimensions align, creating pathways that enable individuals to access, progress through, and benefit from vocational learning. However, these structures are neither fixed nor universally equitable—they evolve with economic, political, and cultural shifts, and they may reinforce or challenge existing inequalities.

In light of these new insights, we can examine how opportunity structures function within VET systems, and what factors enable or constrain them. The conceptual figure below provides a way to visualize the overlapping dimensions of opportunity structures, illustrating how vocational pathways are shaped by the interplay of individual, institutional, and societal forces. At the core of any discussion on opportunity structures is individual agency—the choices people make regarding education, training, and employment. However, these choices are embedded in social and economic realities that either enable or restrict access to vocational pathways. While individuals have agency in navigating opportunities, their ability to do so is conditioned by the institutional and structural contexts in which they operate. Furthermore, they are shaped by the institutional arrangements that organize education and employment pathways. VET systems are embedded within national policies, training standards, and apprenticeship models that can either facilitate or restrict progression. A third layer involves macroeconomic trends, labor market policies, and cultural attitudes toward vocational education. While individual agency and institutional structures shape immediate access to VET, broader societal forces determine the long-term sustainability and success of vocational pathways.

Figure 16.1 illustrates how these three dimensions intersect to shape the opportunities available in vocational education and training. While strong institutional policies and positive economic conditions can expand opportunity, misalignment between these layers can create barriers, limiting who can benefit from vocational learning. Attempting to understand some of these dynamics helps us see VET as a system of interest alignment, mobility, inclusion, and transformation.

Rethinking Opportunity Structures in VET

As shown, the concept of opportunity structures can refer to the structures shaped by policy frameworks, institutional arrangements, and collaborative networks that create the conditions under which learners can progress, thrive, and contribute meaningfully to society. By examining how these opportunity structures emerge and evolve, we can gain insights into what makes vocational education successful. We see that it is not just about delivering job-ready skills, but about building systems that are flexible, inclusive, and responsive to the changing needs of both individuals and economies.

One key lesson from this book is that there is no single blueprint for successful VET. The cases presented here highlight a range of approaches, each tied to specific national contexts. In Norway, we see how a strong welfare state and tripartite collaboration have shaped an inclusive and

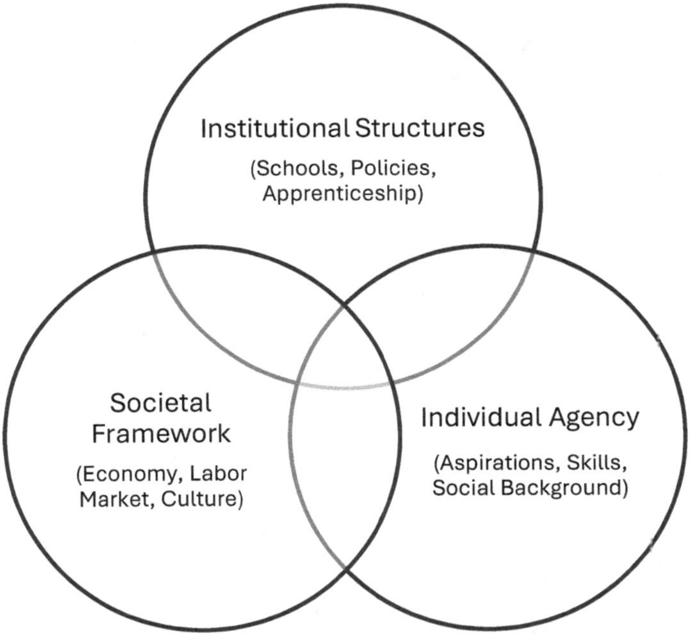

Fig. 16.1 The Overlapping Dimensions of Opportunity Structures in VET

adaptable VET system (Smeplass, 2025). In South Korea, innovative Meister High Schools demonstrate how targeted reforms can elevate the status and effectiveness of vocational education in a credential-driven society. (Jeong, 2025) In Ohio, a well-developed and articulated delivery system is accessible to all residents of the state (Zirkle, 2025). Meanwhile, in Zimbabwe, we witness the dual challenge of overcoming resource constraints while maintaining the cultural and community-oriented foundations of vocational training (Mupinga & Chakamba, 2025). Yet, despite their differences, these chapters reveal recurring themes that cut across borders. The integration of work-based learning, the importance of industry partnerships, the emphasis on lifelong learning, and the need for social equity all emerge as crucial factors. Successful VET systems are those that do not merely respond to labor market demands but actively shape them—helping individuals navigate economic uncertainty, access meaningful careers, and contribute to the social fabric.

While VET has long been a cornerstone of many national education systems, it faces unprecedented challenges. The ever-accelerating pace of technological change, the push for greener economies, and the need to address persistent inequalities are forcing us to rethink the very foundations of vocational education. Can VET systems keep pace with the rise of artificial intelligence, automation, and new industry demands? Can they provide technical skills but also the adaptability, creativity, and critical thinking needed in an era of continuous disruption? Moreover, how can VET institutions maintain inclusivity in the face of widening social and economic divides? For many, VET remains a powerful vehicle for upward mobility. However, structural barriers—such as gender inequalities, regional disparities, and systemic underfunding—still prevent too many learners from realizing the full potential of vocational pathways. Addressing these inequities will require targeted policy interventions, and a broader cultural shift that recognizes and values the unique contributions of VET to society.

As this book demonstrates, the concept of opportunity structures provides a powerful lens for understanding how vocational education operates in different contexts. But it also invites us to think about what a more unified, global vision of VET might look like. How can countries learn from one another's successes without losing their distinct strengths? What would it mean to develop shared frameworks for quality assurance, mobility, and recognition of skills that transcend national borders? Developing a global vision for VET does not mean imposing a one-size-fits-all model. Rather, it calls for deeper dialogue, informed international partnerships, and a commitment to mutual learning. By fostering collaborative connections in VET research, we can move beyond the traditional binary of academic versus vocational, of national versus global. We can begin to see VET as a field that bridges these divides—enabling individuals to build meaningful lives in conjunction with the economy.

Looking Forward—Developing New Tools to Learn from Each Other

A significant challenge facing the international VET research community is not simply gathering more data but engaging with each other's findings in ways that enrich our collective understanding of the field. Rather than focusing solely on methodological advances, the development of shared theoretical concepts and deeper intellectual dialogue offers a meaningful path forward. The opportunity structures concept explored throughout this book is a step in this direction, showing how a unifying framework can facilitate both critical analysis and productive exchange. By emphasizing theoretical concepts that resonate across national and regional contexts, we can foster conversations that go beyond comparisons of surface-level outcomes. When researchers align their analyses around shared conceptual frameworks, they create common ground for discussing the challenges, successes, and ongoing challenges and reforms within VET systems. These frameworks also make it possible to connect seemingly disparate cases, drawing out patterns and insights that might otherwise remain obscured.

However, simply adopting shared concepts is not enough. We must also consider how we engage with one another's work. Scholarly exchange should aim to bridge divides: both the geographic and economic differences between countries, but also the disciplinary and institutional divides that can fragment the field. Meaningful engagement means reading across contexts, applying a critical lens to each other's findings, and treating differences as possibilities for reflection rather than as barriers to consensus. It also means recognizing the strengths in every system—what can be learned, more than what needs to be fixed.

Such a focus on theoretical alignment and mutual engagement provides a way of understanding those challenges within a larger, interconnected landscape. As more scholars turn to shared concepts and genuinely engage with diverse perspectives, we can deepen our understanding. This approach enriches individual research projects and strengthens the field as a whole, allowing us to move forward together with a clearer sense of purpose and direction.

This book can be seen as a call to action for what could be. For policymakers, educators, researchers, and community members, to identify the opportunity to celebrate existing achievements and invest in VET systems that are inclusive, forward-looking, and deeply connected to the communities they serve. It means rethinking opportunity structures as frameworks for lifelong development, societal well-being, and collective progress.

References

Abbott, A. (2010). *Chaos of disciplines*. University of Chicago Press.

Avis, J. (2012). Global reconstructions of vocational education and training. *Globalisation, Societies and Education, 10*(1), 1–11. https://doi.org/10.1080/14767724.2012.646876

Biesta, G. J. (2015). *Good education in an age of measurement: Ethics, politics, democracy*. Routledge.

Billett, S. (2025a). Contemporary purposes for vocational education. In E. Smeplass & C. Zirkle (Eds.), *Opportunity structures in vocational education and training* (pp. 255–278).

Billett, S. (2025b). Localized capacity to realize the potential of vocational education. In E. Smeplass & C. Zirkle (Eds.), *Opportunity structures in vocational education and training* (pp. 303–323).

Dewey, J. (1916). *Democracy and education: An introduction to the philosophy of education*. Macmillan.

Fraser, N. (2008). *Scales of justice: Reimagining political space in a globalizing world*. Polity Press.

Gonon, P., & Deissinger, T. (2021). Towards an international comparative history of vocational education and training. *Journal of Vocational Education & Training, 73*(2), 191–196. https://doi.org/10.1080/13636820.2021.1912945

Jeong, S. J. (2025). Advancing secondary vocational education system: A case study of Korea's Meister High School. In E. Smeplass & C. Zirkle (Eds.), *Opportunity structures in vocational education and training* (pp. 97–125).

Láscarez Smith, D. (2025). Challenges and contradictions between the reduction of youth unemployment and the implementation of dual apprenticeship in Costa Rica. In E. Smeplass & C. Zirkle (Eds.), *Opportunity structures in vocational education and training* (pp. 23–46).

Lyckander, R. H., Skålholt, A., & Hovdhaugen, E. (2025). Improving adult students' access to vocational education and training in Norway. In E. Smeplass & C. Zirkle (Eds.), *Opportunity structures in vocational education and training* (pp. 47–70).

Mupinga, D. M., & Chakamba, J. (2025). The benefits, challenges, and opportunities for technical & vocational education & training in Zimbabwe. In E. Smeplass & C. Zirkle (Eds.), *Opportunity structures in vocational education and training*.

Palomares-Montero, D. (2025). Second chance schools in Spain: A bridge towards continuity of training and employment for young people at risk of social exclusion. In E. Smeplass & C. Zirkle (Eds.), *Opportunity structures in vocational education and training* (pp. 201–226).

Peirce, C. S. (1905). Issues of pragmatism. *The Monist, 15*(4), 481–499.

Pilz, M. (2012). International comparative research into vocational training: Methods and approaches. In *The future of vocational education and training in a changing world* (pp. 561–588). VS Verlag für Sozialwissenschaften.

Pilz, M., & Li, J. (2020). *Comparative vocational education research*. Springer Fachmedien.

Posch, K. J., Li, J., & Pilz, M. (2025). Quality indicators for the vocational education and training of refugees: Acceptance of German vocational school principals. In E. Smeplass & C. Zirkle (Eds.), *Opportunity structures in vocational education and training* (pp. 153–175) .

Seitamaa, A., & Hansen, P. (2025). Opportunity structures for who and what? The discursive struggle between educational equality and economization in Europeanized Finnish vocational education and training. In E. Smeplass & C. Zirkle (Eds.), *Opportunity structures in vocational education and training* (pp. 127–152).

Smeplass, E. (2025). Investment in youth through vocational education and training. In E. Smeplass & C. Zirkle (Eds.), *Opportunity structures in vocational education and training* (pp. 279–301).

Swedberg, R. (2014). *The art of social theory*. Princeton University Press.

Zirkle, C. (2025). 100+ years of historical opportunity: Vocational/career and technical education in the state of Ohio. In E. Smeplass & C. Zirkle (Eds.), *Opportunity structures in vocational education and training* (pp. 71–96).

Index

E. Smeplass and C. Zirkle (eds.), *Opportunity Structures in Vocational Education and Training*, Palgrave Studies in Technical and Vocational Education and Training, https://doi.org/10.1007/978-3-032-04410-5

M

N

O

P

Q

FSC
www.fsc.org
MIX
Papier | Fördert gute Waldnutzung
FSC® C083411

Kolibri 360 GmbH
Ferdinand-Jühlke-Straße 7
99095 Erfurt, Deutschland
produktsicherheit@kolibri360.de